INSIDER TRADING
Global Developments and Analysis

INSIDER TRADING
Global Developments and Analysis

Edited by

Paul U. Ali

Greg N. Gregoriou

CRC Press is an imprint of the
Taylor & Francis Group, an **informa** business
A CHAPMAN & HALL BOOK

CRC Press
Taylor & Francis Group
6000 Broken Sound Parkway NW, Suite 300
Boca Raton, FL 33487-2742

First issued in paperback 2019

ISBN-13: 978-1-4200-7401-7 (hbk)
ISBN-13: 978-0-367-38699-3 (pbk)

Library of Congress Cataloging-in-Publication Data

Ali, Paul U.
Insider trading : global developments and analysis / Paul U. Ali and Greg N. Gregoriou.
p. cm.
Includes bibliographical references and index.
ISBN 978-1-4200-7401-7 (alk. paper)
1. Insider trading in securities. I. Gregoriou, Greg N., 1956- II. Title.

HG4551.A5847 2008
332.64--dc22 2008005100

Visit the Taylor & Francis Web site at
http://www.taylorandfrancis.com

and the CRC Press Web site at
http://www.crcpress.com

Contents

Introduction

Insider trading—the illegal use of price-sensitive, nonpublic information to buy and sell securities and other financial instruments—has long been considered an endemic feature of the world's financial markets, despite the almost-universal criminalization of insider trading. It is thus unsurprising that the recent boom in mergers and acquisitions has been accompanied by a resurgence in insider trading and a concomitant increase in the prosecution of insider trading on a scale not seen since the 1980s (Drummond 2007; *Economist* 2007).

The U.S. Securities and Exchange Commission (SEC), for instance, has made clear that it views prosecuting insider trading a priority and has recently devoted considerably more resources to monitoring suspicious trading in securities and security-linked derivatives. SEC Chairman Christopher Cox, in his testimony to the U.S. Congress in early 2007, identified insider trading as one of three major risks affecting the U.S. capital markets (SEC 2007b).

This new focus on insider trading is well illustrated by three major insider trading–related lawsuits brought by the SEC during 2007. In March 2007, the SEC charged fourteen persons, including three hedge funds, with illegally trading shares using information stolen from two investment banks (SEC 2007a). Then, in May 2007, the SEC charged a New York–based investment banker with illegally providing inside information to an accomplice in Pakistan who purchased call options over TXU Corp shares ahead of the buyout of TXU Corp by KKR and Texas Pacific (SEC 2007c). Also, in May 2007, the SEC charged a Hong Kong couple with illegally trading Dow Jones shares ahead of the announcement of New Corp's takeover offer for Dow Jones (2007d).

The prohibition of insider trading and the imposition of criminal penalties for insider trading have a moral dimension. One of the key justifications for this response to insider trading is that it is "unfair" or even "immoral" for securities to be traded (in a public market) when one party has private information which, if it were publicly available, would or would be reasonably expected to affect the price of the securities and other parties in the market are ignorant of that information. Thus, insider trading, if unchecked, would lead to an erosion of confidence in the market and the exit of investors from the market with adverse consequences for the cost of capital. Yet the concept of "unfair" trading alone cannot be a sufficient basis for prohibiting insider trading, for a better-informed investor could always be said to have an advantage—and arguably an unfair advantage—over less-well-informed investors (Leland 1992). The logical extrapolation of this unfairness argument would be to ban all trading using private information.

The economic dimension, however, of insider trading is also murky and is, likewise, a less than satisfactory basis for prohibiting insider trading. Insider trading may reduce market liquidity, cause a widening of spreads, and increase market volatility (Leland 1992; Du and Wei 2004; Cheng et al. 2006) and also reduce the returns to outsiders since they are trading against better-informed insiders (Leland 1992). However, it is equally possible that insider trading may facilitate price discovery, leading to more informative security prices (Cornell and Sirri 1992; Meulbroek 1992; Bhattacharya and Nicodano 2001) and, consequently, reduce the risks of investing in securities for outsiders (Leland 1992). Nonetheless, the economic consequences of insider trading appear to be on a more defensible footing than unfairness for prohibiting insider trading, especially if the presence of insiders deters outsiders from trading, thus impeding price discovery (Fishman and Hagerty 1992), and if insider trading does not differentially affect the prices of securities compared to outsider trading (Chakravarty and McConnell 1999).

Having said this, it is an ineluctable fact of life in most financial markets that insider trading is a criminal offense. The criminalization, however, of insider trading has not served to reduce the incidence of insider trading (Seyhun 1992). Moreover, what evidence there is suggests that insider trading prohibitions have in fact made insider trading more profitable (Bris 2005).

One might, on this basis, be forced to adopt the uncomfortable conclusion that the criminalization of insider trading has less to do with deterring economically damaging conduct and more to do with placating public resentment about insider trading, namely, "the resentment by the investment public that other persons have the good fortune to enjoy something to which the public has no right" (Schroeder 2005, 2027). This particular point is well explored in a separate volume by one of the contributors to this book (Heminway 2007).

This book brings together some of the latest research on insider trading, covering established U.S., European, and Australasian markets as well as the key emerging markets of Brazil and China. The book combines a variety of approaches toward the study of insider trading, with the contributors coming from the fields of accounting, economics, finance, and law. The book is divided into three parts. Part 1 (Chapters 1 to 6) broadly examines the ethics of insider trading and the rationale for criminalizing insider trading. Part 1 also examines insider trading in the context of emerging markets and the new market for credit derivatives. Part 2 (Chapters 7 to 15) is concerned with regulatory responses to insider trading, including the controversial topic of legal insider trading and whether the regulatory response to insider trading should differentiate between positive and negative information and price-increasing and price-decreasing insider trading. Part 3 (Chapters 16 to 20) investigates the economic consequences of insider trading, including market responses to insider trading and the impact of insider trading on equity returns.

—Paul U. Ali and Greg N. Gregoriou

REFERENCES

Bhattacharya, S., and G. Nicodano. 2001. Insider trading, investment, and liquidity: A welfare analysis. *Journal of Finance* 56(3):1141–56.

Bris, A. 2005. Do insider trading laws work? *European Financial Management* 11(3):267–312.

Chakravarty, S., and J. J. McConnell. 1999. Does insider trading really move stock prices? *Journal of Financial and Quantitative Analysis* 34(2):191–209.

Cheng, L., M. Firth, T. Y. Leung, and O. Rui. 2006. The effects of insider trading on liquidity. *Pacific Basin Finance Journal* 14:467–83.

Cornell, B., and E. R. Sirri. 1992. The reaction of investors and stock prices to insider trading. *Journal of Finance* 47(3):1031–59.

Drummond, B. 2007. Insider trading. *Bloomberg Markets,* August.

Du, J., and S. J. Wei. 2004. Does insider trading raise market volatility? *Economic Journal* 114:916–42.

Economist. 2007. When greed is bad. May 10.

Fishman, M. J., and K. M. Hagerty. 1992. Insider trading and the efficiency of stock prices. *RAND Journal of Economics* 23(1):106–22.

Heminway, J. M. 2007. *Martha Stewart's legal troubles.* Durham, NC: Carolina Academic Press.

Leland, H. E. 1992. Insider trading: Should it be prohibited? *Journal of Political Economy* 100(4):859–87.

Meulbroek, L. K. 1992. An empirical analysis of illegal insider trading. *Journal of Finance* 47(5):1661–99.
Schroeder, J. L. 2005. Envy and outsider trading: The case of Martha Stewart. *Cardozo Law Review* 26:2023–78.
Securities and Exchange Commission. 2007a. SEC Charges 14 in Wall Street insider trading ring, March 1.
Securities and Exchange Commission. 2007b. Testimony concerning fiscal 2008 appropriations request, March 27.
Securities and Exchange Commission. 2007c. Litigation release 20105, May 4.
Securities and Exchange Commission. 2007d. Litigation release 20106, May 8.
Seyhun, N. H. 1992. The effectiveness of the insider-trading sanctions. *Journal of Law and Economics* 35(1):149–82.

About the Editors

Paul U. Ali is an associate professor in the Faculty of Law, University of Melbourne, and a member of that Law Faculty's Centre for Corporate Law and Securities Regulation. Prior to becoming an academic, Paul was, for several years, a finance lawyer in Sydney. Paul has published widely on banking and finance law, corporate governance and institutional investment law, securitization law, and structured finance law. His most recent publications include *Secured Finance Transactions* (London, 2007), *Innovations in Securitisation* (The Hague, 2006), and *International Corporate Governance after Sarbane–Oxley* (Hoboken, NJ, 2006). In 2006, Paul was appointed by the Australian Federal Attorney-General as a member of the Personal Property Securities Review Consultative Group. Paul holds an SJD degree from the University of Sydney.

Greg N. Gregoriou is professor of finance in the School of Business and Economics at State University of New York (Plattsburgh). A native of Montreal, Greg received his bachelor of arts in economics from Concordia University in 1988. In 1991, he completed his MBA from UQAM (University of Quebec at Montreal), and graduate Diploma in Applied Management from McGill University. He then completed his PhD in finance at UQAM in the joint doctoral PhD program in Montreal, which pools the resources of Montreal's universities (McGill University, Concordia, and HEC Montreal). He specializes in hedge funds, funds of hedge funds, and managed futures. He has published more than fifty academic articles on hedge funds and managed futures in various peer-reviewed journals, such as the *Journal of Portfolio Management*, *Journal of Futures Markets*, *European Journal of Operational Research*, *Annals of Operations Research*, *and Computers and Operations Research*, and has written twenty book chapters. He is hedge fund editor and editorial board member for *Journal of Derivatives and Hedge Funds*, a London-based academic journal and also editorial board member of the *Journal of Wealth Management* and the *Journal of Risk and Financial Institutions.* Since his arrival at SUNY (Plattsburgh) in 2003, he has published 26 books with John Wiley & Sons, Elsevier Butterworth-Heinemann, McGraw-Hill, Palgrave-Macmillan, Bloomberg Press, and Risk Books.

Contributors

Nihat Aktas
Academic Fellow—Europlace Institute of Finance
CORE & IAG Louvain School of Management
Université catholique de Louvain
Louvain-la-Neuve, Belgium

Paul U. Ali
Faculty of Law
University of Melbourne
Melbourne, Australia

Yee Ben Chaung
Faculty of Law
University of Melbourne
Melbourne, Australia

Blanaid Clarke
Law School
University College Dublin
Dublin, Ireland

Eric de Bodt
Université de Lille 2
IAG Louvain School of Management
Université catholique de Louvain
Louvain-la-Neuve, Belgium

Otavio R. de Medeiros
University of Brasilia
Campus Universitario Darcy Ribeiro
Brasilia, Brazil

Jan de Smedt
Banking, Finance and Insurance Commission (CBFA)
Brussels, Belgium

Anna-Athanasia Dervenis
Macquarie Bank
Sydney, Australia

Jana Fidrmuc
University of Warwick
Coventry, West Midlands, England

Marc Goergen
University of Sheffield Management School
Sheffield, United Kingdom

Kristoffel R. Grechenig
University of St. Gallen
St. Gallen, Switzerland

Philippe Grégoire
Département de finance et assurance
Faculté des sciences de l'administration
Université Laval
Quebec City, Quebec, Canada

Greg N. Gregoriou
School of Business and Economics
State University of New York, Plattsburgh
Plattsburgh, New York

Michael Halperin
Lippincott Library
Wharton School
University of Pennsylvania
Philadelphia, Pennsylvania

Joan MacLeod Heminway
The University of Tennessee College of Law
Knoxville, Tennessee

William Kelting
Professor of Accounting
School of Business and Economics
State University of New York, Plattsburgh
Plattsburgh, New York

Thomas A. Lambert
University of Missouri School of Law
Columbia, Missouri

Zhihui Liu
Faculty of Law
Chinese University of Political Science and Law
Beijing, China

Edward J. Lusk
School of Business and Economics
State University of New York, Plattsburgh
Plattsburgh, New York

Robert W. McGee
Director, Center for Accounting, Auditing and Tax Studies
School of Accounting
College of Business Administration
Florida International University
Miami, Florida

Geraldine Szott Moohr
University of Houston Law Center
Houston, Texas

Sadakazu Osaki
Nomura Institute of Capital Markets Research
Tokyo, Japan

François-Éric Racicot
Department of Administrative Sciences
University of Quebec at Montreal
Montreal, Quebec, Canada

Colin Read
School of Business and Economics
State University of New York, Plattsburgh
Plattsburgh, New York

Luc Renneboog
Tilburg University
Tilburg, The Netherlands

Ilham Riachi
IAG Louvain School of Management
Université catholique de Louvain
Louvain-la-Neuve, Belgium

Raymond Théoret
Department of Administrative Sciences
University of Quebec at Montreal
Montreal, Quebec, Canada

Margaret Wang
Chambers & Co
Melbourne, Australia

About the Contributors

Nihat Aktas is associate professor of finance at the IAG Louvain School of Management, Universite Catholique de Louvain and Academic Fellow of the Europlace Institute of Finance.

Yee Ben Chaung works in Melbourne for one of Australia's leading national law firms. He holds degrees in commerce and in law from the University of Melbourne, and is currently undertaking postgraduate studies in law at the University of Melbourne.

Blanaid Clarke, BCL, MBS (Banking & Finance), Barrister at Law, PhD, is associate professor in corporate law in the Law School, University College Dublin (UCD). Blanaid is one of the founding members of the Institute of Directors' Centre for Corporate Governance at UCD, a member of the Irish Takeover Panel Executive, and Academic Director for the Business and Legal Studies Degree in UCD. In Spring 2006, Blanaid was Visiting Scholar to the University of Queensland and Parsons Visitor to the University of Sydney. Blanaid has published several texts including *Contract Cases and Materials* (coauthored with R. Clark, 3rd ed. Dublin: Gill & Macmillan; 2004) and *Takeovers and Mergers Law in Ireland* (Dublin: Roundhall Sweet & Maxwell; 1999). She has produced numerous articles including most recently "Articles 9 and 11 of the Takeover Directive (2004/25) and the Market for Corporate Control." *Journal of Business Law* (UK) (2006): 355–74.

Eric de Bodt is professor of finance at Université de Lille 2 and at the IAG Louvain School of Management, Universite Catholique de Louvain.

Otavio Ribeiro de Medeiros was born in Rio de Janeiro. He started his academic career with a university degree in mechanical engineering from Pontifícia Universidade Católica in Rio. After some time, he studied for an MSc in administration with a major in finance at the COPPEAD Institute in Rio. Later on, he moved to the United Kingdom where he received an MSc in economics from Birkbeck College–University of London, and a PhD in economics from the University of Southampton. He worked as a financial executive for a number of large Brazilian companies such as Furnas, BNDES, AD-Rio, Aços Villares, and Etesco. After joining academia in 2002, he is currently full professor of finance at the University of Brasilia, Brazil, teaching in the graduate programs of administration and accounting in that institution. He is also a research fellow of CNPq–National Counsel of Technological and Scientific Development (Brazil). He has written a number of articles and research papers, and his research interests include stock markets, corporate finance, and applied econometrics.

Jan de Smedt is administrative secretary with the Banking, Finance and Insurance Commission (CBFA), Belgium.

Anna-Athanasia Dervenis is a lawyer with the Macquarie Bank Limited Banking and Property Group, Sydney. Anna specializes in corporate and property law within the Real Estate Capital division. She holds degrees in commerce (finance) and in law from the University of New South Wales and was admitted as a solicitor of the Supreme Court of New South Wales in 2006.

Jana Fidrmuc is a graduate of the University of Bratislava with an MSc in physics and of the University of Tilburg with an MSc and PhD in financial economics. She held a post-doc position at Erasmus University in Rotterdam and joined the University of Warwick as a lecturer in finance in 2005. She has published on insider trading in the *Journal of Finance*, and on transition economics in the *Journal of International Money and Finance* and in the *Economics of Transition*. Her research interests are corporate governance, corporate finance, ownership structure, and restructuring in transition.

Marc Goergen has a degree in economics from the Free University of Brussels, an MBA from Solvay Business School, and a DPhil from the University of Oxford. He has held appointments at the University of Manchester Institute of Science & Technology (UMIST), Manchester Business

School, and the ISMA Centre (University of Reading). He currently holds a chair in finance at the University of Sheffield Management School. Marc Goergen's research interests are in corporate ownership and control, corporate governance, mergers and acquisitions, dividend policy, corporate investment models, insider trading, and initial public offerings. Marc has widely published in academic journals such as *European Financial Management, Journal of Business Finance & Accounting, Journal of Corporate Finance, Journal of Finance*, and *Journal of Law & Economics.* He has also contributed chapters to several edited books and written two books on corporate governance (published by Edward Elgar and Oxford University Press). Marc is a research associate of the European Corporate Governance Institute and a fellow of the International Institute for Corporate Governance & Accountability.

Kristoffel R. Grechenig received both his first law degree (Magister) and his second law degree (Doktor) from Vienna University School of Law. He wrote his doctoral thesis on Spanish corporate law in comparison to German and Austrian corporate law and was awarded the Walther-Kastner-Prize and the IVA Stipendien Prize. He enrolled for a postgraduate study at Columbia Law School and received his LLM degree as a Harlan Fiske Stone Scholar. He started working as an assistant professor of law at the Vienna University of Economics and Business Administration, focusing on comparative corporate law. He was a fellow at the FOWI Institute (Institute for Eastern European Business Law) and a clerk at the Austrian Court for Commercial Law. He currently holds the position of Post-Doc Assistant Professor of Law at the Vienna University of Economics and Business Administration, where his main fields of research include comparative corporate law, law and economics, and comparative legal theory. He has published numerous articles on these topics in English, German, and Spanish. He currently teaches a corporate law and economics course at his university and a comparative legal theory course at the Vienna University School of Law.

Philippe Grégoire is a professor in the Faculty of Business Administration, Laval University, Quebec.

Michael Halperin is director of the Lippincott Library of the Wharton School of the University of Pennsylvania. He is an information retrieval expert who has given numerous seminars on the creation of information from the EDT linkages available in the nexus of Compustat and CRSP. Further, he has developed numerous workshops on the search retrieval of text information using ABI-Inform and Business Source Premier.

Joan MacLeod Heminway regularly teaches business associations, corporate finance, representing enterprises (a transaction simulation course), and securities regulation in The University of Tennessee (UT) College of Law James L. Clayton Center for Entrepreneurial Law. She received the University Chancellor's Award for Teaching Excellence in 2006, the college's Marilyn V. Yarbrough Faculty Award for Writing Excellence for 2005, and the college's Harold C. Warner Outstanding Teacher Award for 2004. Professor Heminway's stock merger module for the Representing Enterprises course was recognized by the UT Innovative Technology Center in its September 2002 Best Practices@UT Showcase. She was Visiting Professor at Boston College Law School for the Fall 2005 semester, where she taught mergers & acquisitions and securities regulation, and is visiting at Vanderbilt University Law School to teach a short course on Animals & the Law for the Spring 2007 semester. Professor Heminway edited and contributed to a recently released book, *Martha Stewart's Legal Troubles*, that includes chapters focusing on the white collar crime, securities regulation, and corporate law aspects of Martha Stewart's recent legal entanglements. Her research agenda principally focuses on securities disclosure law and policy, with a special focus on insider trading regulation. Recent articles authored by Professor Heminway have appeared in (among other publications) the *American University Law Review*, *University of Cincinnati Law Review*, *Fordham Journal of Corporate & Financial Law*, *Maryland Law Review*, and *Texas Journal of Women and the Law*. Before starting her teaching career in 2000, Professor Heminway spent fifteen years practicing law in the Boston office of Skadden, Arps, Slate, Meagher & Flom LLP, where she specialized in mergers and acquisitions and securities regulation matters. Leveraging off this experience, Professor Heminway has coauthored (with several Tennessee practitioners) a series of annotated merger and acquisitions agreements that have been published in the Spring 2003, 2004, 2005, and 2006 issues of *Transactions: The Tennessee Journal of Business Law* (http://www.law.utk.edu/centers/entrep/claytontransactions.htm). A new article in this series has been published in the Spring 2007 issue of *Transactions*.

William R. Kelting is emeritus professor of accounting, State University of New York, Plattsburgh. He has been teaching for almost 40 years at the institution. Before joining the college, he worked for several accounting firms including Arthur Andersen, and Telling, Kelting, & Potter, CA (Founding partner). Dr. Kelting earned his BA in economics from Washington & Lee University, and his MBA in public administration from Rutgers University, and a PhD in accounting with minors in economics and psychology from University of Arkansas (1988). He is also a CPA. Dr. Kelting's research has appeared in refereed journals. Recently, his articles appeared in *The Accounting Educator's Journal, Journal of Financial Crime,* and *Pension: An International Journal.* Dr. Kelting also coauthored a computerized case study in governmental accounting and a study guide for a major textbook in auditing. In addition, Dr. Kelting has several publications in professional journals and has made several presentations at academic and professional organizations. He is a member of the American Institute of Certified Public Accountants, Association of Certified Fraud Examiners, and the New York State Society of Certified Public Accountants.

Thomas A. Lambert, associate professor of law at the University of Missouri, graduated with highest honors from Wheaton College (Illinois). He then worked as an environmental policy analyst at the Center for the Study of American Business at Washington University in St. Louis before attending the University of Chicago Law School. While at Chicago, Lambert was a Bradley Fellow and served as comment editor of the Law Review. After graduating with honors in 1998, he clerked for Judge Jerry E. Smith of the U.S. Court of Appeals for the Fifth Circuit and spent a year as the John M. Olin Fellow at Northwestern University Law School. He then joined the Chicago office of Sidley Austin LLP, where he practiced antitrust and securities litigation. In 2003, he joined the law faculty at Missouri, where he teaches business organizations, contracts, antitrust law, and environmental law. Lambert has published numerous scholarly articles on regulatory theory and business law topics. In addition, he is a member of the advisory board of the eSapience Center for Competition Policy and a regular contributor to *Truth on the Market* (www.truthonthemarket.com), a weblog devoted to "academic commentary on law, business, economics, and more."

Zhihui Liu is associate professor of Faculty of Law and an associate director of Civil Law Institute, at China University of Political Science and Law in Beijing, China. She specializes in civil and commercial law and has published extensively in these areas. Her latest publications include *Comments on Complex Issues Relating to China's new Property Law,* published by Jiangsu People's Press, and *Fundamental Principles of Possessory System*, published by Renmin University Press.

Edward J. Lusk is currently professor of accounting, the State University of New York, College of Economics and Business, Plattsburgh, New York, and Emeritus: The Department of Statistics, The Wharton School, The University of Pennsylvania, Philadelphia, Pennsylvania. From 2001 to 2006 he held the Chair in Business Administration at the Otto-von-Guericke University, Magdeburg Germany. He has also taught in China at the Shanxi University of Finance and Commerce and the Chulalongkorn University Bangkok, Thailand. He as published more than 145 articles in peer-reviewed journals including *The Journal of Political Economy, Statistics and Probability, The Accounting Review, Management Science, The Journal of the Operational Research Society, Gender, Work and Organizations*, and *Decision Support Systems.*

Robert W. McGee is an accounting professor at Florida International University in Miami, Florida. He has published more than 400 articles and more than 50 books in the areas of accounting, taxation, economics, law, and philosophy. His experience includes consulting with the governments of several former Soviet, East European, Asian, African, and Latin American countries to reform their accounting and economic systems. Dr. McGee is an attorney, certified public accountant, and economist and holds doctorates in several fields, including accounting, economics, law, and philosophy. The *Journal of Business Ethics* ranked him number 1 in the world for business ethics scholarship. The Social Science Research Network ranks him in the top one-tenth of 1 percent in terms of downloads from its Web site, out of more than 83,000 academics worldwide.

Geraldine Szott Moohr joined the University of Houston Law Center as an associate professor of law in 1995. In 2001, she was awarded a George Butler Research Chair, a rotating professorship honoring excellence in scholarship and in 2005 she was awarded the Alumnae Law Center Chair. Professor Moohr has held visiting appointments at the Washington and

Lee University School of Law and the University of Toledo College of Law. In the autumn of 2006, she served as the John J. Sparkman Visiting Professor of Law at the University of Alabama School of Law. A graduate of the University of Illinois with an MS from Bucknell University, Professor Moohr received her law degree in 1991 from the American University College of Law. She graduated first in her class and was editor-in-chief of the American University Law Review. Following graduation she clerked for the Honorable James M. Sprouse of the U.S. Court of Appeals for the Fourth Circuit. She joined Covington & Burling's Washington office in 1992 as a litigation association; her practice focused on commercial fraud and included arbitrations as well as trials. Professor Moohr has published numerous articles in law journals. Her areas of expertise are federal criminal law, white collar crime, fraud offenses, government corruption, and the emerging criminal laws that govern various forms of information. Her work on the federal mail and wire fraud offenses and intellectual property crimes has been widely cited. Another area of interest is arbitration, specifically the arbitration of employment disputes. Professor Moohr frequently speaks at conferences and symposia on these topics, as well as on post-Enron criminal initiatives and intellectual property crimes. She is the author of a casebook, *The Criminal Law of Intellectual Property and Information* (West, 2007). She is also coauthor of the sixth edition of *Criminal Law* with Joseph Cook, Linda Malone, and Paul Marcus (Lexis, forthcoming 2008).

Sadakazu Osaki is Executive Fellow, Nomura Institute of Capital Markets Research, Tokyo and Visiting Professor, Graduate School of Asia-Pacific Studies (MBA), Waseda University, and Visiting Associate Professor, Graduate School of Law, University of Tokyo. He holds an LLB degree from the University of Tokyo, an LLM (economic law) from University College, London, and an LLM, from the Europa Institute, University of Edinburgh. He has published extensively on Japan's securities markets.

François-Éric Racicot, PhD, holds a joint doctorate in business administration (finance) from the University of Quebec at Montreal (UQAM). He also holds an MSc in economics (econometrics) from University of Montreal where he also received his BSc in economics (quantitative economics). He is associate professor of finance at the Department of Administrative Sciences of the University of Quebec, Outaouais (UQO). He was professor of finance at the Department of Strategic Business of ESG-UQAM. He is a permanent member of the Laboratory for Research in Statistics and Probability (LRSP) and a research associate at the Chaire d'information financière et organisationnelle located at ESG-UQAM. He is also a consultant in financial engineering for various financial institutions in Quebec. His research fields include the theory of fixed income securities, the theory of derivative products, the empirical analysis of hedge funds, and financial engineering. His research focuses on the development of new econometric techniques for correcting and detecting specification errors in financial models, especially in the context of estimating the alpha and conditional beta of hedge funds. This research should be useful, especially for improving the selection of hedge funds used in the construction of a fund of funds. Professor Racicot has published many books on quantitative finance used at the graduate levels in universities and also in financial institutions. He has also published several articles on empirical finance in international journals.

Colin Read is the dean of the School of Business and Economics at State University of New York (SUNY), Plattsburgh. He holds a PhD from Queen's University in economics, a juris doctor from the University of Connecticut, and a master's of taxation from the University of Tulsa. He has published numerous articles on economic theory, location theory, and the microeconomic underpinnings of information and taxation.

Luc Renneboog is professor of corporate finance at Tilburg University, and a research fellow at the Center for Economic Research and the European Corporate Governance Institute (ECGI, Brussels). He is a graduate of the Catholic University of Leuven with degrees in management engineering (MSc) and in philosophy (BA), of the University of Chicago with an MBA, and of the London Business School with a PhD in financial economics. He held appointments at the University of Leuven and Oxford University, and visiting appointments at London Business School, European University Institute (Florence), HEC (Paris), Venice University, and CUNEF (Madrid). He has published in the *Journal of Finance, Journal of Financial Intermediation, Journal of Law and Economics, Journal of Corporate Finance, Journal of Banking and Finance, Journal of Law, Economics & Organization, Cambridge Journal of Economics, European Financial Management*, and others. He has coauthored and edited several books on corporate governance, dividend policy, and venture capital with Oxford University Press. His research interests are corporate finance, corporate governance, dividend policy, insider trading, law and economics, and the economics of art.

Ilham Riachi is a research assistant at the IAG Louvain School of Management, Universite Catholique de Louvain.

Raymond Théoret, PhD, holds a doctorate in economics (financial economics) issued by the University of Montreal. He is professor of finance at l'École des Sciences de la Gestion (ESG) of the University of Quebec, Montreal (UQAM). He was previously professor in financial economics at l'Institut d'Économie Appliquée located at HEC Montreal. He was an economic and financial consultant at various financial institutions in Quebec and secretary of the Campeau Commission on the improvement of the situation of financial institutions in Montreal which gives way to the foundation of Institut de Finance Mathématique de Montreal. He has published many articles and many books on financial engineering, especially in the fields of numerical methods, computational finance, and asset pricing. Moreover, he is the founder of DESS (finance) issued by UQAM and a cofounder of the Maîtrise en finance appliquée at the same university. He teaches portfolio management theory and computational finance. His research focuses on modeling hedge fund returns, especially on its link with specification errors. He is an associate member of the Chaire d'information financière et organisationnelle located at ESG-UQAM.

Margaret Wang is senior associate at Chambers and Company International Lawyers, the Australian affiliated law firm of Chadbourne & Parke LLP, a global law firm headquartered in New York, specializing in corporate and commercial law. Prior to taking up her current post, she was a lecturer in law at Victoria University, Melbourne, Australia. She has published extensively in her areas of specialization, including in the *European Business Law Review.*

Acknowledgments

We thank the anonymous referees for the selection of papers for this book. We also thank Sunil Nair, Jay Margolis, Jessica Vakili, Sarah Morris, and Carol Shields of Taylor & Francis/CRC Press. Paul Ali acknowledges the assistance of the Australian Research Council (DP0557673). Each of the chapters in this book is the original work of the relevant author(s). The publisher and editors are not responsible for the accuracy of the individual chapters.

Part 1

The Taxonomy of Insider Trading

CHAPTER 1

Market Inefficiencies and Inequities of Insider Trading—An Economic Analysis

Colin Read

CONTENTS

1.1 INTRODUCTION

Insider information can be described as trading on a security based on asymmetric information the inside trader has but which has yet to be reflected in the security price. There is no advantage to such closely held information until the information is fully capitalized in the security price. Hence, insider trading is primarily an information-timing issue. One might argue that because insider trading is simply taking early advantage of information before the market ultimately incorporates this information, insider trading does no harm. Indeed, this information will likely benefit other early traders who can act on the information before it is fully incorporated in the market price. Under this reasoning, insider trading merely allows an insider to profit earlier than these other early traders. I argue that this simplistic approach neglects other important economic consequences.

In Section 1.2, I model the economic losses that arise for risk-averse traders. I then model the diversion of economic surpluses from all traders to insiders as a consequence of insider trading and discuss the social welfare consequences if inside traders consume valuable resources to garner such information. In Section 1.4 and Section 1.5, I discuss the depressing effect such insider trading has on the rate of return for securities, with the implication of reduced capital formation.

In Section 1.6, I discuss the implications of the creation of a market for lemons, along the lines of George Akerlof's seminal paper. In Section 1.7, I make the distinction between the benefits and value created by the creation, analysis, and distribution of information (and the market efficiencies they create) and the inefficiencies and inequities that result from the information hording insider trading requires. I discuss some public policy and regulatory implications of insider trading in Section 1.8, and conclude in Section 1.9.

1.2 RISK AVERSION

Consider two otherwise identical, risk-neutral traders with an identical wealth W_l who own all outstanding shares of a security in equal amounts. Let us assume that a piece of information could increase the value of the security by $1,000,000, resulting in a wealth W_h. Absent insider trading, each would benefit by $500,000, resulting in an equal average wealth W_{av}. If one trader can benefit from insider trading, she could benefit by the full $1,000,000 gain, while the other will not benefit at all. Figure 1.1 demonstrates that risk-averse traders would prefer the average wealth W_{av} (in this example, the $500,000

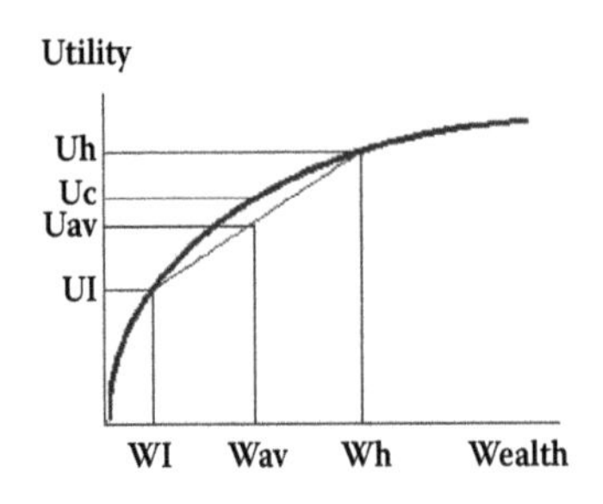

FIGURE 1.1. The costs of risk if traders are risk averse.

with certainty) over the 50 percent chance at $1,000,000. The horizontal axis denotes the level of wealth for the otherwise identical traders, and the vertical axis represents the corresponding level of satisfaction or utility. The figure shows that utility rises with increased wealth, but rises at a decreasing rate (begins to flatten out) because greater and greater wealth produces correspondingly smaller increases in utility.

Under this analysis we can plot the level of utility U_h for the inside trader who secures all the gains of information, the status quo utility U_l for the trader that does not, the utility U_c if both split the gains evenly, and the average utility U_{av} if there is a 50 percent chance either trader will be the winner (or loser).

This translates in a utility or enjoyment U_h for the agent with the new wealth and a utility U_l for the agent without the insider information. Note that the average utility between the two individuals, at the halfway point between U_l and U_h on the vertical axis, is lower than the utility U_c each would had received had they evenly divided the average wealth W_{av}. This is a consequence of diminishing marginal utility—the additional gains in wealth accruing to the inside trader are less satisfying than the losses in wealth sacrificed by the agent exploited in the transaction. The difference in utility $U_c - U_{av}$ is a social sacrifice as a consequence of insider trading.

1.3 DEADWEIGHT LOSSES

For a given security, the demand and supply can be represented by the diagram shown in Figure 1.2. Demand is downward sloping, implying more will purchase the security at a lower price, all else being equal. Supply is upward sloping because different owners of a security have different beliefs of the underlying potential for the security and hence different stop-loss or reservation prices, for instance. Of course, additional issues of stock or stock repurchases either dilute or concentrate the value of the original stocks but have no net effect on the value to existing shareholders.

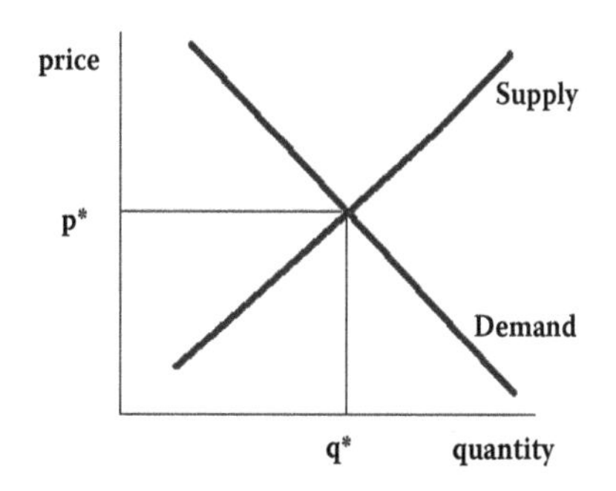

FIGURE 1.2. Demand and supply of a security as a function of security price.

Let us assume an individual (or group of individuals, without loss of generality) has some private information that would in effect increase the demand by an amount x (Figure 1.3). Note that only those inframarginal sellers would be affected along the ray a–b that valued the security above a price p* but below the price p* + x before the information became publicly available. The increased trade activity (in excess of the usual trade activity q*) is a consequence of the insider trading that attempts to profit from information not yet known by the broader market.

Once this information becomes available, the supply of the security is also ratcheted up by the information that has becomes available. The usual level of market activity is restored, and the market is again at equilibrium, as shown in Figure 1.4. In other words, those that held the private information can usurp the entire triangle of the gains *a–b–c* that would normally have been evenly divided by the sellers and buyers both. Ultimately, sellers and buyers should be indifferent and the holders of the private information profit.

These potential gains could actually be quite a bit larger, however. Let us now permit a parallel futures market in the stock. In such a case, there could be multiple puts and calls, resulting in gains (and losses) much larger than the disputed surplus triangle. One may argue, however, that these exchanges are simply transfers of wealth from one willing agent to another. If transactions costs are essentially zero, society would be indifferent to such exchanges because we cannot conclude that the losses of one risk-neutral agent are worse than the gains of another. For this reason, economists typically are unwilling to compare equity or fairness

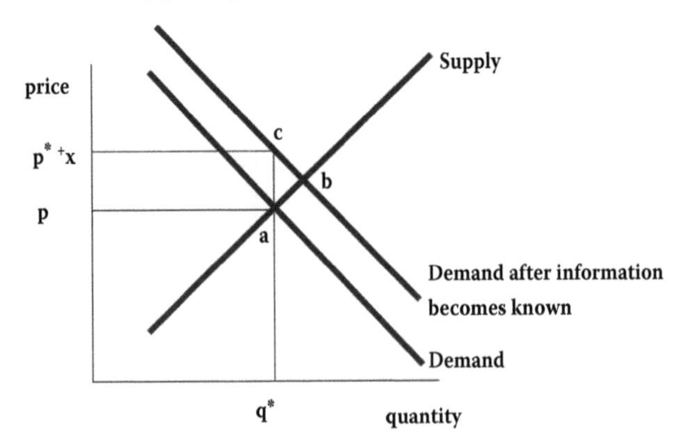

FIGURE 1.3. Demand for the security as the inside traders act on inside information.

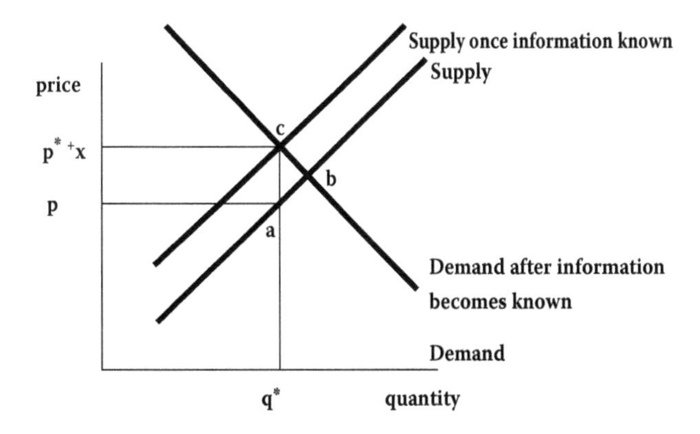

FIGURE 1.4. Equilibrium once the insider information affects both sides of the market.

with regard to the final allocations, and instead confine statements to overall efficiencies.

1.4 THE PURSUIT OF UNPRODUCTIVE ACTIVITIES

Although economists are unwilling to draw equity conclusions, there remain a number of significant and negative consequences of such insider trading. The first and the most obvious is that those who may profit from insider information may indeed use productive resources to garner such information, thereby diverting productive resources from actual value-creating activities. Efficiency is lost if resources are used to try to divide up the existing pie rather than create a larger pie. Equity considerations aside, there is a social loss to this diversion of productive resources to unproductive enterprises.

1.5 SPLITTING THE PIE

The second, more subtle, and perhaps more significant negative consequence also stems from the fact that while new wealth is neither created nor consumed (except if productive resources are used to discover the inside information), existing returns are diverted from the traditional market to the insider market. This depresses the returns in the traditional market and reduces the ability of an economy to mobilize capital. Again, absent any costs to discover insider information, this reduced return is exactly proportional to the share of wealth creation within the insider trading activity vis-à-vis the wealth creation in the entire market. Furthermore, because the insider traders also receive a return as participants in the legitimate market, the return to those participating only in the legitimate market is further diluted.

1.6 THE MARKET FOR LEMONS

There is a third effect that also comes into play. In George Akerlof's seminal paper, "The Market for Lemons," we observe that if insider information were sufficiently widespread, honest traders would be forced out of the marketplace. The strength of the marketplace, and the raison d'être for securities markets, is in the artifact that it creates a market and hence a return on good analysis and good information. Any artifact that frustrates this important function of the marketplace will ultimately hamper the ability to form capital and innovate.

While Akerlof's paper treats the market for used cars, some of which may be defective (lemons), the car market is an analogy to any market in which there is an asymmetry of information between traders. In this case, market participants assume the price of a given security incorporates all known information.

Of course, insider information could be on the upside or the downside, encouraging buys or short sells on the part of those holding the insider information. Let us simply assume that there could be such information in equal proportion. As a consequence, although the market price is about right over time, we know that insiders are capitalizing on profits as the security moves in either direction. The Akerlof analogy in the market for lemons results in a similar conclusion here as would Gresham's law with regard to good versus counterfeit money—the bad drives out the good. Nobody wants to invest in a currency if he believes there is a reasonable chance the bill he has is counterfeit. In the case of insider trading, the greater amount of insider trading there is in a security, the greater uncertainty an investor would have with regard to its true value. As with money, the prudent investor would simply shift her portfolio to an instrument that is more reliable or transparent, even if on average the security prone to insider trading is priced correctly on average over time.

1.7 TOWARD A MORE COMPETITIVE MARKET

Before we leave the impression that information gathering is unhelpful, let us make the distinction that the activity of collecting and profiting from information is an essential element of the marketplace. Indeed, it is the sophistication of this function of information creation and dissemination that creates the maturity of well-developed markets. A well-functioning market will invest in the proper level of information collection and dissemination to the point where the cost on the margin of accumulating or analyzing additional information is just compensated by the increased value of that information in better and more efficient trading.

There even exists a positive externality in this activity. Those who create and profit from this information that is eventually factored into the security price make the entire market function better, even for those who ultimately take the cue from others. This positive externality benefits the market overall. If such rewards were not provided for those who create the good information flows in the market, this (and most other markets in which there exists costly information) tends to degenerate toward the monopolistic solution in which only a few extract such monopoly profits. See, for instance, Read (1994, 1997) for examples of the social welfare consequences and reduced competitive effects of imperfect information.

It is this value of transparent information and information dispersion (and the value such activities create) that creates market value. This is in contrast to the hording of information that underpins insider trading. The former effect results in increased market efficiency and improved equity, whereas the latter effect arising from illegal market timing unambiguously results in decreased market efficiency and inequities.

1.8 PUBLIC POLICY RAMIFICATIONS—THE APPROPRIATE PENALTY FOR INSIDER TRADING

The inefficiencies arising from insider trading are the primary rationale for the regulation of securities markets and the criminalization of insider trading. However, the cost of insider trades, while relatively easy to quantify in theory, is very difficult to quantify in practice because they are necessarily secretive. Insider trading relies on horded information within a close-knit circle of traders who could benefit from the information.

For instance, let us assume the inefficiencies arising from insider trading amount to only 1 percent of the value of the market. Such an inefficiency (or deadweight loss) would actually be considered reasonably low. However, the value of trades of stocks on the New York Stock Exchange can approach $100 billion on some days. This inefficiency can result in a billion dollars of losses on a given day.

Because detection rates are low and the costs so high, fines and penalties for insider trading are likewise high. Let us further assume that only 10 percent of all insider trading is detected. While these trades may then only induce $100 million of inefficiencies on a given day, these same traders may have also participated in the $900 million of inefficiencies that went undetected. As a consequence, fines and sentences would have to be a multiple of damages equal to the reciprocal of the detection rate to sufficiently deter insider trading. In other words, in this numerical example, the deterrence should be 1/10 percent or ten times the estimated inefficiencies.

Note also that the deterrence and fines should be proportional to the damages to market efficiency, not to the profits incurred from the act of insider trading. The relationship of damages to the actual gains is an interesting and open area for future research into this important dimension. Ultimately, while some may observe that the fines and sentences for insider trading seem disproportionately high when compared to the typical gains, they are likely not high once one factors in the relatively low probability of detection and conviction. This effect is further worsened by the likelihood that all who are in a position to observe the insider trading are also in a position to benefit, and are quite possibly part of a conspiracy to obscure the insider trading act. Such a conspiracy of silence, combined with the difficulty to detect insider trading in the first place and the difficulty to convict (beyond a reasonable doubt) in criminal court, further biases upward the appropriate level of damages in such cases.

1.9 CONCLUSION

Insider trading is insidious for a number of reasons. However, although most would argue their objection to insider trading is because it is simply unfair, perhaps the greatest consequence is that insider trading makes the market less efficient. Although there are many other competing avenues for market inefficiency, new financial instruments and insights increasingly allow us to solve these competing inefficiencies. The secretive nature of insider trading makes detection difficult, conviction more difficult, and the huge sums involved difficult to deter.

A simplistic analysis of insider trading might conclude that trading earlier than the market on information merely rewards the early bird. Indeed, it can be shown that such early trading on information will not affect the ultimate market price. However, I establish there are at least four additional consequences of insider trading that damage the market place or reduce the overall level of economic welfare that would otherwise arise from fair trading of securities. These explanations allow us to focus on negative efficiency effects of insider trading even if one does not accept the legitimacy of concerns over equity or fairness.

These conclusions are drawn based on a number of premises. First, insider trading creates riskiness regarding who will benefit from the early information, as opposed to inherent uncertainty that is typically and naturally priced into the security in any regard. Risk aversion aside, the second type of loss arises because the profits from early and private information create an incentive for agents to employ valuable resources in an effort to secure such profits. However, the ultimate value of the security is unchanged and no value is created from capitalizing on insider information. Third, insider trading is a form of seeking to split the pie in this way that results in a lower return for the traditional trader and a higher return for those engaging in the illegal activity. This depresses the rate of return in the securities market and makes it more costly to raise capital. Finally, asymmetric information results in a flight of good money from the market with insider trading and into markets with greater transparency. Again, this flight of capital hinders the ability of the market to mobilize capital for legitimate purposes.

Of course, this corroding effect of insider trading is the primary rationale for the regulation of securities markets and the criminalization of insider trading. However, insider trades are by their very nature necessarily secretive—and hence extremely difficult to detect. Because detection rates are low and the costs so high, fines and penalties for insider trading are likewise high. Indeed, the fines and sentences for insider trading may even seem disproportionately high when compared to the typical gains, but are not high once one factors in the relatively low probability of detection and conviction. This effect is further worsened by the likelihood that all who are in a position to observe the insider trading are also in a position to benefit, and are quite possibly part of a conspiracy to obscure the insider trading act.

Finally, I close with a discussion of the value of transparent information and information dispersion (and the value such activities creates) in contrast to the hording of information that underpins insider trading. The former effect results in increased market efficiency and improved equity while the latter effect of using market timing illegally unambiguously results in decreased market efficiency and inequities.

REFERENCES

Akerlof, G. 1970. The market for lemons: Quality uncertainty and the market mechanism. *Quarterly Journal of Economics* 84(3):488–500.

Read, C. 1994. General equilibrium in a simple economy with imperfect information. *Canadian Journal of Economics* 27(3):393–407.

Read, C. 1997. Development effort in speculative real estate competitions. *Journal of Housing Economics* 6(1):1–16.

CHAPTER 2

Securities Fraud and Its Enforcement

The Case of Martha Stewart

Geraldine Szott Moohr*

CONTENTS

* This chapter is adapted from a previously published article, "What the Martha Stewart Case Tells Us about White Collar Criminal Law," *Houston Law Review* 43 (2006): 591, and is used here with permission of the *Houston Law Review*. Full citations to sources and authorities are provided in the law review article.

THE MARTHA STEWART CASE attracted attention for many reasons. An author, television personality, and entrepreneur who had built a large, profitable company, Stewart ran afoul of the securities laws. The tragedy, as it unfolded over a long investigation and hotly contested trial, was gripping. As it turns out, her story is gripping for another reason: the light it casts on the federal crimes of securities fraud and insider trading and the government policies for enforcing those laws. Insights from the case apply generally, not only to insider trading and securities fraud, but also to other federal white collar crimes. This chapter analyzes those issues by tracing the case against Stewart from the investigation stage through sentencing. We begin by reviewing briefly the statutory framework and the facts of the Stewart case.

2.1 BACKDROP TO THE CASE

2.1.1 The Statutory Framework

Securities fraud is a special type of deceptive conduct that applies to misrepresentations that are made when issuing or trading securities. The type of securities fraud dealt with here is prohibited by the Securities Exchange Act of 1934, which governs trades in secondary markets and sales of already-issued securities. The securities fraud provision, 15 U.S.C. § 78j(b), makes it unlawful to "use or employ, in connection with the purchase or sale of any security ... any manipulative or deceptive device or contrivance" in contravention of rules established by the Securities Exchange Commission (SEC). Congress authorized the SEC to promulgate rules that implement the statutory prohibition. The relevant rule, 17 C.F.R. § 240.10b-5, is hardly more specific. Rule 10b-5 makes it unlawful for any person "to employ any device, scheme, or artifice to defraud," to make "any untrue statement of a material fact," or "to engage in any act which operates as a fraud or deceit ... in connection with the purchase or sale of any security."

Under the criminal provision, 15 U.S.C. § 77ff, the Department of Justice (DOJ) may bring criminal charges. To establish criminal liability, the government must prove, beyond a reasonable doubt, all the elements of the civil fraud and that the defendant acted "willfully."

This statutory framework generally prohibits any deception in the course of buying or selling securities and has been applied to a more specific kind of securities fraud, insider trading. The statutes provide for three types of enforcement. Injured investors may bring private civil suits alleging negligence and civil fraud to obtain compensation for the loss. The SEC may bring administrative civil actions, in which defendants face disgorgement of profits, fines, and orders barring service as officers and directors of public companies. The SEC may also refer cases to the DOJ for criminal investigation. A conviction can result in fines of up to $5,000,000 or imprisonment for up to twenty years, or both.

2.1.2 Summary of Stewart's Case

A few facts will refresh your memory of Stewart's case. Unless noted, the facts were taken from reported decisions: *United States v. Stewart*, 323 F. Supp. 2d 606 (S.D.N.Y. 2004); *United States v. Stewart*, 433 F.3d 273 (2d Cir. 2006); *United States v. Stewart*, 305 F. Supp. 2d 368 (S.D.N.Y. 2004). On December 27, 2001, Stewart's friend Sam Waksal sold his holdings in ImClone, a technology company he had founded and ran. Shortly after Stewart was told about Waksal's sales by her Merrill Lynch broker, she sold all of her shares in ImClone, Inc. On the following day, when ImClone announced that the Food and Drug Administration (FDA) had not approved the company's cancer drug, the value of ImClone shares fell by 18 percent. By selling when she did, Stewart had avoided a loss of approximately $45,000.

Regulators soon suspected that sales of ImClone stock were inside trades, based on information about the firm's failure to obtain FDA approvals, and they began to investigate whether sellers had violated the securities law. During the initial investigation, Stewart was interviewed twice by federal investigators. On both occasions she offered information about her reasons for selling that a jury later found to be untrue.

Eventually, Waksal pleaded guilty to that offense and is now serving a seven-year prison term. Stewart and her broker, Peter Bacanovic, were indicted and tried. Bacanovic's assistant, Douglas Faneuil, testified against them. Stewart was not charged with insider trading, but was accused and acquitted of securities fraud. Stewart and Bacanovic were charged with several other offenses and found guilty of cover-up crimes. Stewart received a ten-month prison sentence. In January 2006, the case officially closed when the Second Circuit rejected her appeals.

2.2 THE INVESTIGATION

When a criminal investigation becomes public, several side effects, or collateral consequences, of the investigative process are likely to occur. Most tellingly, financial and business markets react to news of the investigation—and to leaks regarding its progress. Stewart's company, Omnimedia, felt serious repercussions that were intensified because she was so closely identified with the company's main product lines. Between January 2002, when the investigation began, and March 2004, when the verdict was announced, the company's overall revenues fell 17 percent and earnings from publications fell 68.5 percent. In addition to falling revenue, the strength of her brand, as measured by a consumer loyalty index, dropped from a high of 120 (out of a possible 150) in June 2002 to just 83 by June 2003. As a result of falling share prices and general uncertainty about Stewart's future role, corporate shareholders filed thirteen civil suits against Stewart and her company.

The investigation of her trade had personal and political consequences for Stewart. As the investigation was beginning in February 2002, she was nominated to the board of the New York Stock Exchange. Seven months later, still during the investigation period, she resigned after her broker's assistant, Douglas Faneuil, pleaded guilty and agreed to testify against her. To add insult to injury, her company's shares fell 8.7 percent on the news to $6.21; the high for 2002 had been $20.01 in March. On the political front, the House Energy and Commerce Committee investigated Stewart's ImClone sale, and asked the DOJ to investigate whether Stewart had lied to investigators.

Eighteen months into the investigation, Stewart stepped down as CEO and chairwoman of her company. The SEC filed a civil administrative action, alleging violation of the insider trading laws, to which we shall return. In addition to all of this, the Stewart case and Stewart herself were subject to intense public scrutiny throughout the eighteen-month investigation. Stewart probably exacerbated the media attention by posting her side of the story on her Web site, an act said to be akin to poking a prosecutor in the eye.

What does this account tell us about the enforcement of securities law? Note first that federal officials were in charge of this criminal matter: administrative agency personnel from the SEC, agents from the Federal Bureau of Investigation, and lawyers from the DOJ New York office. The securities laws are largely a matter of federal criminal law. There is significant negative commentary on proliferation of federal crimes and the tendency of members of Congress to show they are tough on crime and to support new criminal laws as an election tactic. Nevertheless, there is little controversy about treating securities crimes as a federal offense. Congress has constitutional jurisdiction under the Commerce Clause to enact laws that regulate capital markets, and the federal interest in protecting the national securities markets is robust.

Even though Stewart's transaction was known to authorities almost immediately, an entire year and a half passed between the initial inquiry and her indictment. This fact illustrates a second characteristic: investigations of white collar crimes can take a long time. Lengthy investigations are due, at least to some extent, from the private and secret nature of many fraud offenses. Hampered by the absence of witnesses, investigators painstakingly follow paper trails.

Third, during such a lengthy period, an investigation can set off harmful collateral consequences, such as those experienced by Stewart and her company. These consequences are not, strictly speaking, legal in nature. Rather, most are "extra legal"—they do not result from a legal determination, but from market and social forces that are stimulated by an investigation. Investors sell, shareholders sue, clients withdraw, advertisers bail, and targeted firms soon encounter financial difficulties, phenomena best illustrated by the Arthur Andersen case. By the time that accounting firm was tried, it was only a shadow of its former self.

In addition to the financial consequences illustrated by Stewart's case, a social stigma attaches to the subject of an investigation, and brings suspicion of colleagues and friends as well as of the government, with consequent loss of respect and status. Thus, the stigma of conviction begins long before a finding of guilt. Such serious collateral consequences should, at least in theory, deter criminal behavior, especially among respectable middle-class business people who are presumed to care about what their neighbors think of them, their status in the broader community, and the effect on their careers and businesses. Although the stigma may be justified upon conviction, it is important to realize that the side effects of a criminal investigation apply to the innocent as well as to the guilty, and it is often impossible to recover from them.

2.3 THE INDICTMENT

The pertinent point here is that Stewart was indicted. The decision to indict—or not—rests in the discretion of the federal prosecutor in charge of the case. Nominally, under the Constitution, the grand jury investigating the case decides whether a criminal charge is warranted. But a prosecutor has great influence over the grand jury, and in practice the decision to indict rests with the government, not its citizens.

Stewart's celebrity status and the media attention that was given to the investigation posed an initial dilemma that argued for and against indictment, and several commentators debated whether she was targeted for reasons other than her conduct. But her notoriety made it very difficult for prosecutors not to charge Stewart and, at the same time, provided an incentive for them to charge her. If the government had not indicted Stewart, prosecutors would be seen as giving the rich and famous a break. Failing to indict would also indicate that the laws at issue were not worth enforcing. On the other hand, if she was charged, prosecutors would be seen as treating all offenders equally and all laws seriously.

Martha Stewart's celebrity status also provided the government with a bonus. When someone rich and famous is charged with a crime, the case attracts greater media attention than cases against ordinary defendants. Simply stated, media attention increases the deterrent effect of a case. Indicting a celebrity sends a message that everyone pays attention to because the message attaches to someone who is famous. What message was sent in the Martha Stewart case?

2.4 THE CHARGES

The message sent by a criminal case depends on the offenses that are charged, a decision that is also in the discretion of the prosecutor. Stewart was charged with lying to investigators, obstructing justice, conspiracy, and one count of securities fraud. Thus, the messages here are that law-abiding citizens must be honest with investigators and absolutely truthful with shareholders. Rather stunningly, Stewart was not charged with insider trading. What do the charges—and the noncharge—tell us about the securities laws?

2.4.1 The Absence of an Insider Trading Charge

Why wasn't Stewart indicted for the crime of insider trading? Although we can only speculate, we know that the prosecutor handling the case referred to an insider trading charge as "unprecedented." According to the original indictment, Stewart sold her ImClone stock after she was told that Waksal was trying to sell the ImClone stock in his Merrill Lynch account. Thus, she had traded while in possession of nonpublic information that was arguably material to an ImClone investor. Why would indicting Stewart for insider trading be unprecedented?

The Supreme Court, in three cases decided over seventeen years, endorsed insider trading applications that had been developed by lower courts. Under the first case, *Chiarella v. United States*, 445 U.S. 222 (1980), only insiders who have a fiduciary obligation to those with whom they trade violate the statute when they buy or sell stock of their corporation. Insiders who trade on the basis of material nonpublic information deceive the buyer or seller by failing to disclose the information to those with whom they trade. This classic theory protects investors who buy as well as those who sell and counts parties such as lawyers and accountants as insiders. But Stewart was not an insider who owed fiduciary duties to ImClone shareholders, and could not be charged under the classic theory of insider trading.

In the second case, *Dirks v. SEC*, 463 U.S. 646 (1983), the Court endorsed liability of tippers and tippees, but insisted on a relationship between the tipper and investors in the tipper's firm. Thus, the insider/tipper breached a fiduciary duty, in order to obtain some benefit, and the tippee must have known those facts. At first appraisal, this theory seems to fit because Stewart was a tippee who had learned information from Bacanovic. But neither Bacanovic nor his assistant owed fiduciary duties to ImClone shareholders.

Both classic insider trading and tippee liability apply only when the actor has a fiduciary duty to investors in the firm. This produced a gap in the law because an outsider—one who does not have such a duty—remained free to trade in the stock. In the third and most recent case, *United States v. O'Hagan*, 521 U.S. 642 (1997), the Court accepted the misappropriation theory of insider trading that closed this gap. An individual, even an outsider who does not owe fiduciary duties to investors, may not trade if that person makes unauthorized use of information to trade in securities. A prerequisite is that the person used the information in breach of a confidential relationship owed to the source of the information. Again, Stewart could not be charged. She had not betrayed any confidential relationship to the source of her information, in this case her broker or his employer, Merrill Lynch.

It could be argued, however, that the broker had misappropriated information that "belonged" to Merrill Lynch, thus betraying a confidential relationship with, and obligations to, the firm. The "unprecedented" issue was whether Stewart had committed insider trading when she traded on information that had been misappropriated by someone else. In other words, she was the tippee of a misappropriator. An insider trading charge on this theory combines two related but distinct doctrines: tippee liability, a species of classic insider trading; and misappropriation liability.

There is profound disagreement about whether this conduct violates insider trading law and, if it is construed to do so, whether it should be adopted. For instance, Professor Jeanne Schroeder argues that Bacanovic was not a misappropriator under *O'Hagan* because he merely breached a contract with Merrill Lynch. Further, the information about Waksal's sale was not the property of Merrill Lynch. In contrast, Professor Kelly Strader sees no impediment under Supreme Court precedents to tippee liability when the tipper is a misappropriator. The SEC apparently accepts this view, as the civil complaint is based on the theory.

Two federal courts have thus far accepted the theory, but with reservations. In *United States v. Falcone*, 97 F. Supp. 2d 297 (E.D.N.Y. 2000), the trial court affirmed a conviction of a tippee who traded on misappropriated information. The court nevertheless expressed "real reluctance," "dismay," and concern with the "boundless expansion of the misappropriation theory." In a civil case, *SEC v. Yun*, 327 F.3d 1263 (11th Cir. 2003), an appellate court rejected the SEC position that the standard in classic tippee cases was not relevant when the tipper had misappropriated information.

Thus, the prosecutor had good reasons for declining to charge Stewart with insider trading. It was reasonable to avoid an expansive charge in such a high-profile case. Second, significant issues regarding Stewart's culpability may have led to a finding of not guilty, even had the trial court accepted this new type of insider trading. Third, incorporating tippee liability into misappropriation doctrine clearly expands misappropriation theory, and the prosecutor might have had reservations about a criminal charge when the defendant had no notice that this specific conduct was a crime. Finally, the prosecutor may have hesitated to establish a new cause of action in a criminal case.

Regulatory agencies like the SEC often test new causes of action in civil cases, slowly developing precedents on which future criminal cases may rely. The agency pursued the tippee/misappropriation theory in its civil enforcement action against Stewart. Stewart settled the SEC suit, neither admitting nor denying any wrongdoing. The terms of the settlement forbid Stewart from serving as a company executive or director of her own or any public company for five years, and she was fined $195,000. Stewart is still listed on company publications as "founder and editorial director," and has yet to assume her former positions at the company.

The SEC enforcement action against Stewart tells us that civil regulatory actions and criminal prosecutions often overlap. This is not unique to securities law, and almost every regulatory regime has an omnibus criminal provision. It is typically a simple statement that any willful violation of the civil law shall be subject to criminal penalties. Environmental laws have criminal kickers, so do antitrust offenses, so do labor laws, so do food and drug offenses. This overlap means that the same conduct may give rise to civil or criminal charges, or, as in Stewart's case, both. The only distinctions between civil and criminal liability in many statutes are the defendant's felonious intent and the prosecution's burden to prove all elements of the crime beyond a reasonable doubt. In short, if the government believes it can prove that the defendant acted with criminal intent—as defined in the relevant statute—a civil violation can be treated as a crime.

The convergence of civil and criminal law has two troubling aspects. First, treating the same conduct as the basis for civil liability and for criminal punishment tends to blur the distinction between the standards for each cause of action. The insider trading laws are a good example of this tendency. There is no definition of the culpability element of "willful" in the criminal statute. In this vacuum, courts have applied standards that are strikingly similar to the civil standard to criminal cases. The merger of civil and criminal standards in securities fraud cases means that the only distinction between civil and criminal liability is the standard of proof in a criminal case, beyond a reasonable doubt. Finally, in some cases, very harsh administrative penalties raise constitutional due process issues and confuse the standards of civil liability and punishment.

The dual remedies, civil and criminal, may also lead to an expansion of liability. Professor Lawrence Solan's analysis shows that expansive judicial interpretations are more likely when a government agency is charged with civil enforcement. In an effort to satisfy the mandate of a civil remedy for harm suffered by individual investors, courts broadly interpret statutory terms. When criminal courts use these standards, they expand the scope of the criminal provision.

The second troubling aspect of the overlap is the possibility of parallel civil and criminal proceedings against defendants, or the threat of them. In the Stewart case, the civil and criminal cases are not strictly parallel. The SEC took on insider trading and the DOJ focused on Stewart's other conduct. Nevertheless, Stewart faced three proceedings that emanated from her infamous trade: civil suits by her shareholders, regulatory action by the SEC, and criminal action by the DOJ. A three-front defense effort is very challenging.

Commercial practitioners must be very careful not to implicate clients in the criminal matter when representing clients who may have violated civil provisions. Providing information to administrative authorities or civil parties may result in a forfeiture of attorney–client and work-product privileges in the criminal case. And the threat of criminal charges obviously strengthens the government's position in negotiations over the civil matter. Yet parallel or multiple proceedings are not unusual in insider trading, securities fraud, or other white collar crimes. Indeed, many of those indicted in Enron-related criminal actions were also the subject of parallel SEC proceedings.

2.4.2 The Securities Fraud Charge

The government passed on a criminal charge of insider trading. But it was not so cautious when it indicted Stewart for another form of securities fraud. The reasons for not indicting Stewart for insider trading seem to apply equally to this charge. It too is an untested theory that posed a risk to prosecutors. On the other hand, because of alleged investor losses of $400 million, the securities fraud charge exposed Stewart to a ten-year prison term.

Thus, the decision to charge securities fraud illustrates first the power of the prosecutor to decide what charges to file. Stewart was charged with fraudulently misleading investors in Omnimedia—not ImClone investors, but shareholders of her company. In June 2002, after news reports disclosed Stewart's sale and Waksal's arrest, Stewart made public statements aimed at her company's shareholders. Among other things, she denied any wrongdoing and explained her reason for selling: the price of ImClone had fallen to $60 per share, a price that she and Bacanovic had previously decided would trigger selling her stock in the company. Prosecutors charged that Stewart's statements had affirmatively misrepresented material facts in order to manipulate the price of Omnimedia stock. The government's implicit theory was that Stewart was motivated to do so because she was the majority shareholder, was concerned about the company's share value, and acted to protect her own wealth. According to the government, her deceptive statements fraudulently misled investors and affected their decisions to buy or sell shares of her company.

Commentators almost immediately identified several problems with this unusual—many said "novel"—charge. A former SEC enforcement lawyer stated he was not aware of any other case where the government alleged a manipulation of stock prices by claiming innocence. The reason for surprise was that her conduct fell outside the heartland of securities fraud. Fraud requires more than deception; at a minimum, the deception must be material. Stewart's statements were arguably not material because they referred to a personal transaction that did not concern the business of or at Omnimedia. In addition, the statements were no different from press reports about her reasons for selling, so investors would already have known about Stewart's explanation. Finally, securities law requires that deception is connected to the purchase or sale of a security. Here, the connection between Stewart's statements and an investor's decision to buy shares of Omnimedia as particularly attenuated.

In addition, the charge is contrary to principles that are basic to our criminal justice system and seem to interfere with her right to defend herself. Because a basic tenet of American law is that a person is innocent until proven guilty, our intuition is that everyone has a right to say "I am innocent." It is perverse to charge someone with a crime for exercising that right because the charge implicitly rejects the basic premise of innocence. One might argue that Stewart had a right to deny guilt, but did not have a right to add false information. The key information she provided was apparently true, however, because the jury did not find Stewart had lied about the prearranged decision to sell. The securities fraud charge also raised First Amendment concerns relating to free speech. Her lawyers argued these and other points in a motion to dismiss, which the trial court rejected.

The securities charge tells us first that Congress prefers to write federal criminal laws with broad, undefined terms that can be applied to a wide range of conduct. The securities fraud charge was based on the same statute that bars insider trading—a person may not willfully make material, untrue statements that would defraud investors. But neither the statute nor the pertinent administrative rule defines key terms, such as "deceptive device," "willfully," and "material," or explains how tight a link is required between a deception and the ultimate transaction. Instead, Congress—purposefully or inadvertently—delegated the task of defining these terms to the courts. One result is that some crimes continue to evolve, as the judicial treatment of insider trading illustrates. Another result of continual interpretation is that definitions and standards can vary circuit by circuit. Appellate courts also revisit precedents and reframe definitions. Thus, the law of securities fraud does not articulate a clear standard.

Broad, open-ended laws like this may be so vague that they offend the Constitution. The Constitution requires that statutes provide notice to citizens and guidelines to prosecutors, judges, and juries about the conduct that has been prohibited. The constitutional standard for vagueness, articulated in *Connally v. General Construction Co.*, 269 U.S. 385 (1926), is whether citizens "of common intelligence must necessarily guess at its meaning and differ as to its application." The standard is particularly pertinent in white collar crimes such as those involving securities because not all unethical or immoral business conduct is criminal. As the insider trading allegation against Stewart illustrates, the line between acting in one's self-interest or the interest of a firm is not clearly distinct from criminal conduct.

Stewart invoked the vagueness doctrine to argue that she had no notice that general statements regarding her innocence could violate the securities statute. This seems like a reasonable position. An intelligent reading of the statute would not have disclosed that the conduct she was about to undertake was a crime, and even experts disagreed about the charge. In Stewart's case, whether the government's expansive interpretation was indeed encompassed by the statute was not known until the judge refused to dismiss the charge. Had she been indicted for insider trading, she would have made the same arguments.

The vagueness doctrine also requires that the written law provide enforcement standards to guide the decisions of prosecutors, judges, and juries. Legislators must provide guidelines that prevent executive and judicial branch officials from pursuing their own predilections with regard to enforcement and conviction. Prosecutors need to know when and what to charge, judges need to advise jurors about the law, and juries need to measure factually proven conduct against a legal standard that they understand. Providing such standards prevents arbitrary enforcement and leads to consistent, uniform application, properties now missing from the crimes of securities fraud and insider trading.

There is also a policy reason to avoid vague criminal laws. Vague laws may not effectively deter criminal conduct. When the law is unclear, persons who are considering some action may not realize that they are in danger of violating criminal laws. In those circumstances, people do not stop to weigh the benefit of the conduct against the risk of being caught and punished. The point is particularly relevant in the white collar context where conduct is often based on ethical lapses, betrayals of trust, and deceptions that are not always crimes.

Vague laws can be interpreted so that the statute applies to facts of a given case. Over time, the interaction among the regulatory agency, executive branch prosecutors, and the courts expands the meanings of undefined terms, and the law becomes ever broader. The expansion process typically begins, as here, when the prosecution presses for a new application of the statute. In criminal cases, courts tend to defer to the mandate of the executive branch to enforce the laws, and so they often entertain the charge and adopt the new meaning advocated by prosecutors. By rejecting Stewart's motion to dismiss this charge, the trial judge accepted the government theory—a defensive statement about a personal transaction can operate as a fraud on investors. Thus, a new application of the statute, a new theory of criminality, a new form of fraud, was established.

Finally, vague laws increase the discretion of prosecutors in several ways. First, as the securities fraud charge against Stewart illustrates, vague laws enhance the authority of prosecutors to choose what charges to bring. Indeed, some commentators argue persuasively that the federal criminal law has become a menu of possible charges from which prosecutors may choose. Second, vague laws allow prosecutors to urge applications of the law to conduct that the law did not initially seek to prevent. In this way, prosecutors, members of the executive branch, with the cooperation of the judiciary, assume a legislative role. In an adversarial system such as ours, based on the rule of law, a democratically elected legislature is responsible for defining criminal conduct. In the criminal realm, the justification for punishment emanates first from statute, rather than from the defendant's conduct or the enforcer's values. Finally, vague laws also expand the authority of prosecutors to negotiate plea bargains and cooperation agreements, an authority, which we take up in a moment, that has disadvantages as well as benefits.

Ironically, Stewart was ultimately acquitted of securities fraud. The trial court held that the government had not produced enough evidence for the charge to go to the jury. No reasonable juror could have found her guilty beyond a reasonable doubt without undue speculation that she had acted willfully. The decision, based on the court's evaluation of the evidence, is not subject to appellate review. Nevertheless, a precedent has been established: protesting one's innocence may operate as a fraud.

2.4.3 The Cover-Up Charges

As we know, Stewart maintained that she had sold her ImClone stock because its price had fallen to her predetermined $60 limit. Based on these and other statements, Stewart was charged with making false statements to investigators, obstructing justice, and conspiring to commit those offenses. Aptly referred to as "cover-up" crimes, these free-standing, independent offenses are not related to the target offense under investigation.

As here, rather than relating directly to either insider trading or securities fraud, cover-up crimes originate in the investigation, as defendants seek to avoid charges on the target offense. Cover-up crimes encompass a wide range of actions and include lies and other acts involving evidence of and witnesses to the crime under investigation. They are often easier to prove than the underlying crime, and some observers believe there is an increasing tendency to use such charges. A cover-up offense may be brought even when the target offense, such as insider trading or securities fraud, is not a crime or could not be proven.

Cover-up crimes are another example of the breadth of federal statutes. The false statement statute prohibits "knowingly and willfully" saying something that is not true to virtually any government official. Unlike a perjury charge, this crime does not require that the statement be made under oath or in formal circumstances. As Justice Ginsburg has pointed out, the statute gives extraordinary authority to the executive branch.

Similarly, the obstruction statutes are also broadly written to capture a wide range of conduct. The omnibus clauses of the obstruction statutes authorize criminal punishment when a person "corruptly … influences, obstructs, or impedes, or endeavors to influence, obstruct, or impede" justice. The term that defines culpability, "corruptly," has no uniform meaning and has been interpreted broadly. The Supreme Court has issued two opinions that slightly restrict application of obstruction statutes. In the *Andersen* case, the Court held that prosecutors must show the defendant was conscious of wrongdoing, and in *Aguilar*, that a nexus exists between the conduct and the proceeding at issue (*Arthur Andersen v. United States*, 544 U.S. 696, 2005; *United States v. Aguilar*, 515 U.S. 593, 1995). Congress enacted new obstruction provisions, which Professor Julie O'Sullivan has roundly critiqued as confusing and unnecessary.

Second, and the point I want to emphasize here, cover-up charges against Stewart illustrate the depth of the federal criminal code. The code reportedly contains at least 100 false statement statutes and more than 325 fraud statutes. This depth gives prosecutors the ability to pick and choose among various offenses and allows charges of more than one crime for the same conduct. In Stewart's case, she sought to conceal the circumstances of her sale of ImClone shares. Although she made false and misleading statements, she engaged in, as it were, a single scheme to deceive investigators. That single scheme gave rise to three sets of charges: false statement, obstruction, and conspiracy. Overlapping crimes like false statements and perjury—by their nature and almost automatically—also obstruct the due administration of justice. Thus, her deceptive statements to federal investigators resulted in charges of separate crimes. These offenses also suggest conspiracy when another person is involved. On the theory that Stewart and Bacanovic had agreed to violate the false statement and obstruction statutes, they were each charged with conspiracy.

The conspiracy statute adds an extra layer of depth. Co-conspirators are liable for the actions of their cohorts when that action furthers the conspiracy, even when a defendant was not present and did not engage in the wrongful conduct. A 1946 case, *Pinkerton v. United States*, 328 U.S. 640 (1946), established that a coconspirator may be convicted of offenses committed by a cohort in the conspiracy as long as those acts are in furtherance of the conspiracy, fall within the scope of the agreement, and could reasonably be foreseen as a necessary or natural consequence of the agreement. Thus, Stewart was also charged with and found guilty of her broker's independent perjury in testimony before the SEC.

An additional aspect of the depth of federal criminal law is that a single lie can lead to multiple counts of each offense, a practice referred to as "piling-on." Every time a lie is repeated during an investigation, a separate count of the crime may be charged. If a misrepresentation of the truth is repeated in two meetings, prosecutors can charge two counts of false statements and two counts of obstruction. If the statement is made a third time, prosecutors may add another count of each crime. In Stewart's case, prosecutors charged two false statement counts with eight misstatements making up one count and three making up the other. She could have faced eleven charges of making false statements and eleven charges of obstructing justice.

"Piling-on" does not necessarily aid in truth-finding during trial. Jurors who assume that "where there is smoke, there must be fire" are likely to reach compromise verdicts, finding defendants guilty of some but not all of the charges. In addition, multiple counts are used to leverage plea agreements, as prosecutors "give up" some counts in return for the plea. Although the sentencing guidelines have mitigated the effect of multiple counts on the length of the prison term, the practice exposes the limited due process rights of defendants during the plea bargaining stages of a criminal matter.

Multiple counts of overlapping crimes would seem to violate the Double Jeopardy Clause of the Constitution, which protects people from being tried twice for the same crime. However, the restrictive test for double jeopardy claims focuses on the elements of the statutes at issue. Under

Blockburger v. United States, 284 U.S. 299 (1932), double jeopardy does not exist if the elements of one crime require proof of a fact that the second crime does not. In Stewart's case, the elements of false statement and obstruction are different. The false statement charge requires that the defendant made a false statement; obstruction does not. Obstruction requires that the defendant, at a minimum, endeavored to influence, obstruct, or impede the proper administration of the law; false statement does not. Thus, Stewart was not being tried twice for the same crime.

A trial or plea on cover-up charges obscures the substantive crime that is under investigation. Using cover-up charges allows the government to avoid proving that the defendant committed the crime that gave rise to the investigation. Because the defendant is not charged with the substantive charges that gave rise to the investigation, the government need never explain why the conduct that was "covered up" was a crime. The public is not well served by this. We are left, as in Stewart's case, wondering whether a court would agree with the SEC that trading on a tip of misappropriated information is a crime.

Committing obstruction, false statement, and perjury are serious matters. Without them, investigative efforts necessary to uncover and prosecute white collar crimes would be significantly hampered. Lying or withholding information makes it more difficult for the government to enforce our laws; cover-up offenses serve a bona fide function, to deter actors from misleading investigators. Such actions harm the criminal justice system and thus society generally.

Having said that, the threats to due process rights of defendants and the long-term harm to the criminal justice system are also serious matters. Over time, using such charges may offend the community's sense of fairness and lessen respect for the entire criminal enforcement system. The use of cover-up charges betrays a win-at-any-cost attitude on the part of the government that can devalue the criminal justice system in the eyes of the public. As an editorial in the Fort Wayne *News-Sentinel* put it, "In the Martha Stewart case, doesn't it look like the prosecutors are angry because they can't prove insider trading, and are being vindictive in pursuing these other charges?" Enforcement decisions like those in the Stewart case can cause citizens to lose respect for the criminal justice system, which would have a pernicious effect on informal peer-to-peer enforcement and law-abiding behavior.

These long-term effects raise the question of whether there is some middle ground—some constraint on using overlapping charges and multiple counts—that would preserve the integrity of government investigations without impugning the rights of defendants and devaluing the criminal justice system. Two professors, Michael Seigel and Christopher Slobogin, suggest that all the counts that deal with the same conduct or transgression be merged before trial and plea bargaining discussions; Professor Ellen Podgor urges that guilt of false statement requires proof that the statement was material.

2.5 THE TRIAL

In some senses, Stewart's trial was a typical illustration of the enforcement of white collar crimes. In another sense, Stewart's trial was not at all typical: she went to trial.

2.5.1 Typical Aspects of Stewart's Trial

Because of the conspiracy charges, Stewart was tried jointly with her broker, Peter Bacanovic. The strategy of a joint trial shows that the trial strategies applied in federal drug cases have migrated to white collar cases. Following the shocking conduct at some of the nation's most well-known business firms, the executive branch put money and personnel into punishing the "bad apples" who disgraced themselves and American businesses. Stewart's case shows that prosecutors are now as committed to winning white collar cases as they are to winning cases in the war on drugs.

Stewart's trial also illuminates the crime of conspiracy. The substantive law, in Justice Jackson's words, is an "elastic, sprawling and pervasive offense ... that is so vague that it almost defies definition" (*Krulewitch v. United States*, 336 U.S. 440, 1949). On a procedural level, a joint trial on conspiracy charges offers the government several procedural advantages that disadvantage defendants. Rules regarding joinder, venue, and statutes of limitations give the government an edge before the trial even begins. The advantages continue throughout the trial. For instance, testimony that recounts statements made by a third party is usually not allowed because the defendant cannot cross-examine the absent third party. But this rule does not apply when the testifying witness is an admitted co-conspirator. Thus, Faneuil was able to testify about statements Bacanovic made in conversations with him. Moreover, it is very difficult for a jury to view defendants as individual actors when they are linked together in the courtroom, and testimony as to one defendant can "stick to" a codefendant. As mentioned earlier, individuals may be guilty of the acts of co-conspirators. In Stewart's case, she was found guilty of the perjury committed by Bacanovic.

In trying Stewart, the government relied on Faneuil as a cooperating witness who testified for the government. A person who "cooperates" has pleaded guilty to a (usually) lesser charge and agreed, in return, to help the government in its investigation and subsequent trials of other individuals. In this case, Faneuil pleaded guilty to a misdemeanor charge of withholding information from authorities: see *United States v. Faneuil,* 02-CR-1287 (S.D. N.Y. Oct. 2, 2002) (Misdemeanor Information charging a violation of 18 U.S.C. § 830). After testifying against Stewart and Bacanovic, he was sentenced to a $2,000 fine.

The use of a cooperating witness raises a troubling issue, even though the person provides crucial evidence for the government. Consider that Faneuil is almost as liable as Stewart or Bacanovic on the trading issue and the subsequent cover-up. By his own testimony, he was aware of the insider trading issue at the time of the trade and yet he agreed to and facilitated the sale. He supported Stewart and Bacanovic's story for six months. The cynical trade of small fish for big fish that privileges guilty parties is not lost on the public. For those who believe in the moral basis of criminal law, allowing an offender like Faneuil to escape punishment is far from satisfying. Those who justify punishment because it deters others may question the efficacy of the tactic if it leads to widespread cynicism about criminal enforcement.

2.5.2 The Atypical Aspect of Stewart's Trial

By one index Stewart's case is unique: she did not plead guilty, but actually went to trial. In the federal criminal justice system, around 95 percent of defendants plead guilty. Several factors converge to explain why so many white collar defendants plead. Some defendants are contrite and willing to accept punishment. Others are willing to forgo trial in hopes of a lesser or certain sentence. Like Faneuil, some defendants can trade information about others for the probability of a lesser sentence.

Another reason that white collar defendants plead guilty relates to a characteristic of crimes like insider trading and securities fraud. Most white collar cases involve some kind of organizational entity, usually the defendant's workplace. Under federal case law, firms may be found guilty of the conduct of employees who acted within the scope of their authority and on behalf of the firm. Recall that the Arthur Andersen firm was essentially dissolved by the time the trial began. The message was loud and clear, and firms are now anxious to avoid indictment and the collateral consequences that flow from an official investigation.

The threat of indictment and trial encourages firms not only to plead guilty but also to cooperate with the government in prosecuting their employees. Firms may abandon employees by firing them, by refusing to pay their attorney's fees, and by providing the government with suggestive and incriminating evidence. In the Stewart case, Merrill Lynch suspended Bacanovic and Faneuil and eventually dismissed them. Actions such as these enhance a firm's chances of avoiding indictment and securing a deferred prosecution agreement. Often DOJ policies, which include expecting firms to waive attorney–client and work-product privileges, meant that firms, in the words of a former prosecutor, were to "cooperate with the government in convicting their former employees." Former employees are caught between their former firms, who may release incriminating documents and refuse to pay legal fees, and prosecutors, who threaten them with prison terms. Many opt for a plea agreement. The DOJ policy regarding waiver has recently been altered, and time will tell whether that formal response can undo the informal expectations that have been developed.

What does plea bargaining tell us about the enforcement of white collar crimes? Is it a good thing? It cannot be a good thing if individuals who plead are not guilty of a crime. Yet a rational defendant may accept a plea to a lesser charge—and a certain prison term—rather than risk trial on a more serious charge, even if the person is innocent. Under current sentencing policy, prosecutors may ask the judge to depart from a guideline sentence when defendants "substantially assist" the government. Neither the judge nor the defense attorney may initiate this departure from the guideline sentence. A defendant who declines to plead guilty and cooperate thus risks alienating a person who can influence the length of the prison term.

Plea bargains of those who are in fact guilty may be an expeditious and efficient way of settling the matter, saving taxpayers the expense of trial. But of course there are other considerations. One is that fewer cases go to trial. In giving up public adjudication, we also incur a loss, although it is not immediately obvious. Trials create a public record and educate the public, and I would suggest they are critical to deterrence. At trial, details of the conduct are fully revealed in a convincing way that explains what happened and why it was a crime. Everyone understands after a trial what conduct constituted the crime and that it was wrong, whereas information contained only in an indictment and plea bargain are not generally known and do not have this effect.

Trials also affirm shared social values and reinforce law-abiding behavior of people who do not "skate too close to the wind" in order to benefit themselves. This is the kind of behavior that we want to encourage. We do not want people to go around thinking, "I'm a chump because I obeyed the law." We want business executives to see that people who fail to obey the law are tried and punished. A trial accomplishes this in a visceral way that is more effective than a plea bargain, even one that includes headlines and perp walks. As a result of Stewart's trial and conviction, citizens understand that they are obliged not to lie to federal investigators.

2.6 THE SENTENCE

Federal prison sentences depend on the prison term provided in the statute and on the Federal Sentencing Guidelines. The guidelines were devised, in part, to constrain judicial discretion in the interests of providing uniform and certain sentences. Overall, punishment of federal white collar offenses became harsher and more certain after 1989, when the guidelines were adopted. As noted, prosecutors may request a downward departure from a guideline sentence, a privilege that does not extend to defense attorneys or even to the judge.

Stewart was sentenced as a first offender convicted of nonviolent crimes that did not cause financial harm. She received a five-month prison term, a five-month term of house arrest, two years of probation, and a $30,000 fine. The ten-month term of incarceration was split between prison and home confinement, and Stewart spent five months in a federal prison. Bacanovic received the same term of incarceration, and a $4,000 fine. As previously mentioned, Faneuil avoided a prison sentence. The sentences in this case tell us several things about the enforcement of crimes like insider trading and securities fraud.

First, federal prison terms for similar conduct can be inconsistent. Faneuil, at a minimum, conspired in the trade and the subsequent cover-up. Yet he was indicted on a lesser charge and, on the prosecutor's recommendation that he receive "extra" leniency for his cooperation, avoided both a prison term and probation. One reason some defendants serve less time in prison than others for the same offense is that prosecutors can readily evade and influence guideline sentences. They may choose the offenses that are charged, negotiate plea and cooperation agreements, and exercise their authority to request downward departures.

Inconsistent sentences are not a good thing. First, they are contrary to the purpose of modern sentencing reform, which is to ensure uniform sentences. They also undermine a major reason for subjecting individuals to criminal punishment in the first place. Inconsistent sentences impair the deterrent function of criminal law because people are more likely to desist from crime when sentences are more certain. Finally, inconsistent sentences raise troubling fairness issues and erode citizens' respect for the criminal justice system.

2.7 NEW DIRECTIONS

A few additional points that are not directly suggested by Stewart's case bear mentioning. Responding to public outrage over corporate misconduct in 2002, Congress enacted the Sarbanes–Oxley Act. Among other provisions, the Act increases prison terms for several white collar crimes by as much as 400 percent; in frauds involving pensions, penalties were increased 1,000 percent, from one to ten years. Criminal penalties for securities fraud, which includes insider trading, now include fines of up to $5,000,000, imprisonment for up to twenty years, or both. This is a significant increase; previous penalties had rested for many years at maximum penalties of fines of $1,000,000 and ten years in prison.

Congress apparently believed that long prison terms would deter potential offenders. Evidence suggests, however, that long prison terms do not increase deterrence—even for white collar criminals who presumably plan and consider the risk of criminal behavior. Armed with this knowledge, it would be wise to consider shorter sentences. The money saved by shorter sentences could be used to strengthen another source of deterrence, the likelihood that white collar offenders will be apprehended.

Congress also added an independent securities fraud statute to the federal criminal code which is broader than the existing provision. The new provision, 18 U.S.C. § 1348, prohibits the knowing execution of (or attempt to execute) "a scheme or artifice to defraud any person in connection with any security" of a registered or reporting company. The maximum prison term for violating this provision is twenty-five years. In passing this provision, Congress intended to give the DOJ greater enforcement flexibility. Not only does it include attempts, but it also applies to frauds "in connection with any security" rather than those that occur in the course of sales or purchases, and requires only that defendants acted knowingly, a lower standard than willful conduct.

At the same time, Congress has restricted use of civil complaints of securities fraud that may be brought by private investors. The statutory scheme allows injured investors to bring private civil suits alleging negligence and civil fraud. In private suits, guilty defendants pay damages and disgorge profits. The Private Securities Litigation Reform Act of 1995, however, generally handicapped the use of private lawsuits by making it more difficult to bring class action suits, limiting joint and several liability, expanding safe harbors for certain company statements, and toughening pleading requirements for fraud. In a very recent case, *Tellabs, Inc. v. Makor Issues & Rights, Ltd.*, No. 06-484, 2007 WL 1773208 (U.S. June 21, 2007), the Supreme Court endorsed a rigorous standard for private plaintiffs alleging securities fraud, holding that they must present "cogent and compelling" inferences of deceitful intent in order to survive defendants' motions to dismiss the case. In addition, in 1993 the Court had reduced the incentive to bring private suits by reducing potential damages.

The combined effect of these developments is more reliance on government enforcement, either through the SEC civil administrative actions or through criminal charges. Restricting the private cause of action while expanding the criminal sends a puzzling and inconsistent message to the business community. If private investors cannot effectively bring actions in tort, it seems odd, to say the least, to be subject to criminal penalties for the same conduct.

2.8 CONCLUSION

The Martha Stewart case exposes features of insider trading and securities fraud that were once apparent only to those who closely followed the case law. Insider trading law continues to evolve and seems to be in a constant state of development. Securities fraud is in similar flux, with little agreement about the type of conduct and culpability that constitute a crime. The case also illustrates significant issues about the substance of these crimes, namely, the consequences of their vagueness and breadth and the merger of civil and criminal standards.

Stewart's case also reveals enforcement issues that cause concern: reliance on cover-up charges, collateral consequences of investigations, parallel and multiple proceedings, use of cooperating witnesses, plea bargains, and sentencing policies. From lawmaking through sentencing, the enormous discretion given federal prosecutors stands out. In sum, the Martha Stewart case discloses problematic issues that damage the integrity of federal criminal laws and their enforcement.

CHAPTER 3

An Economic and Ethical Look at Insider Trading

Robert W. McGee

CONTENTS

3.1 INTRODUCTION

Insider trading, the trading on nonpublic information, has been going on for hundreds of years. Aquinas (1225–1274) spoke about it in the thirteenth century (Aquinas n.d.; McGee 1990). He stated that there is no inherent moral duty for a newly arrived grain merchant to tell the people in the city of his arrival that there are other grain merchants one day behind him. Failure to disclose this information means he will be able to fetch a higher price for his grain but there is nothing morally wrong about keeping the information to himself. In other words, there is nothing inherently wrong about profiting from nonpublic information. The fact that the grain merchant did little or nothing to acquire that information is apparently irrelevant.

But some commentators would disagree with the view that keeping inside information private is morally acceptable. Some authors have concluded that insider trading is always or almost always unethical (Ferber 1970; Werhane 1989, 1991) while other authors have reached the opposite conclusion (Machan 1996; Manne 1966a, 1966b, 1970). Some authors would decide on the ethics of insider trading based on whether the result is the greatest good for the greatest number (Bentham 1997; Mill 1993).

The two main sets of ethical principles that have been applied to the analysis of insider trading ethics have been utilitarianism and rights theory. Insider trading studies have focused on a few major issues, such as fairness, the level playing field argument, overall welfare, fiduciary duty, misappropriation, and property rights in information.

This chapter does not go into a detailed analysis of all those studies. That work has been done elsewhere (Bainbridge 2003). The purpose of the present study is to construct a framework that can be applied to analyze any insider trading issue. The two main ethical systems that have been used to address the ethics of insider trading are utilitarian ethics and rights theory. This chapter applies both of these ethical theories to determine when, and under what circumstances, insider trading is ethical and when it is not.

3.2 REVIEW OF THE LITERATURE

Manne's early work (1966a, 1966b) was published before many empirical studies of market efficiency were conducted. Thus, he was not able to cite empirical research to support his *a priori* arguments. Manne's work is commonly regarded to be the classic work in the field, as evidenced by the numerous citations it has received in the finance, legal, and business ethics literature. The main thrust of his argument is that we should not view insider trading from the perspective of ethics. He takes the position that insider trading is good because it increases economic efficiency. It causes prices to move in the correct direction quicker than would be the case in the absence of insider trading. His argument is utilitarian in nature.

Another Manne argument is that insider trading can be viewed as a form of executive compensation. If corporate officers are permitted to trade on inside information, there will be less pressure to pay them big salaries. Paying smaller salaries is good for the corporation because it reduces its compensation costs. What is good for the corporation is good for shareholders.

This argument is good as far as it goes. But not all insiders are corporate officers. Some insiders may be clients or customers of the corporation. Allowing these groups to trade on inside information may be good business at times. But allowing other noninsiders to trade on inside information may be to the detriment of the corporation. Or maybe not.

Hartikainen and Torstila (2004) surveyed 230 finance professionals and solicited their views on various ethical issues in finance, including the ethics of insider trading. They analyzed results to determine if there were differences in ethical attitudes based on gender or age. Young (1985) discussed insider trading from the perspective of benefits and harms, which is utilitarian based. His article also discusses market efficiencies that result from insider trading, which is also a utilitarian argument.

Leland (1992) also takes a utilitarian approach. He used a rational expectations model and concluded that, where insider trading is permitted: (1) stock prices better reflect information and will be higher on average, (2) expected real investment will rise, (3) markets are less liquid, (4) owners of investment projects and insiders will benefit, and (5) outside investors and liquidity traders will be hurt. He then goes on to discuss whether total welfare increases or decreases, an approach that has often been taken in the literature (Carlton and Fischel 1983; Dennert 1991; Fishman and Hagerty 1992; Fried 1998; Leland 1992; Manne 1966a, 1966b; Medrano and Vives 2004).

Chakravarty and McConnell (1999) analyzed the trading activity of Ivan Boesky, a convicted inside trader, and found that their tests were not able to distinguish the price effect of Boesky's informed purchases from the effect of noninsider purchases. Their goal was to determine whether insider trading moves stock prices. Various other studies have also attempted to determine the effect that insider trading has on stock prices and whether it causes them to move in the right direction sooner than would otherwise be the case (Chakravarty and McConnell 1997; Cornell and Sirri 1992; Meulbroek 1992), which is a utilitarian approach, since it deals with measuring market efficiency. Other studies have also examined insider trading to determine its effect on market efficiency (Seyhun 1986).

Bernardo (2001) takes a utilitarian approach when he analyzes the welfare effects of permitting firms to contractually negotiate the right to allow corporate insiders to trade shares in the firm on private information. In his study he examines the informational efficiency of stock prices and the welfare of all investor types. He also investigates the effectiveness of various compensation schemes on mitigating conflicts of interest between shareholders and managers. He concludes that shareholders generally choose not to allow managers to engage in insider trading and that this decision is socially optimal.

O'Hara (2001) looks at the legality, ethics, and efficiency of insider trading. He dismisses the view that all traders are entitled to equal information and states that there is no real evidence to allow us to conclude that legalizing insider trading would lead to increased instability in markets. Bhattacharya and Nicodano (2001) did a welfare analysis of insider trading, investment, and liquidity and concluded that insider trading is beneficial, which is a utilitarian approach.

Bhattacharya and Daouk (2002) studied 103 countries that have stock markets and found that the cost of equity does not change after the introduction of insider trading laws but decreases significantly after the first insider trading prosecution. Jeng, Metrick, and Zeckhauser (2003) measured gains and losses to various groups and found that (1) insider purchases earn abnormal returns of more than 6 percent, (2) insider sales do not earn significant abnormal returns, and (3) expected costs of insider trading to noninsiders are about 10¢ for a $10,000 transaction.

Kara and Denning (1998) concluded that the U.S. securities market did not fit the strong form of the efficient market hypothesis, which made room for insider trading profits. Bushman, Piotroski, and Smith (2005) looked at the relationship between insider trading restrictions and the degree of interest analysts had in following stocks. They found that interest increased in cases where there were restrictions on insider trading and that the effect of restrictions was more pronounced in emerging economies.

The accounting literature also addresses insider trading from an ethical perspective (Keenan 2000a, 2000b; Ronen 2000; Walker 2000; Williams 2000). However, the discussion is mostly limited to a determination of whether insider trading increases or decreases efficiency. They mention Manne's so-called ethical nihilism on the issue, not realizing that Manne's argument is based on utilitarian ethics. What appears to be nihilistic is actually nothing more than a cold analysis of winners and losers.

The Boatright book (1999) devotes part of a chapter to insider trading. The basic thrust of his argument centers on breaking laws, fairness, or breaching fiduciary duties. He also discusses inside information as property, the level playing field argument, and the O'Hagan case.

Bear and Maldonado-Bear (1994) devote a few pages to the discussion of insider trading. Almost the entire discussion is from a legal perspective. However, some mention is made of efficiency. Property rights are discussed only briefly and only one subset of property rights, the property rights of a trader who obtains information through hard work.

Padilla (2002) looks at agency theory to determine whether the regulation of insider trading is justified. He looks to the separation of ownership and control topic that Berle and Means discussed a few generations ago (Berle and Means 1932). Berle and Means also discussed property rights in their book, a topic that is not discussed as often as utilitarian theory. Padilla is one of the few authors to discuss property and contract rights in conjunction with insider trading, which places him in the same category as McGee (1988), McGee and Block (1992), and McGee and Yoon (1998).

3.3 UTILITARIAN ETHICS

The vast majority of economists are utilitarians. The legal system of the United States and of many other countries also has a strain of utilitarianism running through it. The general welfare clause of the U.S. Constitution is only one of many examples that may be given. Thus, any complete discussion of insider trading must at least discuss the utilitarian perspective.

Basically, utilitarian ethics classifies an act as good if the result is the greatest good for the greatest number (Bentham 1997; Mill 1993) or if the majority benefits (Goodin 1995; Quinton 1988; Scarre 1996). Economists would say that an economic act or policy is good if the result is a positive-sum game. Another variation of this approach is that something is ethical if it is efficient. The American jurist Richard Posner has been an advocate of this approach (1983, 115, 205; 1998).

Not all ethicists agree with the ethics as efficiency argument. Egger (1979) comes down quite firmly against it, arguing that efficiency is not a substitute for ethics. Hoppe (1993, 195–201) argues that only the institution of a strong property rights regime can lead to maximizing efficiency and, thus, that economic efficiency can be justified, but on the basis of property rights rather than utilitarianism.

There are several problems with the utilitarian perspective. Part of the problem is that there are several utilitarian perspectives. Sometimes the greatest good and the greatest number do not go together. Situations can exist where the result is the greatest good but only a few benefit, whereas in other cases the vast majority might benefit but the result is not the greatest good.

But there are more fundamental flaws with the utilitarian ethics approach. One structural problem is that it is impossible to accurately measure gains and losses (McGee 1994; Rothbard 1970). At best, one may only estimate. That does not stop economists from trying, of course, but obtaining accurate calculations of gains and losses is impossible if for no other reason than the inconvenient fact that different individuals have different preferences. Also, individuals make economic decisions based on ranking choices, not by calculating relative benefits. If some consumer has to decide between competing hamburger chains, one might say that he prefers McDonald's over Burger King, but one might not say that he prefers McDonald's over Burger King by 13.7 percent.

Another insurmountable flaw in the utilitarian ethics approach is that it totally ignores rights. For a utilitarian, rights are not important. If the result is a positive-sum game, that is all that counts. The fact that grandma's ancestral home must be confiscated so that developers can use the land to build a shopping mall that would employ hundreds of people is all that counts.

Manne (1966a, 1966b, 1970) was one of the first scholars to apply utilitarian ethics to insider trading but he is not the only one. Most of the scholars cited in the literature review section also applied utilitarian ethics to at least a certain extent, although others relied mostly on emotional arguments (Ferber 1970; Werhane 1989, 1991), which is rather anti-intellectual. However, emotional arguments do influence courts and legislators, so emotional arguments cannot be totally dismissed because they have had an effect on how insider trading law has evolved. But there is no need to discuss or examine emotional arguments in any detail because such arguments have no logical or intellectual basis to stand on.

Economic studies have had mixed results when it comes to measuring gains and losses from insider trading. Some studies have concluded that insider trading causes the market to become more efficient while other studies have found that insider trading has negative effects on markets. At times, it is not possible to determine whether a particular inside trade has a positive or negative effect on the market until long after the trade, which makes good policy making difficult, if not impossible.

Part of the problem with any utilitarian economic analysis of insider trading is that it is difficult to come up with the numbers. It also is not always possible to determine which groups are affected and by how much. Manne (1966a) goes so far as to say that it is impossible to identify a single individual who loses as a result of insider trading. That is because stock trades are conducted anonymously through the stock exchanges. The person buying shares does not know who the seller is, and vice versa. The person who bought from an insider trading would have purchased the shares anyway, so she is no worse off after the trade than she would have been if she had bought from someone who is not an inside trader.

Then there is the philosophical problem of how to deal with people who do not trade on insider information. This happens all the time when an insider merely sits on the shares that he already owns when he has information that suggests the stock price is about to climb. No one has suggested that individuals should be punished for not trading.

The utilitarian approach to analyzing insider trading can be summed up in the following flowchart shown in Figure 3.1 (McGee 2007).

Next we examine an alternative to utilitarian ethics.

3.4 RIGHTS THEORY

The two main problems with utilitarian ethics are the inability to measure gains and losses and the total disregard of rights. One advantage of the rights approach is that there is never any need to measure gains and losses. All that needs to be done is to determine whether anyone's rights have been violated. If they have, then the act is automatically unethical. Rights trump majorities and positive-sum games.

The difference between the rights approach and the utilitarian approach may be highlighted by the following example. Let's say that two wolves and one sheep vote on what's for dinner. A utilitarian would conclude that voting to eat the sheep would be an ethical result because there are two winners and only one loser. A rights theorist would conclude that the act

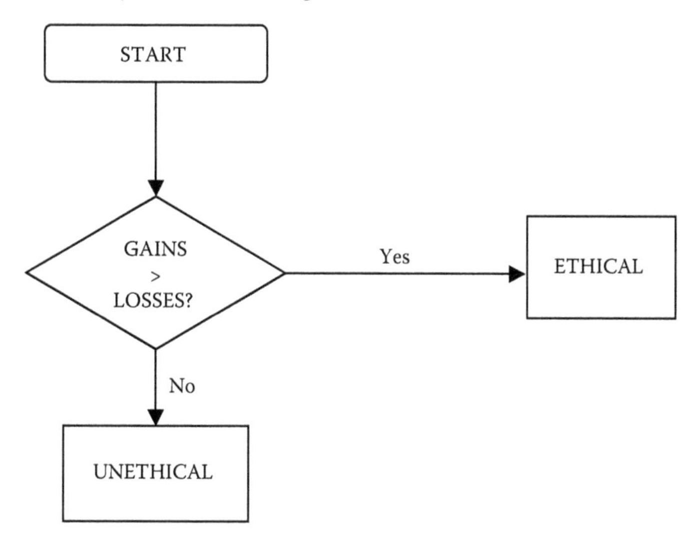

FIGURE 3.1. Utilitarian ethics.

of voting in this case would be unethical because the rights of the sheep would be violated if the wolves were allowed to carry out their intent.

Here is another example, taken from medical ethics. Let's say that three terminally ill patients are sitting in the hospital cafeteria pondering their fate when a young, healthy looking man walks by their table. Because the three terminally ill patients are also philosophers, they see this healthy individual differently than would a group of nonphilosophers. They start plotting to kidnap him and harvest his organs. One would take his heart. Another would take his lungs. The third would take his kidneys. The young guy would die, of course, but his death would make it possible for three philosophers to live.

A utilitarian would conclude that their plotting and subsequent act was perfectly ethical, since the result was the greatest good for the greatest number, a positive-sum game. A rights theorist would disagree, pointing out that in order to carry out their scheme they would have to violate the rights of the young guy.

The flowchart in Figure 3.2 summarizes the rights-based position (McGee 2007). The flowchart points out that violating someone's rights makes the act unethical automatically. However, acts that do not violate anyone's rights are not necessarily ethical. They may be ethical but it is

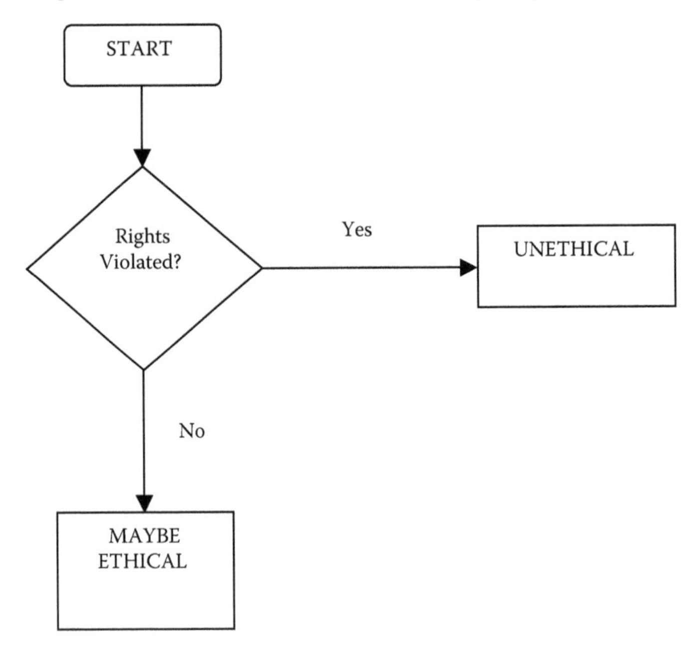

FIGURE 3.2. Rights-based ethics.

not a foregone conclusion. Whether non-rights-violating activity is ethical depends on the act. Having sex with dead animals, as is the ritual in some religions, may not be ethical, but it does not violate anyone's rights. Dead animals do not have rights, and neither do live animals, in the opinion of the author. Paying for sex or for illegal drugs may or may not be an ethical act but neither of these acts violates anyone's rights. Whether engaging in victimless crimes is ethical or unethical cannot be determined on the basis of rights theory. So rights theory can be used to determine whether some acts are ethical but it cannot be used to determine whether non-rights-violating activities are ethical.

Applying rights theory to insider trading makes it unnecessary to calculate gains and losses or identify which groups or individuals are affected by an insider trade. All that need be done is determine whether anyone's property or contract rights are violated. If rights are violated, then the act of trading on insider information is unethical. If no one's property or contract rights are violated, then probably there is nothing ethically improper about the trade. There is no need to talk about level playing fields, asymmetric information, or "fairness," whatever that is. Breaches of fiduciary duty violate contract rights, and so are unethical. Trading on information that is owned by someone else is unethical if the trader obtained the information without the owner's permission.

One problem that might be encountered when applying the rights approach to insider trading is trying to determine who owns the property in question. The rights approach works best when property rights are clearly defined, which is not always the case for insider trading.

3.5 CONCLUDING COMMENTS

The finance literature on insider trading uses a variety of methodological approaches. Manne (1966a, 1966b) had to use *a priori* reasoning because there were not many empirical studies of insider trading at the time he wrote his classic book on the topic. Since Manne, a number of empirical as well as *a priori* studies have been made of insider trading, something Manne was not able to do. Some of those studies have confirmed Manne's initial thesis: that insider trading causes capital markets to operate more efficiently.

The finance and business ethics literature that discusses ethical issues does so mostly from the perspective of utilitarian ethics—positive-sum and negative-sum games, gains or losses to shareholders or to some stakeholder community, and so forth. However, as has been shown above, utilitarian ethics has basic flaws that cannot be eradicated by rigor. One needs to apply another ethical system—rights theory. But rights theory operates best when rights are clearly defined, which is not always the case for insider trading. What needs to be done is more clearly define who has rights in information.

This chapter has discussed the two main approaches to examining issues related to insider trading but it does not examine all aspects of insider trading. Much of that has been covered previously (Bainbridge 2003). However, a few words can be said about some of the various subtopics that relate to insider trading.

One argument that has been used to criticize insider trading is the fairness argument. This argument is sufficiently squishy that one can massage it in practically any direction without fear of criticism. Who can criticize the advocacy of fairness?

The problem with the fairness argument is that ten people can have twelve different opinions about what constitutes fairness. One underlying premise of the fairness argument is often envy. There is the belief that rich people should not be able to make huge amounts of money with apparently little effort. It just isn't fair. But who is being treated unfairly?

The little guy, who is on the short end of asymmetric information, is usually identified as the one being treated unfairly. But, as Manne points out, it is difficult or impossible to actually find these little guys who have been treated unfairly because the anonymous mechanism of market trades produces the same result whether they bought their shares from an insider or a noninsider. Besides, many of these so-called little guys, blue collar workers or lower or mid-level white collar workers, own their shares through pension funds, which are usually big institutional investors that sometimes have access to nonpublic information.

Another problem with the fairness argument is fairness to whom? Is it fair to force hardworking, knowledgeable traders to disgorge information that they have spent many hours to obtain just so some less worthy individuals can have access to the same information that they have? Is it fair to prevent them from trading on information they have legally obtained? I think not. But this side of the fairness argument is seldom analyzed.

An argument that is a close cousin of the fairness argument is the level playing field argument. The main problem with the level playing field argument is that it is being misapplied. A valid application would be to sporting events. Both teams in a competition should have identical rules to abide by. Basketball teams whose name ends in a vowel should not be able to take extra foul shots. Football teams that have more than two Puerto

Ricans should not be able run downhill for the entire game. Teams should have to change sides of the field every quarter or every half.

Injustice results when the level playing field argument is applied to business situations. Those who have obtained property rights in information should not be forced to give that information to people who have done nothing to earn it. It is an abuse of the legal process. It is also bad policy. If individuals cannot benefit by trading on nonpublic information, they will not bother to invest the time and energy to uncover such information, which would cause the market to work less efficiently, to the detriment of the general public.

Regulating publishers of financial newsletters violates their right to free speech and press. If some government agency prohibits them from saying whatever they want, to whomever they want, it automatically violates their rights. It is improper to apply some balancing test in such cases because balancing tests can only be justly applied where *interests* diverge. Interests can conflict but rights cannot. Rights never need to be balanced. Besides, real rights, in the negative sense of that term, can never conflict. My right to free speech does not violate your right to free speech. My right to property does not conflict with your right to property.

Part of the problem with the current state of insider trading law and regulation is that property rights are not clearly defined. The misappropriation of property is a violation of property rights. The problem is that property rights are not yet clearly defined in the case of insider trading. Furthermore, Congress abdicated its responsibility by deliberately failing to define exactly what constitutes insider trading in the insider trading law it passed two decades ago, hoping that courts would define it for Congress. In the meantime, individuals do not always know whether engaging in certain kinds of activity will cause them to have legal problems down the road. As a result, they do not act as aggressively as they otherwise would, which makes markets operate less efficiently.

Ideally, insider trading laws should clearly define what constitutes acceptable and unacceptable conduct, so that everyone knows what the law is. The law should also clearly define property rights and allow insider trading that does not violate property rights, while punishing insider trading that does violate property rights.

Actually, there may not be any need for any insider trading law. We already have contract laws to protect contract rights. We already have laws to protect property rights. All that really needs to be done is to more clearly define property rights in information. There is no need to have special laws on insider trading. Having good laws that protect contract and property rights would be sufficient.

REFERENCES

Aquinas, T. n.d. *Summa theologica*, II-II, Q.77, Art. 4.

Bainbridge, S. M. 2003. The law and economics of insider trading: A comprehensive primer. www.ssrn.com.

Bear, L. A., and R. Maldonado-Bear. 1994. *Free Markets, finance, ethics, and law.* Upper Saddle River, NJ: Prentice-Hall.

Bentham, J. 1987. Anarchical fallacies: Being an examination of the Declaration of Rights issued during the French Revolution. In *Nonsense upon stilts: Bentham, Burke and Marx on the rights of man,* ed. J. Waldron, 46–76. London: Methuen.

Berle, A. A., and G. C. Means. 1932. *The modern corporation and private property.* New York: Macmillan.

Bernardo, A. E. 2001. Contractual restrictions on insider trading: A welfare analysis. *Economic Theory* 18:7–35.

Bhattacharya, S., and G. Nicodano. 2001. Insider trading, investment, and liquidity: A welfare analysis. *Journal of Finance* 56(3):1141–56.

Bhattacharya, U., and H. Daouk. 2002. The world price of insider trading. *Journal of Finance* 57(1):75–108.

Boatright, J. R. 1999. *Ethics in finance.* Malden, MA: Blackwell Publishers.

Bushman, R. M., J. D. Piotroski, and A. J. Smith. 2005. Insider trading restrictions and analysts' incentives to follow firms. *Journal of Finance* 60(1):35–66.

Carlton, D. W., and D. R. Fischel. 1983. The regulation of insider trading. *Stanford Law Review* 35:857–95.

Chakravarty, S., and J. J. McConnell. 1997. An analysis of prices, bid/ask spreads, and bid and ask depths surrounding Ivan Boesky's illegal trading in Carnation stock. *Financial Management* 26:18–34.

Chakravarty, S., and J. J. McConnell. 1999. Does insider trading really move stock prices? *Journal of Financial and Quantitative Analysis* 34(2):191–209.

Cornell, B., and E. R. Sirri. 1992. The reaction of investors and stock prices to insider trading. *Journal of Finance* 47(3):1031–59.

Dennert, J. 1991. Insider trading. *KYKLOS* 44(2):181–202.

Egger, J. B. 1979. Comment: Efficiency is not a substitute for ethics. In *Time, uncertainty and disequilibrium,* ed. M. J. Rizzo. Lexington, MA: Lexington Books.

Ferber, D. 1970. The case against insider trading: A response to Professor Manne. *Vanderbilt Law Review* 23:621–27.

Fishman, M. J., and K. M. Hagerty. 1992. Insider trading and the efficiency of stock prices. *RAND Journal of Economics* 23(1):106–22.

Fried, J. M. 1998. Reducing the profitability of corporate insider trading through pre-trading disclosure. *Southern California Law Review* 71:302–92.

Goodin, R. E. 1995. *Utilitarianism as a public philosophy.* Cambridge: Cambridge University Press.

Hartikainen, O., and S. Torstila. 2004. Job-related ethical judgment in the finance profession. *Journal of Applied Finance* 14(1):62–76.

Hoppe, H. H. 1993. *The economics and ethics of private property.* Boston: Kluwer Academic Publishers.

Jeng, L. A., A. Metrick, and R. Zeckhauser. 2003. Estimating the returns to insider trading: A performance-evaluation perspective. *Review of Economics and Statistics* 85(2):453–71.

Kara, A., and K. C. Denning. 1998. A model and empirical test of the strong form efficiency of U.S. capital markets: More evidence of insider trading profitability. *Applied Financial Economics* 8:211–20.

Keenan, M. G. 2000a. Insider trading, market efficiency, business ethics and external regulation. *Critical Perspectives on Accounting* 11:71–96.

Keenan, M. G. 2000b. Inefficiency, immorality and insider trading: A reply to my critics. *Critical Perspectives on Accounting* 11:123–28.

Leland, H. E. 1992. Insider trading: Should it be prohibited? *Journal of Political Economy* 100(4):859–87.

Machan, T. R. 1996. What is morally right with insider trading? *Public Affairs Quarterly* 10:135–42. Reprinted in T. R. Machan and J. E. Chesher, *A primer on business ethics.* Little Falls, NJ: Rowman & Littlefield, 2003.

Manne, H. G. 1966a. *Insider trading and the stock market.* New York: Free Press.

Manne, H. G. 1966b. In defense of insider trading. *Harvard Business Review* 44(6):113–22.

Manne, H. G. 1970. A rejoinder to Mr. Ferber. *Vanderbilt Law Review* 23:627–30.

McGee, R. W. 1988. Insider trading: An economic and philosophical analysis. *Mid-Atlantic Journal of Business* 25(1):35–48.

McGee, R. W. 1990. Thomas Aquinas: A pioneer in the field of law and economics. *Western State University Law Review* 18(2):471–83. www.ssrn.com.

McGee, R. W. 1994. The fatal flaw in NAFTA, GATT and all other trade agreements. *Northwestern Journal of International Law & Business* 14(3):549–65.

McGee, R. W. 2007. A flow chart approach to analyzing the ethics of insider trading. Andreas School of Business Working Paper, Barry University.

McGee, R. W., and W. E. Block. 1992. Insider trading. In *Business ethics & common sense,* ed. Robert W. McGee, 219–29. Westport, CT: Quorum Books.

McGee, R. W., and Y. Yoon. 1998. Insider trading and finance law after O'Hagan: A discussion from the perspectives of law, economics & ethics. *Journal of Accounting, Ethics & Public Policy* 1(3):400–13.

Medrano, L. A., and X. Vives. 2004. Regulating insider trading when investment matters. *Review of Finance* 8:199–277.

Meulbroek, L. A. 1992. An empirical analysis of illegal insider trading. *Journal of Finance* 47(5):1661–99.

Mill, J. S. 1993. *On liberty and utilitarianism.* New York: Bantam Books.

O'Hara, P. A. 2001. Insider trading in financial markets: Legality, ethics, efficiency. *International Journal of Social Economics* 28(10, 11, 12):1046–62.

Padilla, A. 2002. Can agency theory justify the regulation of insider trading? *Quarterly Journal of Austrian Economics* 5(1):3–38.

Posner, R. A. 1983. *The economics of justice*. Cambridge, MA: Harvard University Press.

Posner, R. A. 1998. *Economic analysis of law*. 5th ed. New York: Aspen Law & Business.

Quinton, A. 1988. *Utilitarian ethics*. LaSalle, IL: Open Court.

Ronen, J. 2000. Insider trading regulation in an efficient market, a contradiction. *Critical Perspectives on Accounting* 11:97–103.

Rothbard, M. N. 1970. *Man, economy and state*. Los Angeles: Nash.

Scarre, G. 1996. *Utilitarianism*. London: Routledge.

Seyhun, H. N. 1986. Insiders' profits, costs of trading, and market efficiency. *Journal of Financial Economics* 16(2):189–212.

Walker, M. 2000. Civil law versus criminal law remedies for insider trading: The case for the plaintiff. *Critical Perspectives on Accounting* 11:105–10.

Werhane, P. H. 1989. The ethics of insider trading. *Journal of Business Ethics* 8:841–45.

Werhane, P. H. 1991. The indefensibility of insider trading. *Journal of Business Ethics* 10:729–31.

Williams, P. F. 2000. Loosening the bonds: A comment on "Insider trading, market efficiency, business ethics and external regulation." *Critical Perspectives on Accounting* 11:111–21.

Young, S. D. 1985. Insider trading: Why the concern? *Journal of Accounting, Auditing & Finance* 8(3):178–91.

CHAPTER 4

Martha Stewart

Insider Trader?

Joan MacLeod Heminway

CONTENTS

4.1 INTRODUCTION

In the spring of 2002, Martha Stewart learned that she was under investigation for insider trading in connection with her December 2001 sale of shares of common stock of ImClone Systems Incorporated. On October 8, 2004, Stewart went to jail to serve out a five-month sentence for crimes committed in connection with the investigation—but not for insider trading. Instead, she served time for obstructing justice and making false statements to government officials in connection with the insider trading investigation. Although the Securities and Exchange Commission (SEC) did pursue Stewart for insider trading in a civil case, the parties settled that case before trial. So, despite public opinion to the contrary, Stewart, a prominent publishing and media executive of her own public company (Martha Stewart Living Omnimedia, Inc.), has never been found guilty of or liable for insider trading in connection with her ImClone stock sale.

Yet, the public is justifiably confused. Most who know the story believe that Stewart sold her ImClone shares while in possession of nonpublic information that had the potential to move the market in ImClone shares. This type of unfairness in stock trading commonly is thought to constitute insider trading; those who have an inside track to market-sensitive information should not be able to trade on it.

The applicable law is more complicated than that, however. U.S. insider trading prohibitions are not grounded in unfairness. Rather, they are grounded in the breach of a duty, and violations are dependent on the satisfaction of numerous elements. The breach of duty requirement and these elements sometimes present significant roadblocks to recovery for plaintiffs and prosecutors. It is unclear, for example, whether Stewart's stock sale actually did constitute a violation of U.S. insider trading law.

This chapter looks at that question. In the pages that follow, the basic law of insider trading is described and applied to the facts about Stewart's stock sale, to the extent that the public now knows them. This analysis shows that it ultimately is difficult (albeit not impossible) to label Stewart an insider trader under existing U.S. law.

4.2 INSIDER TRADING LAW IN THE UNITED STATES

Insider trading is a violation of Section 10(b) of the Securities Exchange Act of 1934, as amended (known as the 1934 Act), and Rule 10b-5, which was adopted by the SEC under and in accordance with Section 10(b). Because Section 10(b) and Rule 10b-5 leave significant room for interpretation, the law of insider trading in the United States is largely judge-made law. In general, to be civilly liable for or criminally guilty of insider trading, a person trading in securities must either (1) have and breach a duty to someone by (a) trading while in possession of nonpublic information or (b) divulging information to someone who trades or (2) trade while in possession of nonpublic information that was obtained from someone who breached a known duty to someone else. Moreover, the possessed nonpublic information must be "material" (a word that has a legally specified meaning), and the violator must act with a required state of mind, known as scienter. Accordingly, the mere fact that someone trades on the basis of information that the public does not have is insufficient to establish an insider trading violation. Although we might deem that trading unfair, fairness is not the policy basis underlying federal insider trading regulation.

4.2.1 The Principal Statute and Rule

The principal U.S. law of insider trading comes from judicial interpretations of Section 10(b) and Rule 10b-5. In relevant part, Section 10(b) prohibits a person from using "in connection with the purchase or sale of any security … any manipulative or deceptive device or contrivance in contravention of such rules and regulations as the Commission may prescribe." (15 U.S.C. § 78j(b)) Rule 10b-5, adopted by the SEC under and in accordance with Section 10(b), makes it unlawful for "any person … [t]o employ any device, scheme, or artifice to defraud, [or] … to engage in any act, practice, or course of business which operates or would operate as a fraud or deceit upon any person, in connection with the purchase or sale of any security." (17 C.F.R. § 240.10b-5)

4.2.2 First Key Case—*Chiarella*

Three Supreme Court cases frame the existing doctrine in this area. The first case, *Chiarella v. United States*, 445 U.S. 222 (1980), establishes the threshold importance of duty in establishing an insider trading violation under Section 10(b) and Rule 10b-5. This case involves an intelligent and resourceful employee at a legal and financial printing firm who correctly guessed the identity of tender offer and merger target companies and bought stock in those target companies in the public trading markets before the transactions were announced. He then sold that stock after the tender offer or merger was announced and made a substantial profit. He made no public announcement of the information he had amassed before making his purchases or sales.

The Court finds that the printing firm employee did not violate Section 10(b) and Rule 10b-5 in making his stock purchase because he had no duty to publicly disclose the information that he had pieced together before using it for his own private benefit. "When an allegation of fraud is based upon nondisclosure," the Court states, "there can be no fraud absent a duty to speak. We hold that a duty to disclose under § 10(b) does not arise from the mere possession of nonpublic market information." (445 U.S. at 235) More specifically, the Court notes that

> "the element required to make silence fraudulent—a duty to disclose—is absent in this case. No duty could arise from petitioner's relationship with the sellers of the target company's securities, for petitioner had no prior dealings with them. He was not their agent, he was not a fiduciary, he was not a person in whom the sellers had placed their trust and confidence. He was, in fact, a complete stranger who dealt with the sellers only through impersonal market transactions." (445 U.S. 232-33)

The form of primary insider trading liability described and construed in *Chiarella* (i.e., where an insider—a person with a duty of trust and confidence—trades while in possession of material nonpublic information) is termed "classic" or "classical." Under this liability theory, insiders have a duty to publicly reveal the material nonpublic facts in their possession or refrain from trading in the company's securities. This is known as the duty to disclose or abstain.

4.2.3 Second Key Case—*Dirks*

In footnote 12 of the *Chiarella* case, the Court describes the potential for insider trading liability arising out of a trade made by someone who receives nonpublic information ("tippee") from an insider ("tipper").

> "Tippees" of corporate insiders have been held liable under § 10(b) because they have a duty not to profit from the use of inside information that they know is confidential and know or should know came from a corporate insider.... The tippee's obligation has been viewed as arising from his role as a participant after the fact in the insider's breach of a fiduciary duty. (445 U.S. at 230)

A subsequent Supreme Court case, *Dirks v. Securities and Exchange Commission*, 463 U.S. 646 (1983), analyzes in detail the liability of a person for insider trading in this tipper/tippee context. In that case, Raymond Dirks, an officer of a stock brokerage firm, was tipped off about possible fraud at a company; fraud that he was able to verify after he conducted a private investigation that included interviews with company officers and employees. (The original tip came from a former officer of the company.) Dirks then shared the information he had obtained with his clients who owned the company's stock and with the *Wall Street Journal* (in the hopes that it would publicize the fraud more widely, which it declined to do), but not to the public at large. Dirks's clients sold the company's stock before the fraud was, in fact, revealed, and therefore were able to sell at a higher price.

Reaffirming and supplementing its holding in *Chiarella*, the Court fails to find Dirks liable for violating the insider trading prohibitions of Section 10(b) and Rule 10b-5 on these facts. The Court first takes pains to confirm that it is a fiduciary relationship, and not mere access to or possession of material nonpublic information, that gives rise to a duty to disclose or abstain. Accordingly, the court finds that a tippee—a person who receives information from an insider—does not assume the insider's duty to disclose that information or abstain from trading automatically. Rather, the tippee assumes that duty only under certain circumstances.

> [T]ippees must assume an insider's duty to the shareholders not because they receive inside information, but rather because it has been made available to them *improperly*.... Thus, a tippee assumes a fiduciary duty to the shareholders of a corporation not to trade on material nonpublic information only when the insider has breached his fiduciary duty to the shareholders by disclosing the information to the tippee and the tippee knows or should know that there has been a breach. (463 U.S. at 660)

The court goes on to explain how to determine the existence of a breach by the insider, since (as the Court notes) "[a]ll disclosures of confidential corporate information are not inconsistent with the duty insiders owe to shareholders." (463 U.S. at 661–62)

Specifically, the *Dirks* Court establishes a "personal benefit test" as a means of determining the existence of a breach of duty by an insider who shares material nonpublic information with a noninsider.

> Whether disclosure is a breach of duty … depends in large part on the purpose of the disclosure. This standard was identified by the SEC itself …: a purpose of the securities laws was to eliminate "use of inside information for personal advantage." Thus, the test is whether the insider personally will benefit, directly or indirectly, from his disclosure. Absent some personal gain, there has been no breach of duty to stockholders. And absent a breach by the insider, there is no derivative breach. (463 U.S. at 662)

Lest we misunderstand, the *Dirks* Court clarifies that the personal advantage or benefit obtained by the tipper need not be pecuniary, but rather may be reputational. Moreover, the court notes that objective evidence may give rise to a presumption of personal benefit.

> For example, there may be a relationship between the insider and the recipient that suggests a *quid pro quo* from the latter, or an intention to benefit the particular recipient. The elements of fiduciary duty and exploitation of nonpublic information also exist when an insider makes a gift of confidential information to a trading relative or friend. The tip and trade resemble trading by the insider himself followed by a gift of the profits to the recipient. (463 U.S. at 664)

The Court finds that the insider who tipped Dirks did not do so for any personal advantage or benefit. Instead, he shared the information with Dirks in an attempt to expose corporate wrongdoing. Therefore, Dirks did not assume a duty to disclose or abstain, and he could not be liable for insider trading for sharing the information he obtained with his clients and the *Wall Street Journal.*

A strong and important undercurrent in the Court's opinion in *Dirks* is the value to the public trading markets of analyst inquiries into company affairs for the benefit of their clients. The former executive of the issuer and Dirks, as his tippee, were, in the Court's view, performing desirable activities that, in aggregate effect, protect public company shareholders and support the integrity of the securities markets by encouraging public revelations of corporate fraud that otherwise might remain under wraps. The Court's personal benefit test and its application in this case both seem to be designed to encourage this desirable behavior.

The *Dirks* Court also notes a separate limitation on insider trading actions under Section 10(b) and Rule 10b-5. To violate Section 10(b) and

Rule 10b-5, an insider must trade on the basis of, or (as was alleged in *Dirks*) provide or act on a tip of, information that is both nonpublic and material. "[I[t may not be clear," the Court notes, "either to the corporate insider or to the recipient ... whether the information will be viewed as material nonpublic information. Corporate officials may mistakenly think the information already has been disclosed or that it is not material enough to affect the market." (463 U.S. at 662) Neither trading on nor tipping information that already is public or that is immaterial contravenes insider trading prohibitions.

Finally, Justice Blackmun's dissent in *Dirks* notes the applicability in the insider trading context of the requirement that a Rule 10b-5 violator have scienter—the intent to deceive a corporation's investors or manipulate the market for a corporation's shares. Where an insider "does not intend that ... inside information be used for trading purposes to the disadvantage of shareholders," scienter does not exist (463 U.S. at 674 n.11). The facts supporting or undercutting the existence of scienter often overlap with those that establish other elements of an insider trading claim. For example, "if the insider in good faith does not believe that the information is material or nonpublic, he also lacks the necessary scienter." (463 U.S. at 674 n.11) The Blackmun opinion notes that a primary function of the scienter requirement is to protect those disclosing information in good faith in the best interest of the corporation and its shareholders.

4.2.4 Third Key Case—*O'Hagan*

The Court's third key insider trading case, *United States v. O'Hagan*, 521 U.S. 642 (1997), stems from wholly different facts. In actuality, it is a misnomer to term *O'Hagan* an insider trading case; as many have noted, it truly relates to *outsider*, rather than *insider*, trading liability under Section 10(b) and Rule 10b-5. The *O'Hagan* case does not involve trading by a person with a relationship of trust and confidence to the issuer's shareholders, but rather involves trading by a person with a relationship of trust and confidence to the source of material nonpublic information about an issuer.

The defendant, James O'Hagan, was a partner in a law firm that represented an acquiror (offeror) in connection with an upcoming tender offer for the shares of another corporation (target). Although he was not working on the proposed transaction, he was aware of it. Without informing his fellow partners, he bought options and stock of the target before commencement of the tender offer and then sold off his positions in the target's securities at a profit after the announcement of the tender offer.

Although this behavior certainly was unfair to other investors in and shareholders of the target, O'Hagan owed them no duty. However, the Court found that it was sufficient that O'Hagan owed and breached a duty of confidentiality to his law partners and his client, thereby endorsing what had become known as the "misappropriation" theory of insider trading.

> "Under this theory," the Court states "a fiduciary's undisclosed, self-serving use of a principal's information to purchase or sell securities, in breach of a duty of loyalty and confidentiality, defrauds the principal of the exclusive use of that information. In lieu of premising liability on a fiduciary relationship between a company insider and a purchaser or seller of the company's stock, the misappropriation theory premises liability on a fiduciary-turned-trader's deception of those who entrusted him with access to confidential information." (521 U.S. at 652)

The Supreme Court's endorsement of the misappropriation theory in *O'Hagan* finalizes the trilogy of key cases that define insider trading regulation in the United States.

4.3 MARTHA STEWART AND INSIDER TRADING

And so, with that background, Martha Stewart enters the plot line. At the end of December 2001, having tried (but failed) to sell all of her shares in ImClone Systems in a tender offer a month earlier, Stewart sold off her remaining shares in ImClone in a market transaction. This transaction occurred, by her own account, as a result of ImClone stock price reaching $60 per share under a preexisting "stop loss order" with her brokerage firm. It was one of a number of stock disposition transactions that Stewart made at the end of 2001.

A government investigation revealed, however, that Stewart had information that the public did not have when she made her December 2001 sale of ImClone stock. Specifically, on the day of her stock trade, Stewart had learned from her stock broker at Merrill Lynch, Peter Bacanovic (as conveyed through his assistant, Douglas Faneuil, who turned state's evidence), that ImClone Chief Executive Officer Sam Waksal (a personal friend of Stewart) was attempting to sell his shares in ImClone and also was selling off family interests in ImClone stock. She also learned from Bacanovic (in a telephone message taken at Stewart's office) that he expected ImClone stock to trade down and later from Faneuil that ImClone stock was in fact trading down. Stewart's receipt of this information became the lynchpin of government enforcement activity against Stewart for possible criminal and civil violations of Section 10(b) and Rule 10b-5.

Initial government action focused on the possibility of an insider trading violation under Section 10(b) and Rule 10b-5 based on Stewart's access to nonpublic information. The SEC eventually did bring a civil claim against Stewart for insider trading on that basis. As indicated above, that action was settled before trial. But the federal government also pursued Stewart criminally for defrauding shareholders of Martha Stewart Living Omnimedia, Inc., the public company that Stewart founded and led, and for lying to government officials (in each case, by not fully informing those constituencies of the reasons for her ImClone stock trade). The trial court judge acquitted Stewart of the federal securities fraud charge before sending the case to the jury, leaving only the obstruction of justice and false statement charges to the jury for decision. The jury's guilty verdict on these charges landed Stewart in jail.

4.3.1 Overall Analysis of the Insider Trading Claim against Stewart

When Stewart learned of the Waksal family's stock trades and the downward pressure on ImClone stock, she was not an insider of ImClone; she had no duty of trust and confidence to its shareholders. Accordingly, she is not a classical insider trader. Moreover, Stewart had not been afforded nonpublic information by someone who owed a duty of trust and confidence to ImClone shareholders, so she is not a classical tippee. Finally, Stewart did not breach a fiduciary duty to an information source when she traded in ImClone shares while in possession of nonpublic information, so she is not a misappropriator. However, Stewart may have been tipped off about the Waksal trading transactions and the market for ImClone shares by someone (Bacanovic, alone and through Faneuil) who may have breached a duty to either the source of the information (Waksal) or someone else (Bacanovic and Faneuil's employer, Merrill Lynch). If so, Stewart may be the tippee of a misappropriator. It is this theory on which the SEC proceeded in bringing its insider trading case against Stewart. The Supreme Court has not yet ruled on a case of this kind.

4.3.2 Application of the Misappropriation Theory to Bacanovic and Faneuil

The *O'Hagan* opinion does not expressly contemplate the possibility of a misappropriator tipping rather than trading. In *O'Hagan*, the Court notes that

> "[t]he misappropriation theory targets information of a sort that misappropriators ordinarily capitalize upon to gain no-risk profits through the purchase or sale of securities. Should a misappropriator put such information to other use, the statute's prohibition would not be implicated. The theory does not catch all conceivable forms of fraud involving confidential information; rather, it catches fraudulent means of capitalizing on such information through securities transactions." (521 U.S. at 656)

Accordingly, it is not clear that *tipping* misappropriated nonpublic information, as opposed to *trading* on misappropriated nonpublic information, constitutes insider trading under Section 10(b) and Rule 10b-5. So, even if Bacanovic and Faneuil are misappropriators of nonpublic information and they shared that information with Stewart and Stewart traded on the basis of that information, Stewart may not have violated insider trading prohibitions.

However, to the extent that tipping by a noninsider *is* an actionable fraudulent means of capitalizing on nonpublic information through securities transactions under an extension of the *O'Hagan* case, Bacanovic and Faneuil presumably would be misappropriators if they shared nonpublic information with Stewart in breach of a duty of trust and confidence to Waksal, as the source of the information. Neither the SEC nor the National Association of Securities Dealers (NASD), the regulatory body governing the activity of stock brokers, imposes any such duty on its members. Stock brokers do have an obligation "to serve … customers with honesty and integrity by putting their interests first and foremost" under guidance for brokers published by the NASD on its Web site (http://www.nasd.com/RegistrationQualifications/BrokerGuidanceResponsibility/RegisteredRepresentatives/ObligationsToYourCustomers/index.htm). This description of a broker's obligation implies a duty of loyalty is owed by brokers to their customer. This is unsurprising; brokers are agents of their customers, and under agency law principles, agents owe their principals fiduciary duties of loyalty, care, and candor.

Based on publicly available facts, the disclosure made to Stewart by and for Bacanovic is not apparently contrary to Bacanovic's duty as a broker. It cannot be said definitively that the disclosure was contrary to Waksal's interests. In fact, because Waksal and Stewart were friends (and Bacanovic was aware of this fact), Waksal may have wanted Stewart to have (or at least may not have objected to Stewart having) information about the stock trades being made by his family members; Bacanovic even may have told Waksal that he planned to share the information with Stewart. Moreover, there is no indication that Bacanovic or Faneuil assumed an express or implied confidentiality duty (e.g., by contract or a course of activity) to Waksal. Finally, any confidentiality duty that Bacanovic or Faneuil may have owed to Waksal may have been effectively waived by Waksal.

In addition, Bacanovic had a competing duty of disclosure that may undercut or outweigh any duty of confidentiality owed by Bacanovic to Waksal. Under NASD guidance, brokers have a duty to disclose to their customers all information reasonably relevant to their investment decisions. Bacanovic was a broker for Stewart as well as Waksal. Arguably, Bacanovic had a duty to inform Stewart about the Waksal family ImClone stock dispositions if he deemed it was reasonably relevant to Stewart's investment decisions. Bacanovic's behavior in informing Stewart may have been consistent with his obligation to Stewart to act in her best interest.

On the other hand, NASD guidance for brokers also indicates (somewhat ambiguously) that "[i]t is illegal to use or pass on to others material, nonpublic information or enter into transactions while in possession of such information" (http://www.nasd.com/RegistrationQualifications/BrokerGuidanceResponsibility/RegisteredRepresentatives/SamplePracticesThatViolateRegulations/index.htm). Also, those in the industry generally acknowledge that it is a breach of professional ethics (or at least bad form), if not illegal, to share information about client trades with other clients (Cohen 2001).

Competing with these notions of broker confidentiality on client trades is the general federal securities law requirement that corporate affiliates (including insiders) who are selling more than 500 shares of stock or shares of stock with an aggregate sale price in excess of $10,000 publicly file a notice of proposed sale on Form 144 with the SEC at the time an order is placed with a broker (17 C.F.R. § 230.144(h)). So, insiders (like Waksal) are subject to different rules, and their personal trades often are subject to public scrutiny. (Insiders also have posttransaction reporting obligations under Section 16(a) of the 1934 Act, but these obligations are not relevant to the insider trading claims against Stewart.) Waksal apparently was able to avoid these disclosures because neither Merrill Lynch nor another broker he attempted to engage in later December 2001 would make trades for Waksal's own account when he requested that the trades be made. Accordingly, evidence is inconclusive regarding whether Bacanovic and Faneuil are insider-trader misappropriators based on this analysis.

However, the SEC theory of the case against Stewart rested on a breach of duty owed not to Waksal, as the source of the information, but instead to Merrill Lynch, Bacanovic and Faneuil's employer. The SEC notes in its complaint against Stewart that "Merrill Lynch policies specifically required employees to keep information about . . . [the Waksal stock trades] confidential" (http://www.sec.gov/litigation/complaints/comp18169.htm). The SEC reliance on a breach of duty to a party other than the source of the nonpublic information is not specifically contemplated in the Court's holding in *O'Hagan*. In fact, in its *O'Hagan* opinion, the Court expressly poses and answers in the affirmative the narrower question: "Is a person who trades in securities for personal profit, using confidential information misappropriated in breach of a fiduciary duty *to the source of the information*, guilty of violating § 10(b) and Rule 10b-5." (521 U.S. at 647) (emphasis added) The Court's description of the misappropriation theory in its *O'Hagan* opinion also references breach of a duty owed to the source of the nonpublic information.

> The "misappropriation theory" holds that a person commits fraud "in connection with" a securities transaction, and thereby violates § 10(b) and Rule 10b-5, when he misappropriates confidential information for securities trading purposes, in breach of a duty owed *to the source of the information*.... Under this theory, a fiduciary's undisclosed, self-serving use of a principal's information ..., in breach of a duty of loyalty and confidentiality, defrauds the principal of the exclusive use of that information. In lieu of premising liability on a fiduciary relationship between company insider and purchaser or seller of the company's stock, the misappropriation theory premises liability on a fiduciary-turned-trader's deception of those who entrusted him with access to confidential information. (521 U.S. at 652) (emphasis added)

Although dicta and some of the reasoning in the *O'Hagan* opinion may be applied more broadly to deception conducted through any breach of duty, it is not clear that the Court intended or would endorse this broad reading. However, since the breach of duty underlying the SEC case against Stewart involved the duty of a securities broker to his brokerage firm (which, in turn, is an agent for its customers), the breach is, in effect, an indirect breach of duty to the information source.

In addition, an SEC rule adopted in the wake of the *O'Hagan* case may help to establish the requisite duty on these facts. Rule 10b5-2 under the 1934 Act provides that, for purposes of misappropriation cases under Rule 10b-5, "a 'duty of trust or confidence' exists … [w]henever a person agrees to maintain information in confidence (17 C.F.R. § 240.10b5-2(b))." Therefore, under Rule 10b5-2, Bacanovic and Faneuil's agreement with Merrill Lynch to keep customer information confidential appears to establish the requisite duty.

4.3.3 Application of the Tipper/Tippee Analysis to the Stewart Facts

Of course, in cases where an alleged insider trader tips rather than trades, the need for a tipper/tippee analysis under *Dirks* is apparent. Under *Dirks*, tippee liability depends on the existence and breach of a duty owed by the tipper of which the tippee has knowledge. To determine a breach of duty in this setting, the *Dirks* Court uses the personal benefit test. This test enables the Court to determine whether the information shared by the putative tipper with the putative tippee was shared improperly.

On some level, the personal benefit test makes sense in the *Dirks* classical tipper/tippee environment, since the fiduciary duty of a director or officer of a corporation requires that he act not in self-interest but rather in the interest of the shareholders of the corporation that he serves. It therefore is relatively clear, in classical tipper/tippee cases, that the sharing of information is improper (or at least appears improper) when the tipper receives a personal benefit in violating the corporation's trust. Moreover, as earlier noted, the personal benefit test, as applied in *Dirks*, supports desirable objectives under the securities laws—namely, the timely public disclosure of corporate fraud.

A tipper who is a misappropriator may not owe a similar type of fiduciary duty to an information source or, as in Stewart's case, to another (including an employer). The same level of loyalty, the same abandonment of self-interest, may not be required in all duty-bound relationships. Accordingly, the personal benefit test may not have as a strong a basis in determining the impropriety of the conveyance by and for Bacanovic to Stewart of information about the Waksal family stock trades.

On the other hand, if the point of the Court's endorsement of the misappropriation theory in *O'Hagan* is that a misappropriator should not be permitted to capitalize fraudulently on nonpublic information through a securities trading transaction, then showing that personal benefit inured to a misappropriator in disseminating material nonpublic information would be one way of determining whether the misappropriator did, in fact, capitalize on the shared information. And, as noted above in the discussion of *Dirks*, the personal benefit test is rooted in the SEC's original concern in insider trading matters that an insider not be permitted to benefit personally from inside information. Specifically, in *Dirks*, the Court states that the fraud in an insider trading case "derives from the 'inherent unfairness involved where one takes advantage' of 'information intended to be available only for a corporate purpose and not for the personal benefit of anyone.'" (463 U.S. 654) (citing *In re Merrill Lynch, Pierce, Fenner & Smith, Inc.*, 43 S. E. C. 933, 936 (1968))

The nonpublic information disclosed to Stewart (the fact of attempted and actual stock sales by Waksal and his family) was not, of course, information with a corporate purpose. As indicated below, the personal (rather than corporate) nature of the information at issue in the Stewart affair is part of what makes her case intriguing from an insider trading perspective. But assuming that the information received by Stewart is information of the type and significance that triggers potential insider trading liability, it is not clear whether Bacanovic and Faneuil disclosed that information to Stewart in order to take advantage of the information for their personal benefit. It is seemingly important, therefore, to analyze the Stewart facts under the personal benefit test.

Who, then, benefited from the selective disclosures made to Stewart—was it Bacanovic and Faneuil, as the duty holders, or Sam Waksal or Merrill Lynch, as the beneficiaries of the potentially applicable duties? The answer to this question is not easy to discern from the facts of the case as we know them. Merrill Lynch, as Stewart's broker, benefited, we may presume, by receipt of a stock sale commission from Stewart's disposition of her ImClone shares. No doubt Bacanovic shared in some way in the proceeds of that commission, since Stewart was his customer. Moreover, Bacanovic and Faneuil, as well as Merrill Lynch and Waksal, may have received a reputational benefit from the disclosures. Their respective reputations may have been enhanced with Stewart and with anyone to whom she would promote the services or products of any of them. Evidence in the press suggests that Stewart and Bacanovic were friends as well as business acquaintances. Moreover, evidence exists of a long-term personal (at times romantic) relationship between Waksal and Stewart. But there is no evidence that the disclosure of information to Stewart was a quid pro quo for some other benefit inuring to Bacanovic, Faneuil, Merrill Lynch, or Waksal.

The question remains: did Bacanovic or Faneuil take advantage of information for a personal benefit when they disclosed the Waksal and ImClone stock trading information to Stewart? Or did they disclose the information to her to better serve their other key customer, Waksal, or their employer, Merrill Lynch? Does this disclosure look more like a secret, indirect gift of securities trading profits to Stewart (which presents insider trading concerns) or an ordinary course revelation of material information by a broker to a customer (which does not)? It is not easy, based on publicly available facts, to say that Bacanovic's disclosures to Stewart were made to enhance Bacanovic's or Faneuil's finances or reputation or that these disclosures were tantamount to a gift of securities trading profits from Bacanovic to Stewart. The disclosures made to Stewart likely benefited *both* the beneficiaries and holders of the duties at issue, much like the disclosures at issue in the *Dirks* case. Accordingly, it is not easy to say that Bacanovic or Faneuil breached a fiduciary duty to Waksal or Merrill Lynch that would subject a tippee of either of them to insider trading liability.

However, even if either or both of them breached the requisite duty in disclosing the Waksal family stock trades to Stewart, it remains unclear whether Stewart in fact knew of the breach. As the above analysis suggests, the nature of a broker's duties and the existence of a breach both are uncertain. Therefore, Stewart's knowledge of the breach must be at least as uncertain. Assuming a clear breach, commentators point to the fact that Stewart was a registered stock broker and that, therefore, she knew or should have known of any breach of duty. This may be true with respect to the brokers' duties to Waksal, since those duties are based on general broker–customer guidance of which Stewart likely was aware. Stewart's actual or imputed knowledge of Merrill Lynch's policies would be less easy to prove. In any case, however, Stewart may have reasonably believed that that the information she was getting already was public or that Waksal had authorized Bacanovic and Faneuil to share the information with her or waived any applicable confidentiality policy of Merrill Lynch established for Waksal's protection. Stewart's knowledge of a breach of duty by Bacanovic or Faneuil is significantly in doubt.

4.3.4 The Nature of the Nonpublic Information Provided to Stewart

As earlier noted, the nonpublic information provided to Stewart (information about the Waksal family stock trades) was not corporate in nature. Rather, it was information about the trading of a corporation's chief executive officer. As such, the information was personal, but corporate-related. Dispositions by corporate executives of the corporation's stock are widely regarded as a negative signaling device to the public markets. Still, many have doubts regarding whether this type of information is or should be considered inside information that triggers insider trading liability, especially to the extent that Stewart's trading, like that of Dirks's brokerage customers, more rapidly moves the market closer to an efficient price. Suffice it to say that in many other countries, information of the kind imparted to Stewart would not trigger potential insider trading liability.

However, even if information about the Waksal family stock trades *is* nonpublic information that may subject a tippee to insider trading liability, that does not end the inquiry. As the *Dirks* Court noted, only trading on the basis of, and tips of, *material* nonpublic information are actionable under Section 10(b) and Rule 10b-5. Under the Supreme Court standard adopted in *Basic Inc. v. Levinson*, 485 U.S. 224 (1988), facts are material when there is a substantial likelihood that a reasonable investor would find them important in making an investment decision (or, stated alternatively, facts are material when there is a substantial likelihood that a reasonable investor would view their disclosure as having a significant impact on the "total mix" of publicly available information) (485 U.S. at 231–32). Without analyzing other facts available in the marketplace at the time the Waksal family's stock trades were revealed to Stewart, it cannot be definitively stated that the dispositions of ImClone stock by members of the Waksal family constituted material information. The information shared with Stewart does, however, have a propensity to affect the market price for ImClone securities, making its materiality likely.

4.3.5 The Existence of Scienter in the Stewart Affair

The existence of scienter in connection with Bacanovic's direct and indirect disclosures of nonpublic information to Stewart also is questionable. Stated more specifically, it is not apparent that Bacanovic or Faneuil intended to deceive ImClone investors or manipulate the market for ImClone shares by sharing information about the Waksal family stock trades with Stewart. As earlier noted, it is possible to view the disclosure of nonpublic information to Stewart as information transmitted in the ordinary course of the brokers' business. If Bacanovic and Faneuil did not know or intend that their conduct breached a duty in violation of insider trading prohibitions, they would not have the requisite scienter. At a minimum, they would have had to have engaged in conduct in reckless disregard of the insider trading laws in order for them to violate Section 10(b) and Rule 10b-5 as insider traders.

Similarly, it is not clear that Stewart intended to deceive ImClone investors or manipulate the market for ImClone shares. In particular, even if Stewart understood that Bacanovic had an obligation to hold Waksal's nonpublic stock trading information confidential, Stewart may not have had an appreciation for the fact that the information she was given was nonpublic. She may have assumed that the information already had been made public, whether through the filing of a Form 144 or otherwise. Moreover, Stewart had recently sold off a number of unprofitable securities in her portfolio, and her motive may have been merely to sell off yet another expected loser. Martha Stewart may not have possessed the state of mind necessary to violate insider trading prohibitions.

4.4 CONCLUSION

Martha Stewart is not an insider trader. She never was found guilty of or liable for a violation of federal insider trading prohibitions. Moreover, had the SEC's civil trial against Stewart for insider trading violations proceeded to judgment, the SEC would have had to overcome many substantial legal obstacles in order to prove that Stewart had violated insider trading laws when she sold the last of her ImClone shares in December 2001. The SEC's theory of the case is untested in Supreme Court jurisprudence, and issues of proof exist at virtually every level of analysis. It is not obvious that Merrill Lynch employees Bacanovic and Faneuil were misappropriators of inside information under an extension of the rule set forth in the *O'Hagan* case, and it is similarly unclear that Stewart was a tippee of inside information under an extension of the tipper/tippee liability rules established in the *Dirks* case. In fact, Stewart may not have had inside information at all; and if she did, she may not have realized that she did, making it unlikely that she had the requisite state of mind to violate Section 10(b) and Rule 10b-5. Because the SEC settled its insider trading action against Stewart before a trial was held, we may never know.

REFERENCES

Cohen, R. 2001. The way we live now: 3-04-01: The ethicist; invested interest. *New York Times.* March 4.

Grzebielski, R. J. 2007. Why Martha Stewart did not violate Rule 10b-5: On tipping, piggybacking, front-running and the fiduciary duties of securities brokers. *Akron Law Review* 40:55–83.

Heminway, J. M. 2003. Save Martha Stewart? Observations about equal justice in U.S. insider trading regulation. *Texas Journal of Women and the Law* 12:247–85.

Heminway, J. M. 2007. *Martha Stewart's legal troubles.* Durham, NC: Carolina Academic Press.

Langevoort, D. C. 2006. Reflections on scienter (and the securities fraud case against Martha Stewart that never happened). *Lewis and Clark Law Review* 10:1–17.

Schroeder, J. L. 2005. Envy and outsider trading: The case of Martha Stewart. *Cardozo Law Review* 26:2023–78.

Statman, M. 2005. Martha Stewart's lessons in behavioral finance. *Journal of Investment Consulting* 7:52–60.

CHAPTER 5

Insider Trading Regulation in Transition Economies

Robert W. McGee*

CONTENTS

5.1 INTRODUCTION

Insider trading is generally perceived as evil or at least unethical. The press and television show people being arrested and led away in handcuffs for engaging in it. The media have nothing good to say about the practice.

* An earlier version of this paper was presented at the Fourth Annual International Business Research Conference, University of North Florida, Jacksonville, FL, 13 February 2004.

Politicians enhance their careers by being against it. Commentators make it sound like all insider trading is illegal. Yet some forms of insider trading are perfectly legal (Shell 2001) and some kinds of insider trading are not unethical. In other words, there is a widespread misperception on the part of the public about insider trading.

This misperception has spread to the transition economies that are in the process of converting from centrally planned systems to market systems. This is unfortunate, since there is evidence to suggest that at least some kinds of insider trading are healthy and beneficial for an economy. Thus, transition economies that blindly outlaw all insider trading are unknowingly harming themselves and doing an injustice to the people they are supposed to represent.

5.2 THE PHILOSOPHICAL BASE

5.2.1 Envy and the Labor Theory of Value

Those who think all insider trading should be illegal think so for a variety of reasons. Some say it is inherently immoral to trade on inside information because making a large profit with such little effort is somehow wrong. Others say that there should be a level playing field, and the playing field cannot be level when some individuals enjoy informational advantages over others. A third group takes the position that insiders have some fiduciary duty not to benefit from the information they have access to as part of their position with the corporation. A fourth group subscribes to some kind of misappropriation theory, which basically holds that the information they are using for personal gain belongs to someone else, and using the information results in a violation of property rights or contract rights.

All of these views have received a wide degree of support. However, upon closer analysis, each of these views has major weaknesses. One weakness is that those who advocate outlawing insider trading resort to emotional appeals rather than sound economic or philosophical analysis. There is often a certain amount of envy or jealousy included in the subtext of their arguments (Schoeck 1987). Many of those who would like to see all inside traders punished have what Ludwig von Mises has called the anticapitalist mentality (Mises 1956). They just don't like the free enterprise system, think it is inherently evil, and think that individuals should not be able to make millions of dollars with so little apparent effort. This latter view is a subconscious application of the labor theory of value, which was subscribed to by both Karl Marx and Adam Smith and, in fact, every other economist prior to the 1870s, when the labor theory of value was replaced by the marginal utility theory and the theory of subjective value (Jevons 1871; Menger 1871; Walras 1874).

The problem with applying the labor theory of value to insider trading is that not all value comes from labor. Things are worth whatever someone is willing to pay. The amount of labor that went into the product or service is completely irrelevant. Thus, the fact that someone can make millions of dollars by trading on information that was obtained with little apparent effort has nothing to do with whether the practice is immoral or whether it should be outlawed.

5.2.2 The Level Playing Field Argument

The level playing field argument has been used to justify any number of economic regulations. Trade cannot be free, it must be fair, whatever that means (Bovard 1991). People who have accumulated a great deal of wealth during their lifetimes must have it confiscated when they die so that those who are less fortunate will be able to compete with the children and grandchildren of the rich, who would otherwise leave their wealth to their children. Such thinking is one of the main reasons some countries have adopted punitive estate and inheritance taxes (Buchanan and Flowers 1975). The level playing field argument has been applied to insider trading to argue that all investors should have the same information at the same time, regardless of what they have done, if anything, to earn the information.

The problem with this level playing field argument is that it is not possible or desirable to ever have a level playing field in the realm of economics. The level playing field argument is appropriate to apply to sporting events but not to economics. It would not be fair for one football team to have to run uphill for the entire game while its opponent can run downhill. It is not fair for one basketball team to have a larger hoop to shoot at than its opponent. But there is nothing unfair about allowing banana farmers in Alaska to compete with banana farmers in Honduras. Alaska banana farmers should not be subsidized so that they can compete more effectively with banana farmers from Honduras, and banana farmers from Honduras should not have to comply with punitive regulations or higher tax burdens to make them less able to compete with banana farmers from Alaska. Likewise, there is nothing unfair about allowing experts who work sixty hours a week to gather financial information as part of their job to profit from that information. What is unfair is to force them to disclose such information to people who have done nothing to earn it.

Ricardo's theory of comparative advantage (1817) is at work here. Some individuals or groups are naturally better at some things than others, and some individuals or groups develop skills that are better than those of their competitors. Penalizing those who are better at something or subsidizing those who are worse at something results in inefficient outcomes and is unfair to some groups.

Comparative advantage works to the benefit of the vast majority of the population. It allows specialization and division of labor, which Adam Smith pointed out in his pin factory example (1776) leads to far greater efficiency, higher quality, and lower prices. Not allowing individuals to use their special talents harms the entire community as well as the individuals who are being held back by some government law or regulation. Forcing a level playing field on people is always harmful because it reduces efficiency and violates rights. Using the level playing field argument to prevent individuals from using their insider knowledge for personal gain does not hold up under analysis. If insider trading is to be made illegal and if inside traders are to be punished, some other justification must be found.

5.2.3 Two Philosophical Approaches to the Issue

There are basically two ways to evaluate economic and public policy issues. The utilitarian approach, which is subscribed to by the vast majority of economists, views an action as being good if the result is the greatest good for the greatest number (Bentham [1781] 1988; Mill [1861] 1979; Yunker 1986). They would call it a positive-sum game if the benefits exceed the costs or if the good exceeds the bad.

One problem with the utilitarian approach is that it is impossible to precisely measure gains and losses (Smart and Williams 1973). One may only make estimates. Another related problem is that when individuals rank their choices, they do not calculate that Option A is 20 percent better than Option B. If a consumer prefers McDonald's hamburgers over Burger King hamburgers, it cannot be said that he likes McDonald's hamburgers 20 percent more than Burger King hamburgers, but only that he prefers McDonald's hamburgers to Burger King hamburgers. Furthermore, after he has consumed a few McDonald's hamburgers, he probably prefers no additional hamburgers to a McDonald's hamburger because he is no longer hungry. Not only can individual preferences not be measured, they also change over time. They are not constant. Thus, any precise measurement is impossible.

Another problem with utilitarian approaches, related to the measurement issue, is that there is no way to precisely measure total gains and losses when some minority of individuals or groups benefit a lot from some rule while the vast majority are harmed (Shaw 1999). For example, can it be determined mathematically whether imposing a $5 tariff on the importation of foreign shirts is a good public policy if doing so protects the jobs of 10,000 textile workers but forces 100 million domestic consumers to pay an extra $5 for a shirt? Many empirical studies have found that imposing tariffs results in a negative-sum game, but scholars cannot agree on how negative the result is. Some studies conclude that two jobs are lost for every job saved by some protectionist measure (Baughman and Emrich 1985; Mendez 1986) while other studies conclude that three jobs are lost for every job saved (Denzau 1985, 1987). Much depends on the assumptions made and the economic methodology employed.

Another problem with the utilitarian approach is that it is not possible to compare interpersonal utilities (Rothbard 1997, 1970). Different individuals place different values on things. We may not automatically assume that the theft of a dollar from a rich man results in less disutility than the theft of a dollar from a poor man, since either could use the dollar to buy a candy bar, which might (or might not) give them both the same amount of pleasure, depending on their personal preferences.

Perhaps the strongest criticism that can be made against a utilitarian approach is that it completely and totally ignores rights (Frey 1984; Rothbard 1970). To a utilitarian, violating someone's rights is irrelevant. All that matters is whether the good outweighs the bad. The end justifies the means for a utilitarian. That is an inherent and structural weakness of all utilitarian approaches.

The other approach to analyzing public policy issues is that of rights. The question to be asked is whether someone's rights are violated. If someone's rights are violated, the act is automatically wrong, even if the result would be a positive-sum game. Dostoevsky provides perhaps the strongest illustration of this view in *The Brothers Karamazov*, when he asks whether it would be acceptable to torture one small child to death if the result would be eternal happiness for every other member of the community (1952, 126–7). Although it is not possible to precisely measure the child's pain and compare it to the happiness of the rest of the community, a utilitarian would probably conclude that such an act is just because it results in the greatest good for the greatest number. A rights theorist would reach the opposite conclusion because of the belief that the violation of anyone's rights makes the act automatically wrong.

Most Western legal systems are a mixture of utilitarianism and rights theory. Welfare legislation is at least partially based on utilitarian beliefs. The General Welfare Clause of the U.S. Constitution and the general welfare clauses of other constitutions are also rooted in utilitarianism. The fact that some individuals must be forced to subsidize the existence of others is utilitarian based and necessarily violates rights. But constitutions and laws sometimes protect individuals rights, however defined. So legal systems are a combination of these two competing and sometimes contradictory philosophies.

Another issue to be considered is whether something that is immoral should automatically be declared illegal. The answer to this question depends on which philosophy of law one subscribes to. In a theocratic state, what is deemed to be immoral is also illegal. The law in such countries is a mirror image of the theology being practiced in the community. One may be burned at the stake for being insufficiently Catholic in Spain during the Inquisition or one may be stoned or beheaded for adultery or for saying something unflattering about Islam if one lives in a theocratic Islamic state.

This philosophy of law does not have widespread support in the developed Western democracies, for a variety of reasons. For one thing, these countries are not theocracies. They are basically secular, although their legal systems may contain some religious based philosophy. Thou shalt not kill and thou shalt not steal are religious values that are shared by every religion to a certain extent. But they are more than just religious values. They are values that are subscribed to by atheists and agnostics as well, so we cannot label them purely religious values.

These countries are also pluralist. In a pluralist state, it is difficult to attempt to impose one set of moral values on the entire group, since the population living within the borders of such a state subscribes to different moral values. One may not outlaw alcohol just because some religious minority thinks that imbibing alcoholic beverages is immoral. One may not outlaw pork or require church attendance just because some religious groups think they are morally bound not to eat pork or to attend services on some regular basis. What is immoral to one individual or group may not be considered immoral to another individual or group. In a pluralist society, allowances must be made for such differences if one is to have domestic peace. Trying to make illegal those acts that are considered immoral only by some segment of the community is not good public policy in a pluralist state (Berlin 1991, 2001).

That being the case, we will not go into a detailed analysis of whether insider trading is immoral, since immorality, in and of itself, is irrelevant in a pluralist state. What is immoral should not necessarily be illegal. Our analysis will be confined to a determination of whether insider trading results in a positive-sum game or whether it violates anyone's rights. The morality of insider trading will be discussed from these two ethical perspectives.

5.3 WHAT'S WRONG WITH INSIDER TRADING?

Lekkas (1998) provides a brief summary of the arguments that have been made for and against insider trading. Bainbridge (2000) also summarizes the pro and con arguments and provides a bibliography as well. Their arguments against insider trading include the previously mentioned level playing field argument; the belief that unequal access to information is somehow unfair; insider trading encourages greed; trading on inside information is theft of corporate property; insider trading is a kind of fraud (Strudler and Orts 1999); insiders who profit from the use of inside information are breaching their fiduciary duty.

The main argument supporting insider trading is efficiency. Trading on inside information causes information to be released into the marketplace sooner rather than later, thus causing stock prices to move in the right direction quicker than would otherwise be the case. Studies by Meulbroek (1992), Cornell and Sirri (1992), and Chakravarty and McConnell (1997) support this position. Another argument in favor of insider trading is that inside information is property, and preventing individuals from trading their property violates their property rights.

Bernardo (2001) sees the right to trade on insider information as a contractual problem of allocating property rights between shareholders and stakeholders. Allowing insiders to deal in insider information has also been viewed as a kind of compensation, a salary supplement, or a bonus to be given as a reward for performance.

Henry Manne (1966) was the first to do a detailed study of insider trading and his study has become a classic. He concluded that insider trading does not result in any significant injury to long-term investors and causes the market to act more efficiently. He has called it a victimless crime (Manne 1985), as there are no identifiable victims. Those who sell their stock anonymously to a broker would have done so anyway, so they are no worse off then they would have been if the inside trader had not traded.

Jeng, Metrick, and Zeckhauser (2003) conducted an empirical study that reached basically the same conclusion. They estimated that the expected cost of insider trading to noninsiders was about 10¢ per $10,000 transaction. Allen (1984), Leland (1992), and Repullo (1994) conducted studies concluding that insider trading was beneficial to other shareholders.

The insider trading law does not consider the possibility that an inside trader may profit from inside information by *not* trading. For example, if the insider knows that the stock price is likely to go up, he can refrain from selling the shares he already owns. Likewise, if he knows the stock price is likely to fall, he can refrain from buying shares. These activities are not prohibited by insider trading laws but they are examples of insiders profiting from nonpublic information.

One conceptual problem with insider trading is determining ownership of the property in question. Information can be viewed as property, but it is not always clear who owns the right to use nonpublic information. The misappropriation theory tries to solve this problem but commentators are not in agreement regarding whether this problem has been solved. Quinn (2003), Weiss (1998), and Seligman (1998) think that it has, whereas Swanson (1997) and a plethora of other commentators (Quinn 2003) think it has not. The property issue is one of the keys to solving the problem of whether insider trading should be outlawed or regulated; yet it is unclear in some cases who can claim an ownership right to the property or when it has been misappropriated.

5.4 INSIDER TRADING IN TRANSITION ECONOMIES

The regulation of insider trading is a relatively recent phenomenon. The United States was the first major country to enact an insider trading law and to place restrictions on insider trading. The roots of the U.S. insider law sprouted from the securities legislation that was enacted in 1934 to prohibit other kinds of stock manipulation (Bernardo 2001). France was the second country to enact an insider trading law but France did not place prohibitions on insider trading until 1967 (Gevurtz 2002). Other countries have followed, but slowly. The United Kingdom, Australia, and Japan have adopted insider trading laws along the American model (O'Hara 2001). As of 1990, only thirty-four countries had laws restricting or prohibiting insider trading, and only nine of them had prosecuted anyone for insider trading. By 2000, eighty-seven countries had passed insider trading laws and thirty-eight had prosecuted at least one insider trading case (Gevurtz 2002). China's insider trading law was not enacted until December 29, 1998 and was drafted with the assistance of the United States (Qu 2001). In 1989, the European Union (EU) passed a directive that required all member countries to pass legislation prohibiting certain kinds of insider trading by 1992. Any country that wants to join the EU must also have an insider trading law on the books.

There is a widespread belief that American laws are the best laws (Gevurtz 2002). This view is prevalent among American lawyers and law students in the United States, partly because of their ignorance of laws in other countries, but it is also widespread in developing countries. As a result, policy makers in developing countries often exhibit little resistance to the adoption of American laws when the opportunity presents itself. Indeed, some bureaucrats and political leaders actively encourage such assistance from the United States.

The U.S. Agency for International Development (USAID) has spent tens of billions of dollars sending U.S. "experts" to dozens of countries to give advice and to help them reform their legal systems by adopting laws that more closely resemble the laws of the United States. The American Bar Association supports programs to send American attorneys to numerous developing countries to give advice and assistance in legal reform as well.

There has been somewhat of a shift away from adopting American-like laws in recent years, especially in the developing countries of Eastern Europe. This shift is partly because many of the countries in this region of the world want to become part of the EU, and the EU has laws that are different from those of the United States. However, many EU laws are not all that different from their American counterparts in terms of substance. The EU laws on antidumping, acquisitions and mergers, antitrust, and insider trading are substantially the same as their U.S. counterparts, although perhaps a bit less friendly toward business. The EU economic system is more socialistic than the U.S. system, and this difference is reflected in EU corporate law. However, many corporate laws adopted by the EU are modeled to a certain extent on U.S. law.

The countries in Eastern Europe that want to become part of the EU and countries in other parts of the world that want to join the World Trade Organization (WTO) or that want to obtain loans from the World Bank, the International Monetary Fund (IMF), the European Bank for Reconstruction and Development (EBRD), or other such lenders of last resort often do not take a critical look at the laws the EU, the World Bank, the

IMF et al. want to impose on them. As a result, there is a tendency to "reform" their legal systems to bring their laws into closer compliance with the laws of the more developed countries without critically analyzing whether the laws they adopt are good laws or are in their own best interests. Thus, countries that are in transition often adopt the bad laws along with the good laws when in fact they should be taking a more cafeteria approach by selecting the laws they find attractive and passing on the laws that do not suit them.

One reason they do not use this approach is because they lack expertise in deciding which laws are good and which ones are bad or defective. One reason they allow USAID or World Bank experts to give them advice is because the local bureaucrats do not have any experience living in or working in a developed market economy. Thus, they rely on advisors who *do* have this kind of background. As a result, they sometimes follow bad advice without knowing that it is bad advice. Just because the advice comes from someone who is perceived as an expert from a developed economy does not mean that the advice is good.

Taking the advice of such experts in the area of insider trading is a case in point. Very few of the experts giving advice to transition economies are advocating that the country in question not adopt any prohibitions against insider trading. Indeed, such experts who are funded by USAID or Tacis, its EU equivalent, would likely be fired for giving such advice. Thus, such advice is not given. Instead, these experts are advising the bureaucrats and legislators in transition economies to enact insider trading legislation that mirrors either the EU or U.S. law. The local bureaucrats and legislators often listen to such advice uncritically and often enact the legislation that these foreign experts draft for them.

5.4.1 OECD Position on Insider Trading

As was mentioned previously, various nongovernmental and quasi-governmental organizations are providing advice on economic restructuring in various transition economies. The Organization for Economic Cooperation and Development (OECD) is one such organization. It has poured a great deal of resources into economic restructuring. It has hosted seminars and conferences on corporate governance issues and has published numerous white papers and other documents on the topic.

It began its program to develop corporate governance standards in the aftermath of the Asian Financial Crisis of 1997. In 1999, it issued the

OECD Principles of Corporate Governance, which has become internationally recognized as a major source of guidance. It has become an important component of the Review of Standards and Codes (ROSC) project undertaken by the International Monetary Fund (IMF) and World Bank. It has been endorsed by the International Organization of Securities Commissions (IOSCO) and by private bodies, including the International Corporate Governance Network. In January 2004, it published its revised *OECD Principles of Corporate Governance: Draft Revised*, which also addresses insider trading. In Section II.B. of the revised draft, it states:

> Abusive self-dealing occurs when persons having close relationships to the company, including controlling shareholders, exploit those relationships to the detriment of the company and investors. Since insider trading entails manipulation of the capital markets, it is prohibited by securities regulations, company law and/or criminal law in most OECD countries. However, not all jurisdictions prohibit such practices, and in some cases enforcement is not vigorous. These practices can be seen as constituting a breach of good corporate governance inasmuch as they violate the principle of equitable treatment of shareholders.
>
> The Principles reaffirm that it is reasonable for investors to expect that the abuse of insider power be prohibited. In cases where such abuses are not specifically forbidden by legislation or where enforcement is not effective, it will be important for governments to take measures to remove any such gaps.

The language in the 2004 revised draft is basically unchanged from the original 1999 document. There are several problems with the language used in the OECD documents. For one, they seemingly advocate outlawing *all* insider trading, which would result in punishing individuals who have not violated any rights or breached any fiduciary duties. Such blanket prohibitions would punish some individuals who have done nothing wrong, but who have merely exercised their right to sell their property or to buy new property with information that has been justly acquired. Such blanket prohibitions would also result in making capital markets work less efficiently, to the detriment of the vast majority of the public.

There is also a problem with the statement: "Abusive self-dealing occurs when persons having close relationships to the company, including controlling shareholders, exploit those relationships to the detriment of the company and investors." The word "exploit" is used pejoratively in this statement. A better word to use would be "use." But a more important error in the statement has to do with the presumption that the company or investors are harmed as a result of the inside trade. Some studies have shown that the company and investors stand to gain as a result of insider trading and that the market in general also benefits by such trades, because insider trading causes prices to move in the right direction sooner than would otherwise be the case.

It is also not at all clear that insider trading entails "manipulation of the capital markets." Manipulation is one thing; insider trading is another. It is difficult to see where insider trading results in the manipulation of the market when only a few shares are sold. It is also not easy to see where the failure to sell a large block of shares when the price is expected to go up results in manipulation of the capital markets. Yet insiders who fail to sell their shares because of their privileged information are "using" their inside information, although it is hard to see how such use harms the corporation or other shareholders.

It is difficult to see how shareholders are not being treated equitably if an insider buys shares when the price is expected to rise. The purchase of shares by an insider helps the stock price to rise sooner than would otherwise be the case as soon as the word gets out that an insider has bought shares. Such a price rise works to the benefit of existing shareholders. If the stock price is expected to decline, it is not always clear that the insider who decides to sell is treating shareholders inequitably. The sale would result in inequitable treatment only if the insider were under some duty to announce the expected price decline to shareholders before making the sale.

The OECD has published several white papers on corporate governance that provide guidance for transition and developing countries in various regions of the world. One such document is its *White Paper on Corporate Governance in South Eastern Europe* (2003). This white paper refers to insider trading at least nine times. Chapter 1: Shareholders Rights and Equitable Treatment, para. 111 (p. 20) states: "Insider trading should be forbidden by legislation or securities regulation and monitoring and enforcement of such abusive practiced reinforced." Some of the other relevant paragraphs state the following:

> *Para. 112*—Frequent cases of market manipulation occur in SEE financial markets, due to insiders trading while in possession of

> confidential information. These abusive practices breach the principle of equitable treatment of shareholders. Moreover, they prevent full market transparency, thus harming the integrity of financial markets and public confidence in securities.
>
> *Para. 113*—When necessary, legislation or securities regulations should be completed to bring about prohibition of insider dealing and market manipulation. Any person in possession of inside information should abstain from trading on the related security. This concerns primarily managers and board members, but also any person who has access to specific information by exercising his/her profession or duties, such as the auditors or professionals from the regulatory authorities as well as any persons who have been tipped off by insiders. They should abstain from trading directly or indirectly, for their own account as well as for the account of a third party.
>
> *Para. 114*—Regulatory authorities should monitor more rigorously insider trading and market manipulation. They should to this effect actively supervise the market and effectively investigate suspicious transactions. Such investigations should include requiring any relevant documentation and data, as well as testimony, and carrying out on-site inspections when necessary. Finally, they should be able to impose sanctions on wrongdoing, by freezing assets, prohibiting professional activity or imposing any other adequate administrative and criminal sanctions, as appropriate in co-operation with the judicial authorities.

Again, the language of these paragraphs would seemingly prohibit *all* insider trading, which goes too far. An outright ban on insider trading would delay the movement of stock prices in the correct direction, to the detriment of the capital market. In cases where the inside information has been acquired justly, it appears to be a violation of property rights for some government to prevent individuals from trading in such property. The fact that the property in question is knowledge rather than something tangible merely obscures the substance of the basic transaction, which involves the trading of property—cash for shares.

Actually, not all insider trading is illegal. The laws in a number of countries allow it, provided that disclosure of the insider trades is made within some short period of time. Furthermore, not all insider trading is considered abusive, even by OECD standards. The *OECD White Paper on Corporate Governance in Asia* (2003) admits as much. At page 27 it states:

> With regard to self-dealing/related-party transactions involving the properly disclosed participation of an insider, it is important to remember that not all self-dealing/related-party transactions are abusive, and that some—e.g. executive-compensation arrangements—are unavoidable. A transaction between the company and its insider(s) is only considered abusive when the price is unfair to the company by reference to the price the company would have received from an unrelated party dealing at arm's length.

At pages 72–73 it summarizes the various insider trading civil and criminal penalties for thirteen Asian countries. Eleven countries provide some kind of civil liability, all thirteen assess fines in some cases, and eleven countries have penalties that include possible imprisonment, ranging from a maximum of two years in Thailand to twenty-one years in the Philippines.

5.4.2 The World Bank and IMF Position

The World Bank and IMF have a joint project to issue Reports on the Observance of Standards and Codes (ROSC). Their reports benchmark the state of corporate governance in several countries against the OECD Principles of Corporate Governance. Their report on the Czech Republic (World Bank 2002) is indicative of the kind of reports they have been issuing on the subject of corporate governance in general, and insider trading in particular. At page 8 it states:

> Self-dealing and insider trading have been reported and appear to be pervasive.... Securities laws prohibit the use of inside information for personal benefit. Breaches of the law are punishable by fines up to CZK 20 million (USD 567,000).

The report on Bulgaria (World Bank 2002) states that the law:

> provides for extensive prohibitions of insider trading and market manipulation, including prohibition against entering into transactions, spreading false rumors and forecasts or other acts with the intent of creating of false perception of the prices or volume

> of traded securities. An insider is defined to include members of management and boards of directors, persons holding ten percent of the shares of a company (directly or through related parties) or someone who due to his profession, activities, duties or relations of connection of a traded company has access to privileged information. Insider trading and market manipulation are subject only to civil sanctions and do not carry criminal liability. However, market participants complain that information regarding tender offers is distributed very slowly, allowing for the potential for insider trading. (p. 8)

The report recommends instituting criminal penalties in addition to the already existing civil penalties once the market becomes more active (p. 9).

The World Bank report on Croatia reveals that the securities law prohibits insider trading and provides for fines and imprisonment. However, the law requires insider trades to be reported to the Securities Commission and to the stock exchange within 7 days, so apparently some insider trading is not illegal (World Bank 2001, 10–11).

The World Bank report on Georgia also addresses the issue of insider trading (World Bank 2002). The rules in Georgia prohibit the use of insider information to: "(1) acquire or dispose of shares, (2) disclose insider information to any third party unless the disclosure is made in the normal course of professional duties, or (3) recommend or procure a third party to acquire or dispose of shares" (p. 10). The Georgia Stock Exchange Code of Ethics prohibits member-brokers from using information regarding security ownership to increase, decrease, or create purchases, sales, or exchanges of securities except when the beneficial owner of the security approves.

The World Bank report on Hungary (2003) states that the Hungarian laws on insider trading largely follow the EU rules. The rules are well defined. Insider trades must be reported within 2 days. Civil penalties are provided for violating the law and are equal to the amount of profit generated by the insider trade. The World Bank recommends that the level of fines should be greatly increased. The criminal law has a slightly different definition of insider trading and has penalties of up to three years in prison for violation of the law.

The World Bank also has a report on the Republic of Korea (2003). Korean law strictly prohibits trading in material nonpublic information. Violators are subject to fines and imprisonment and may be held liable for damages. Short-swing profits must be disgorged and profits earned within

6 months must be returned to the firm. Self dealing involving directors requires board approval.

The World Bank report on Latvia (2002) reveals that the civil law and securities law prohibit insider trading by employees, brokers, the Central Depository, and third persons who have information from inside sources (p. 8). Criminal law provisions were in the draft stage at the time the World Bank report was issued.

The law in Lithuania (World Bank 2002) provides for both fines and imprisonment for insider trading violations. The World Bank recommends that insider trading be required for 5 percent shareholders and that monetary fines should be increased. It also recommended that the evidentiary burden for proving insider trading violations should be reduced and that there should be clear disclosure of and approval for potential self dealing actions (p. 6).

The Slovak Republic prohibits insider trading (World Bank 2003). The law defines inside traders as shareholders, employees, professionals, or other positions or offices authorized to acquire inside information. Inside information is defined as "information which has not been published, but which could significantly influence the price of securities" (p. 8). The report also mentions that, although insider trading is illegal, there appears to be no enforcement or surveillance programs that attempt to prevent or detect it. The World Bank recommends adopting some enforcement authority.

The sixteen World Bank *Report on the Observance of Standards and Codes (ROSC), Corporate Governance Country Assessment* studies that have been completed as of this writing categorized the extent of compliance with the OECD benchmark on insider trading into the following five categories: (1) observed, (2) largely observed, (3) partially observed, (4) materially not observed, and (5) not observed. Table 5.1 shows how closely some countries comply with the OECD benchmark rule on insider trading.

As can be seen, most countries miss the OECD benchmark, some by a considerable degree. One might expect that the more developed countries and the countries that either recently became EU members or that are aiming at near-term EU membership would come closer to the OECD benchmark than the other countries, but such is not necessarily the case. The Czech Republic had one of the lowest rankings. Slovakia ranked only slightly higher. Of the Eastern and Central European countries, Hungary did best, with the highest ranking.

TABLE 5.1. Extent of compliance with OECD benchmark on insider trading

Country	Observed	Largely observed	Partially observed	Materially not observed	Not observed
Bulgaria		X			
Chile			X		
Colombia				X	
Croatia		X			
Czech Republic				X	
Egypt		X			
Georgia				X	
Hungary	X				
Korea			X		
Latvia		X			
Lithuania		X			
Mauritius			X		
Mexico		X			
Philippines		X			
Slovak Republic			X		
South Africa	X				

Source: World Bank ROSC Reports www.worldbank.org.

5.5 CONCLUDING COMMENTS

Several studies show that insider trading results in a positive-sum game. There are more winners than losers. Thus, it is ethically justified from a utilitarian perspective, at least in the cases where the result is a positive-sum game. However, gathering reliable data to conduct such studies is hampered because of the fact that some insider trading activity is illegal (Bainbridge 2000). Also, it is not always possible to know whether the result is a positive-sum game, even after the fact. That is one of the insoluble structural deficiencies of the utilitarian approach. Thus, utilitarian ethics is not a good tool for analysis of insider trading cases.

Not all insider trading results in the violation of anyone's rights. In many cases, insider trading is merely the exercise of property rights. Thus, from a rights perspective, it cannot be said that there is necessarily anything wrong with insider trading. It depends on whether anyone's rights are violated in a particular instance. That being the case, any laws that transition economies adopt that outlaw *all* forms of insider trading are bad laws. There should be no blanket prohibitions of insider trading because such laws violate property rights, the right to sell information.

The governments of transition economies should not be so quick to adopt laws that mirror the laws of developed countries, even if the OECD, the World Bank, the IMF, or other organizations put pressure on them to do so. The main responsibility of the political leaders in these countries is to their people, not to some far-off organization that may or may not have the best interests of the people in mind. Legislators have a fiduciary duty to their constituents to make good laws and not to make bad laws. Any insider trading laws they make should be based on the application of some recognized value system. Rights theory seems to be the superior approach, since utilitarianism has so many insoluble structural defects. But even applying utilitarianism to insider trading legislation is better than relying on emotional appeals to determine what form legislation should take.

The presumption should be that all capitalist acts between consenting adults should be legal and unregulated. The only exceptions should be in cases where someone's rights are violated or where some fiduciary duty has been breached. In cases where rights have been violated, the perpetrators should be punished. There are already laws on the books that prohibit the violation of rights, in most cases. Transition economies need to enact such laws where they do not already exist. In cases where a fiduciary duty has been breached, there are already laws on the books, or should be. There is no need to have a special law for breaches of fiduciary duty that involve insider trading.

REFERENCES

Allen, F. 1984. A welfare analysis of rational expectations equilibria in markets. Manuscript, Wharton School, University of Pennsylvania, cited in Bhattacharya and Nicodano 2001.

Bainbridge, S. 2000. Insider trading. In *Encyclopedia of law and economics III*, 772–812. Cheltenham, UK: Edward Elgar.

Baughman, L. M., and T. Emrich. 1985. Analysis of the impact of the Textile and Apparel Trade Enforcement Act of 1985. International Business and Economic Research Corporation, cited in Destler and Odell 1987.

Bentham, J. [1781] 1988. *The principles of morals and legislation*. Amherst, NY: Prometheus Books.

Berlin, I. 1991. *The crooked timber of humanity*. Princeton, NJ: Princeton University Press.

Berlin, I. 2001. *Against the current: Essays in the history of ideas*. Princeton, NJ: Princeton University Press.

Bernardo, A. E. 2001. Contractual restrictions on insider trading: A welfare analysis. *Economic Theory* 18:7–35.

Bhattacharya, S., and G. Nicodano. 2001. Insider trading, investment, and liquidity: A welfare analysis. *Journal of Finance* 56(3):1141–56.

Bovard, J. 1991. *The fair trade fraud*. New York: St. Martin's Press.

Buchanan, J. M., and M. R. Flowers. 1975 *The public finances*. 4th ed. Homewood, IL: Richard D. Irwin.

Chakravarty, S., and J. J. McConnell. 1997. An analysis of prices, bid/ask spreads, and bid and ask depths surrounding Ivan Boesky's illegal trading in Carnation Stock. *Financial Management* 26:18–34.

Cornell, B., and E. Surri. 1992. The reaction of investors and stock prices to insider trading. *Journal of Finance* 47(3):1031–59.

Denzau, A. 1985. American steel: Responding to foreign competition. Washington University, Center for the Study of American Business, St. Louis, MO.

Denzau, A. 1987. How import restraints reduce employment. Washington University, Center for the Study of American Business, St. Louis, MO.

Destler, I. M., and J. S. Odell 1987. *Anti-protection: Changing forces in United States trade politics*. Washington, DC: Institute for International Economics.

Dostoevsky, F. 1952. *The Brothers Karamazov*. In *Great books of the Western world*, ed.-in-chief, R. M. Hutchins. Vol. 52. Chicago: Encyclopedia Britannica.

European Union. 1989. Council Directive 89/592 of November 13, 1989 Coordinating Regulations on Insider Dealing, 1989 O.J. (L 334) 30.

Frey, R. G. 1984. *Utility and rights*. University of Minnesota Press, Minneapolis.

Gevurtz, F. A. 2002. Transnational business law in the twenty-first century: The globalization of insider trading prohibitions. *Transnational Lawyer* 15:63–97.

Jeng, L. A., A. Metrick, and R. Zeckhauser. 2003. Estimating the returns to insider trading: A performance-evaluation perspective. *Review of Economics and Statistics* 85(2):453–71.

Jevons, W. S. 1871. *The theory of political economy*.

Lekkas, P. 1998. Insider trading and the Greek stock market. *Business Ethics—A European Review* 7(4):193–99.

Leland, H. 1992. Insider trading: Should it be prohibited? *Journal of Political Economy* 100:859–87, cited in Bhattacharya and Nicodano 2001.

Manne, H. G. 1966. *Insider trading and the stock market*. New York: Free Press.

Manne, H. G. 1985. Insider trading and property rights in new information. *Cato Journal* 4:933–43.

Mendez, J. A. 1986. The short-run trade and employment effects of steel import restraints. *Journal of World Trade Law* 20:554–66.

Menger, C. [1871] 1950. *Principles of economics*. New York: American Book-Knickerbocker Press.

Meulbroek, L. 1992. An empirical analysis of illegal insider trading. *Journal of Finance* 47(5):1661–99.

Mill, J. S. [1861] 1979. *Utilitarianism*. Indianapolis, IN: Hackett.

Mises, L. 1956. *The anti-capitalistic mentality*. Princeton, NJ: D. Van Nostrand.

OECD. 1999. *Principles of corporate governance*. Paris: OECD.

OECD. 2003. White paper on corporate governance in South Eastern Europe. (June 19) Paris: OECD.

OECD. 2003. White paper on corporate governance in Asia. (July 15). Paris: OECD.

OECD. 2004. OECD principles of corporate governance: Draft revised text (January). Paris: OECD.

O'Hara, P. A. 2001. Insider trading in financial markets: Legality, ethics, efficiency. *International Journal of Social Economics* 28(10–12):1046–62.

Qu, C. Z. 2001. An outsider's view on China's insider trading law. *Pacific Rim Law & Policy Journal* 10:327–52.

Quinn, R. W. 2003. The misappropriation theory of insider trading in the Supreme Court: A (brief) response to the (many) critics of *United States v. O'Hagan*. *Fordham Journal of Corporate & Financial Law* 8:865–98.

Repullo, R. 1994. Some remarks on Leland's model of insider trading. Working paper, CEMFI (Madrid), cited in Bhattacharya and Nicodano 2001.

Ricardo, D. [1817] 1996. *Principles of political economy and taxation*. Amherst, NY: Prometheus Books.

Rothbard, M. N. 1970. *Man, economy and state*. Los Angeles: Nash.

Rothbard, M. N. 1997. *The logic of action one: Method, money, and the Austrian school*. Cheltenham, UK: Edward Elgar.

Schoeck, H. 1987. *Envy: A theory of social behavior*. Indianapolis, IN: Liberty Fund.

Seligman, J. 1998. A mature synthesis: O'Hagan resolves "insider" trading's most vexing problems. *Delaware Journal of Corporate Law* 23(1):1–28.

Shaw, W. H. 1999. *Contemporary ethics: Taking account of utilitarianism*. Oxford: Blackwell.

Shell, G. R. 2001. When is it legal to trade on inside information? *MIT Sloan Management Review* 89–90.

Smart, J. J. C., and B. Williams. 1973. *Utilitarianism—For and against*. Cambridge: Cambridge University Press.

Smith, A. [1776] 1937. *An inquiry into the nature and causes of the wealth of nations*.

Strudler, A., and E. W. Orts. 1999. Moral principle in the law of insider trading. *Texas Law Review* 78(2):375–438.

Swanson, C. B. 1997. Reinventing insider trading: The Supreme Court misappropriates the misappropriation theory. *Wake Forest Law Review* 32(4):1157–1212.

Walras, L. 1874. *Elements of pure economics*.

Weiss, E. J. 1998. *United States v. O'Hagan*: Pragmatism returns to the law of insider trading. *Journal of Corporation Law* 23(3):395–438.

World Bank. 2001. Report on the observance of standards and codes (ROSC), Corporate governance country assessment, Arab Republic of Egypt, September. World Bank, Washington, DC. www.worldbank.org.

World Bank. 2001. Report on the observance of standards and codes (ROSC), Corporate governance country assessment, Republic of Croatia, World Bank, Washington, DC. www.worldbank.org.

World Bank. 2001. Report on the observance of standards and codes (ROSC), Corporate governance country assessment, Republic of the Philippines, September. World Bank, Washington, DC. www.worldbank.org.
World Bank. 2002. Report on the observance of standards and codes (ROSC), Corporate governance country assessment, Bulgaria, September. World Bank, Washington, DC. www.worldbank.org.
World Bank. 2002. Report on the observance of standards and codes (ROSC), Corporate governance country assessment, Czech Republic, July. World Bank, Washington, DC. www.worldbank.org.
World Bank. 2002. Report on the observance of standards and codes (ROSC), Corporate governance country assessment, Georgia, March. World Bank, Washington, DC. www.worldbank.org.
World Bank. 2002. Report on the observance of standards and codes (ROSC), Corporate governance country assessment, Latvia, December. World Bank, Washington, DC. www.worldbank.org.
World Bank. 2002. Report on the observance of standards and codes (ROSC), Corporate governance country assessment, Mauritius, October. World Bank, Washington, DC. www.worldbank.org.
World Bank. 2002. Report on the observance of standards and codes (ROSC), Corporate governance country assessment, Republic of Lithuania, July. World Bank, Washington, DC. www.worldbank.org.
World Bank. 2003. Report on the observance of standards and codes (ROSC), Corporate governance country assessment, Chile, May. World Bank, Washington, DC. www.worldbank.org.
World Bank. 2003. Report on the observance of standards and codes (ROSC), Corporate governance country assessment, Colombia, August. World Bank, Washington, DC. www.worldbank.org.
World Bank. 2003. Report on the observance of standards and codes (ROSC), Corporate governance country assessment, Hungary, February. World Bank, Washington, DC. www.worldbank.org.
World Bank. 2003. Report on the observance of standards and codes (ROSC), Corporate governance country assessment, Republic of Korea, September. World Bank, Washington, DC. www.worldbank.org.
World Bank. 2003. Report on the observance of standards and codes (ROSC), Corporate governance country assessment, Mexico, September. World Bank, Washington, DC. www.worldbank.org.
World Bank. 2003. Report on the observance of standards and codes (ROSC), Corporate governance country assessment, Republic of South Africa, July. World Bank, Washington, DC. www.worldbank.org.
World Bank. 2003. Report on the observance of standards and codes (ROSC), Corporate governance country assessment, Slovak Republic, October. World Bank, Washington, DC. www.worldbank.org.
Yunker, J. A. 1986. In defense of utilitarianism: An economist's viewpoint. *Review of Social Economy* 44(1):57–79.

CHAPTER 6

Credit Derivatives and Inside Information

Paul U. Ali

CONTENTS

6.1 CREDIT DERIVATIVES AND INSIDER TRADING

The laws restricting insider trading in most markets—regardless of whether those laws are predicated on punishing the misappropriation of confidential information or on enforcing parity of information—have traditionally focused on the sale and purchase of publicly traded equity instruments by "insiders," namely, persons in possession of price-sensitive, nonpublic information relating to those instruments. One of the clearest, and most common, examples of insider trading is the case of an insider who, having nonpublic information about the identity of the target of an upcoming takeover bid, purchases shares in the takeover target ahead of the public announcement of the bid.

Trading shares is not, however, the only—or, even it now seems, the predominant—means of profiting from inside information (Drummond 2007). An insider could equally, in terms of the above example, exploit the information in his or her possession by purchasing call options over shares in the takeover target. The leveraged nature of options makes this a more profitable alternative for those engaged in insider trading and it is unsurprising that one of the three recent examples of insider trading mentioned in the Introduction to this book involved share options. Regulators have long been aware of the potential of share options to be used for insider trading and, accordingly, it is commonplace for trading in exchange-traded options over shares in takeover targets prior to the announcement of takeover bids to be subjected to the same level of regulatory scrutiny as the pre-bid trading of the shares themselves.

More recently, the focus of regulators has extended beyond equity derivatives to other instruments which, due to their being traded exclusively in the over-the-counter markets and not having a direct exposure to equity prices, have not ordinarily been considered as a medium for insider trading. The instruments in question are credit default swaps, the most common type of credit derivative. Following a spate of well-publicized incidents of significant pre-bid trading in credit derivatives linked to the debt obligations of takeover targets, it has become increasingly clear that an insider does not need to trade shares or even derivatives linked to shares in order to exploit inside information about shares profitably. In each of the reported instances, the price of credit default swaps referencing the takeover target (that is, the cost of purchasing credit protection in respect of certain debt obligations of the target) rose sharply in the period immediately prior to the public announcement of the bid (Drummond 2007; Harrington 2006; Ng et al. 2006; Scannell et al. 2006).

These credit default swaps are basically privately negotiated contracts between two parties (a protection seller and a protection buyer) under which the protection seller agrees to assume the credit risk of a third party (the reference entity) in respect of specified debt obligations (the reference obligations) of that party (de Vries Robbé and Ali 2005). The protection buyer, by paying a fee to the protection seller, is able to purchase protection from the protection seller against a material deterioration in the creditworthiness of the reference entity, as exemplified typically by the default or insolvency of that entity. If the reference entity defaults in the performance of the reference obligations or becomes insolvent, the protection seller will be obligated either to make a payment to the protection buyer of an amount calculated by reference to the fall in the value of the reference obligations or to purchase the reference obligations (or substitute debt obligations of the reference entity) from the protection buyer for their full face value. Thus, the greater the likelihood a reference entity will default or become insolvent, the more expensive it will be to purchase protection in respect of the reference obligations of that entity.

Generally, a fall in the price of a company's shares is likely to be mirrored in an increase in the price of credit default swaps linked to that company's debt obligations, as the two events may both represent a worsening of the company's creditworthiness with negative implications for both shareholders and creditors (Byström 2005; Logie and Castagnino 2006). In contrast, in the context of a takeover bid, a run-up in the share price of the takeover target (which benefits the shareholders of the target) may nonetheless be accompanied by an increase in the price of credit default swaps, where the takeover is perceived as being likely to affect adversely the target's ability to service its debt obligations and thus detrimental to the interests of the target's creditors (Berndt and Ostrovnaya 2007). The positive correlation between share and credit default swap prices has been clearest in the case of leveraged buyouts where the assumption of substantial debt by the target to finance the acquisition increases the credit risk of the target (Ng et al. 2006).

The above correlations between share and credit default swap prices mean that an insider can take advantage of nonpublic information affecting the price of shares by entering into a credit default swap, as an alternative to trading the shares or entering into an equity derivative linked to the price performance of the shares (for example, by purchasing credit protection and, following the announcement of a takeover bid for the reference entity, engaging in a reverse trade by selling credit protection over the same reference obligations at a higher price).

This raises important questions for the regulation of derivatives in general and, more particularly, the possible extension of insider trading laws to instruments other than equity instruments (or the broader enforcement of existing laws in markets such as Australia whose insider trading laws already apply to all derivatives) (Brown-Hruska and Zwirb 2007).

6.2 CREDIT DERIVATIVES AND INFORMATION ASYMMETRY

The use of credit risk as a proxy for equity price risk is but one of a number of ways in which credit default swaps (and other credit derivatives) can be used to exploit informational advantages or, in other words, make use of the information asymmetry between the protection seller and the protection buyer.

Commercial banks, in particular, routinely employ credit default swaps (either on a standalone basis or as part of synthetic securitizations) to hedge the credit risk of their loan portfolios and, by so doing, release the risk capital held by them in respect of the reference obligations (Bomfim 2005). These banks, due to the nonpublic information they possess about the creditworthiness of their borrowers (which the borrowers are obligated to convey in a timely manner to the banks in compliance with the reporting covenants contained in the loan agreements between the parties), are in a position to exploit that information profitably by entering into credit default swaps with parties over whom they hold an informational advantage (Acharya and Johnson 2007; Duffee and Zhou 2001). For example, a bank that has nonpublic information about a material deterioration in the creditworthiness of a borrower—and even more so a bank that is in a position to call an event of default or is aware of circumstances that are likely to trigger that right—could profit from that information by purchasing credit protection in respect of the obligations owed by the borrower to the bank more cheaply than it could have had that information been generally available to participants in the credit derivatives market.

6.3 SELF-REFERENCED CREDIT DERIVATIVES

The two instances of insider trading involving credit derivatives identified above have each concerned inside information held by one of the parties to the credit derivative about a third party, the reference entity. However, just as it is possible for an insider to use credit derivatives to exploit nonpublic information about a third party, so too is it possible for that third party to use credit derivatives to profit from the informational advantages it naturally holds over all other parties in relation to the nonpublic information it has about its own affairs.

These so-called self-referenced credit derivatives differ from other credit derivatives in one key respect: the reference entity and the protection seller are the same party. The protection buyer, instead of purchasing protection against the default or insolvency of the reference entity from a separate protection seller, purchases that protection from the reference entity itself.

It is, however, not usual for the sale of credit protection by a reference entity to be encountered in the form of a standalone credit default swap, that is, in an "unfunded" form. This is due to two decisive factors. First, if the reference entity's obligation to make a payment under the swap (see Section 6.1 above) is triggered by the reference entity's insolvency, then any such payment is at risk of being set aside by the liquidator of the reference entity as a preference or a fraudulent conveyance (Ali 2004; Firth 2007). Second, the situation in which the reference entity is obligated to make a payment under the swap—usually, the default or insolvency of the reference entity—is likely to be the very situation in which the reference entity lacks the resources to make that payment in full (Ali 2004).

For these reasons, self-referenced credit derivatives are implemented in a funded form with the reference entity lodging with a third party sufficient collateral to cover its payment obligations in relation to its sale of credit protection. The legal structure of a funded self-referenced credit derivative is the same as that employed in conventional, fully funded synthetic securitizations, save for the fact that there is only a single investor and that investor is the reference entity (Ali 2004).

A special purpose vehicle (SPV) is established to act as the conduit for the transfer of credit risk from the protection buyer to the reference entity and also to hold collateral to support this transfer of credit risk. This is accomplished by the SPV entering into a credit default swap with the protection buyer and issuing debt securities to the reference entity, with the cash proceeds from the issue of securities being invested by the SPV in collateral (for example, treasury securities or certificates of deposit). The credit default swap transfers the credit risk of the reference entity to the SPV, and that risk is passed on to the reference entity by making the SPV obligation to redeem the securities for their face value on the scheduled maturity date conditional upon the SPV not being obligated, during the term of the securities, to make a payment to the protection buyer under the credit default swap. If, during the term of the securities, the reference entity defaults in the performance of the reference obligations or becomes insolvent, the securities held by the reference entity will be immediately redeemed for the amount remaining (if any) after the collateral has been applied in satisfaction of the SPV's payment obligations under the credit default swap. Figure 6.1 depicts a generic self-referenced structure.

The key driver for these self-referenced credit derivatives is the nature of the returns that can be generated for the reference entity. In exchange for its assumption of credit risk, the reference entity receives interest payments on the debt securities issued to it by the SPV, comprising the aggregate of the income generated from the collateral held by the SPV and the

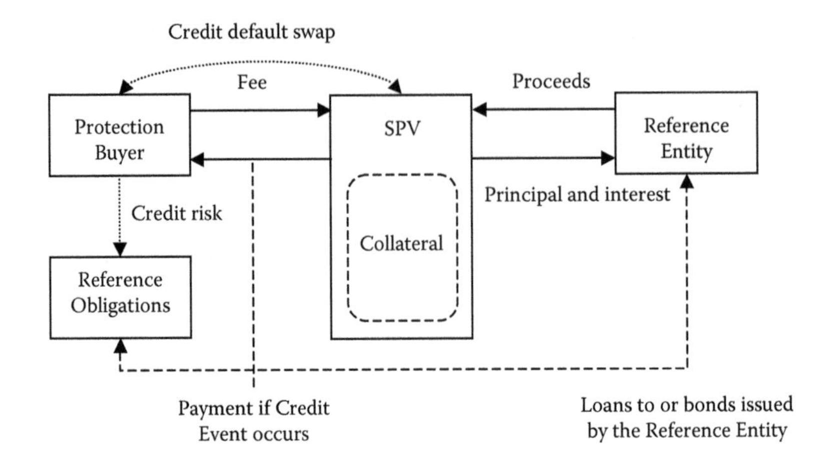

FIGURE 6.1. Self-referenced credit derivative.

fee paid to the SPV by the protection buyer (net of any transaction costs). The risk associated with the returns on the securities is less for the reference entity, due to the information it possesses about its own financial affairs, than it would be for any other party investing in those securities. In addition, the returns on the securities due to their composition are generally higher than the returns on comparably rated, conventional debt securities (Fitch Ratings 2004).

However, self-referenced credit derivatives entail problems not encountered in conventional synthetic securitizations. First, due to the strong, positive correlation between the risk and return profile of the debt securities issued by the SPV and the risk and return profile of the reference entity, the former are likely to be worthless or close to worthless at the very time that the reference entity most needs the securities to retain their value (as the securities will lose value as the creditworthiness of the reference entity deteriorates) (Ali 2004; Fitch Ratings 2004). Second, the liquidity profile of the debt securities will be also be strongly, positively correlated to the liquidity profile of the reference entity (Ali 2004; Fitch Ratings 2004). It is therefore likely to be very difficult, if not impossible, for the reference entity to sell the debt securities, again, at the very time when the reference entity is most in need of cash (as any deterioration in the reference entity's creditworthiness will also impair the liquidity of the debt securities). Finally, the liquidator of the reference entity may be able to unwind a self-referenced structure and thus claw-back any payments made to the protection buyer by the SPV, on the grounds that the reference entity has, by agreeing to the redemption of the debt securities for less than their face value in the event of its insolvency, illegally deprived its creditors of the benefit of the full face vale of those securities (Ali 2004; Farrell 2003; Firth 2007).

6.4 CONCLUDING REMARKS

The inherent flexibility of credit derivatives and the ease with which they can be customized (due in no small measure to the development of standard form credit derivatives documentation by the International Swaps and Derivatives Association) to meet the risk transfer and investment objectives of protection buyers and sellers well explain the widespread use of credit derivatives by participants in the financial markets, ranging from commercial banks desiring to lay off the credit risk of their loan portfolios to hedge funds and other institutional investors seeking specific credit exposures. However, credit derivatives also offer opportunities for market participants to engage in insider trading due, in particular, to the information asymmetry associated with credit derivatives and the relationship between share and credit default swap prices.

This information asymmetry is present in the case of both conventional credit derivatives (where, for example, the protection buyer possesses, as a result of its lender–borrower relationship with the reference entity, price-sensitive, nonpublic information about the reference entity) and self-referenced credit derivatives (since a reference entity has better information about its own financial affairs than anyone else). Moreover, despite the absence of an explicit link to the price performance of shares, the relationship between share and credit default swap prices makes its possible for insiders to use credit derivatives to profit from nonpublic, price-sensitive information about shares. Also, the fact that credit derivatives are linked to debt obligations rather than shares or other equity instruments makes the detection of this type of insider trading considerably more difficult.

Finally, although self-referenced credit derivatives enable reference entities to trade on the basis of inside information about themselves, it is unlikely that parties other than a protection buyer that is already in a borrower–lender relationship with the reference entity would be willing to enter into such transactions (due to the obvious informational advantages possessed by the reference entity). The problem of information asymmetry that would otherwise confront the protection buyer can be addressed by the reporting obligations imposed on the reference entity as part of the lender–borrower relationship. This does not, however, mean that self-referenced credit derivatives have no utility for reference entities. One could, instead, characterize such instruments as analogous to secured finance (where the performance by the reference entity of the reference obligations is supported by collateral) save that the returns on the debt securities issued by the SPV in a self-referenced structure may exceed the returns that could have been obtained from the reference entity itself investing in collateral of a similar credit rating.

REFERENCES

Acharya, V. V., and T. C. Johnson 2007. Insider trading in credit derivatives. *Journal of Financial Economics* 84(1):110–41.

Ali, P. U. 2004. The conundrum of self-referenced credit-linked notes. *Journal of International Banking Law and Regulation* 19:326–30.

Berndt, A., and A. Ostrovnaya. 2007. Information flow between credit default swap, option and equity markets. Working paper, Tepper School of Business, Carnegie Mellon University (December).

Bomfim, A. N. 2005. *Understanding credit derivatives and related instruments.* San Diego, CA: Elsevier Academic Press.

Brown-Hruska, S., and R. S. Zwirb. 2007. Legal clarity and regulatory discretion—Exploring the law and economics of insider trading in derivatives markets. *Capital Markets Law Journal* 2(3):245–59.

Byström, H. 2005. Credit default swaps and equity prices: The iTraxx CDS index market. Working paper, Department of Economics, Lund University (May).

de Vries Robbé, J. J., and P. U. Ali. 2005. *Opportunities in credit derivatives and synthetic securitisation.* London: Thomson Financial.

Drummond, B. 2007. Insider trading. *Bloomberg Markets* (August).

Duffee, G. R., and C. Zhou. 2001. Credit derivatives in banking: Useful tools for managing risk? *Journal of Monetary Economics* 48(1):25–54.

Farrell, S. 2003. Defending the value of self-referenced credit-linked notes. *International Financial Law Review* 22(10):31–33.

Firth, S. 2007. Self-referenced credit derivatives—Are they enforceable under English law? *Capital Markets Law Journal* 1(1):21–31.

Fitch Ratings. 2004. Self-referenced CLNs raise questions and concerns. *Credit Policy Special Report* (13 January).

Harrington, S. D. 2006. Credit-default swap traders anticipated announcements of LBOs. *Bloomberg News* (27 October).

Logie, M. J., and J. P. Castagnino. 2006. Equity default swaps and the securitisation of risk. In *Innovations in Securitisation,* ed. J. J. de Vries Robbé and P. U. Ali. The Hague: Kluwer Law International.

Ng, S., D. K. Berman, and K. Scannell. 2006. Are deal makers on Wall Street leaking secrets? *Wall Street Journal,* July 28, p. C1.

Scannell, K., S. Ng, and A. MacDonald. 2006. Can anyone police the swaps? *Wall Street Journal,* p. C1. August 31, p. C1.

Part 2

Regulating Insider Trading

A. Illegal Insider Trading

CHAPTER 7

Inside Information and the European Market Abuse Directive (2003/6)

Blanaid Clarke

CONTENTS

7.1 INTRODUCTION

The Insider Dealing Directive (1989/592) was the first directive aimed at regulating capital markets in Europe. It was subsequently replaced by the Market Abuse Directive (2003/6), which regulated all aspects of market abuse including market manipulation. The objectives of both directives in regulating insider dealing were to introduce a coordinated legislative response to safeguard the integrity of European financial markets and to enhance investor confidence in those markets. These are based primarily on the perception of insider dealing as unfair in that it gives insiders an unjust advantage over the other market participants. In addition to prohibiting insider dealing, the Market Abuse Directive (MAD) stipulates several preventative measures aimed at reducing the incidence of market abuse and thus reducing the likelihood that the integrity of the market will be undermined. These preventative measures involve a new disclosure regime for issuers and market participants. The regime includes the disclosure and handling of inside information by issuers, the disclosure of dealings by directors and senior management in their own company's shares to the market, and the reporting by firms of suspicious transactions to the competent authorities.

In order for MAD to constitute an effective regulatory mechanism, it must achieve its goals and secure a high level of compliance from member states and market participants (Parker et al. 2004). In this context, the definition of "inside information" is crucial to the efficient application of European insider dealing regulation as it is the term upon which all the duties set out in MAD are based. A clear and workable definition is essential. From a preemptive perspective, market participants must be able to identify "inside information" in order to disclose it and avoid dealing accordingly. From an enforcement perspective, a viable definition facilitates the successful prosecution of breaches and in doing so provides a sufficient deterrent to potential wrongdoers. This chapter examines whether MAD provides such a workable definition of "inside information" and focuses in particular on the interpretation of this term in the United Kingdom and Ireland.

7.2 THE MARKET ABUSE DIRECTIVE

MAD was the first directive to be adopted under the "Lamfalussy format." This format was based on a report by the Committee of Wise Men's on the Regulation of European Securities Markets. The committee was established by the European Council under the chairmanship of Alexandre Lamfalussy. The report, published in February 2001, concluded that the existing regulatory system was too slow and too rigid, failed to distinguish between core principles and detail, and was unevenly implemented. The last problem was attributed both to the use of ambiguous terms and to the lack of coordination by an effective network of European regulators. The report was subsequently endorsed by the European Council (in the Resolution on More Effective Securities Markets Regulation in the EU, March 2001) and the European Parliament (in the Resolution on the Implementation of Financial Services Legislation, February 2002). The Lamfalussy format constitutes a new four-level legislative procedure. The first level is framework principles to be decided by normal European Union (EU) legislative procedures. The second level involves technical implementing measures to be determined by the European Commission with the assistance of two new committees, the European Securities Committee (ESC) and the Committee of European Securities Regulators (CESR). The ESC is composed of high-level representatives from member states and its task is to advise the European Commission on issues relating to securities policy. CESR is an independent advisory body composed of representatives of the national public authorities competent in the field of securities in the different member states. The role of CESR is **to improve coordination among securities regulators**, to advise the European Commission on the technical details of securities legislation, and to **ensure more consistent and timely day-to-day implementation of community legislation in the member states.** CESR thus facilitates the third level of the Lamfalussy format, which is enhanced cooperation and networking among EU securities regulators. Finally, the fourth level involves strengthened enforcement involving "more vigorous action by the European Commission to enforce EU law underpinned by enhanced cooperation between the Member States, their regulators, and the private sector" (Gjersem 2003, 36). A big advantage of the Lamfalussy format approach is that it allows greater flexibility in updating the technical details related to the framework principles. This should ensure that the regulations can be kept up to date with market and supervisory developments.

MAD constitutes a Level 1 measure and sets out framework principles. At Level 2, CESR provided the European Commission, as mandated, with advice regarding Level 2 technical implementing measures for the proposed directive in December 2002 (CESR/02-089d) and in August 2003 (CESR/03-212c). The Commission adopted this advice and introduced

Commission Directive 2003/124/EC implementing MAD as regards the definition and public disclosure of inside information and the definition of market manipulation; Commission Directive 2003/125/EC implementing MAD as regards the fair presentation of investment recommendations and the disclosure of conflicts of interest; Commission Directive 2004/72/EC implementing MAD as regards accepted market practices, the definition of inside information in relation to derivatives on commodities, the drawing up of lists of insiders, the notification of managers' transactions, and the notification of suspicious transactions; and Commission Regulation 2273/2003 implementing MAD as regards exemptions for buy-back programs and stabilization of financial instruments. All these directives have been implemented. At Level 3, CESR released in May 2005 a first set of guidance and information on the common operation of the MAD as regards accepted market practices and the notification of suspicious transactions (CESR/04-505b). In July 2007, CESR released a second set of guidance and information on the common operation of MAD to the market (CESR/06-562b).

As all member states already had existing provisions in place regulating insider dealing, certain member states such as the United Kingdom and Ireland chose to adapt the existing systems in order to comply with MAD while retaining any requirements which went beyond those, but were not inconsistent with those, in the directive. MAD was implemented in the United Kingdom through the Financial Services and Markets Act 2000 (Market Abuse) Regulations 2005 (SI No. 301 of 2005), which amended the Financial Services and Markets Act 2000, and through changes to the Financial Services Authority Code of Market Conduct. In Ireland, MAD was implemented by way of the Market Abuse (Directive 2003/6/EC) Regulations 2005 (SI No. 342 of 2005) and Part 4 of the Investment Funds, Companies and Miscellaneous Provisions Act 2005.

A recent report by the British Institute of International and Comparative Law on the manner of implementation of MAD in five member states—the United Kingdom, Germany, France, Spain, and the Netherlands—highlighted the fact that in implementing MAD, member states tended to copy the provisions of the directive relating to insider dealing verbatim, thus aligning their national laws. (This was in marked contrast to the implementation of the "loosely drafted minimum harmonization" Insider Dealing Directive (1989/592).) The report attributes this mainly to the fact that the Level 2 directives provide so much detail to flesh out MAD principles that little room exists for national variations. It also identifies as causal factors the pressure on member states to adopt a common approach to the implementation of the first directive "to be adopted under the Lamfalussy procedures and the involvement of the national securities regulators through CESR in advising the Commission on the content of the Level 2 Directives." (Welch et al. 2005, 11) However, as the report correctly states, "it remains to be seen whether Member States will interpret similar provisions in similar ways or whether they will retain the approach developed in relation to the Insider Dealing Directive" (1989/592) ("the 1989 Directive").

7.3 DEFINITION OF "INSIDE INFORMATION" IN THE DIRECTIVE

The definition of "inside information" is set out initially in MAD but is expanded on in the Level 2 Commission Directive 2003/124/EC. Article 1(1) of MAD defines the term "inside information" as:

> information of a precise nature which has not been made public, relating, directly or indirectly, to one or more issuers of financial instruments or to one or more financial instruments and which, if it were made public, would be likely to have a significant effect on the prices of those financial instruments or on the price of related derivative financial instruments.

There are two major changes of note from the 1989 Directive. First, the definition in MAD refers to "financial instruments" rather than merely "transferable securities." Article 1(3) of MAD defines the term "financial instruments" as including transferable securities, ucits, options, derivatives, and any other instrument admitted to trading on a regulated market in the EU. This marks an acknowledgment that the scope of instruments affected by inside information is not limited to those of the issuer but also extends to related derivative financial instruments. Options on equity, futures on indices, and so forth will thus fall within the scope of MAD. The second change is that in addition to the general definition of inside information set out above, MAD provides a further two definitions. Article 1(1) of MAD provides a replacement definition for inside information in relation to derivatives on commodities. Such information is defined as:

> information of a precise nature which has not been made public, relating, directly or indirectly, to one or more such derivatives and

> which users of markets on which such derivatives are traded would expect to receive in accordance with accepted market practices on those markets.

This replaces the price-sensitivity aspect of the general definition with an expectation of disclosure in accordance with "accepted market practice." Furthermore, an additional definition of inside information is provided in Article 1(1) for persons charged with the execution of orders concerning financial instruments. For such persons inside information also means:

> information conveyed by a client and related to the client's pending orders, which is of a precise nature, which relates directly or indirectly, to one or more issuers of financial instruments or to one or more financial instruments and which, if it were made public, would be likely to have a significant effect on the prices of those financial instruments or on the price of related derivative financial instruments.

This definition thus focuses purely on information from a client which relates to pending orders from that client.

Because of the more prescriptive nature of the directive itself and partly because, as noted above, member states have moved closer to straight transposition in implementation, the definition of inside information adopted in EU member states has tended to be almost identical. In Ireland, Regulation 2 of the Market Abuse (Directive 2003/6/EC) Regulations 2005 defines insider information almost identically to the MAD as:

> information of a precise nature relating, directly or indirectly, to one or more issuers of financial instruments or to one or more financial instruments which has not been made public and which, if it were made public, would be likely to have a significant effect on the prices of those financial instruments or on the price of related derivative financial instruments.

In the United Kingdom, section 118C(2) of the Financial Services and Markets Act 2000 made one change (discussed below) in providing that in relation to qualifying investments, or related investments, which are not commodity derivatives, inside information is:

> information of a precise nature which is not generally available, relates directly or indirectly, to one or more issuers of the qualifying investments or to one or more of the qualifying investments, and would, if generally available, be likely to have a significant effect on the price of the qualifying investments or on the price of related investments.

The definitions of inside information in relation to client orders and derivatives were incorporated directly in both jurisdictions.

To constitute inside information thus under the MAD, there are five prerequisites. The first three relate to the nature or content of the information. First, the information must be of a "precise" nature. Second, the information must relate, directly or indirectly, to one or more issuers of financial instruments or to one or more financial instruments. Third, in relation to derivatives on commodities, the information must be information which users of markets on which derivatives are traded would expect to receive in accordance with accepted market practices on those markets. The fourth prerequisite is factual and relates to the question of disclosure of the information. The information must not have been "made public." The final point is arguably the trickiest as it involves an assessment of the consequences of its disclosure. The information must be information which, if it were made public, would be likely to have a significant effect on the prices of those financial instruments or on the price of related derivative financial instruments. Each of these points is now considered in turn.

7.3.1 Information of a Precise Nature

Although neither the 1989 Directive nor MAD defined the term "precise," the use of the Lamfalussy format allowed the term to be defined in Article 1(1) of Directive 2003/124. It provides that, for the purposes of MAD, information shall be deemed to be of a precise nature if:

> it indicates a set of circumstances which exists or may reasonably be expected to come into existence or an event which has occurred or may reasonably be expected to do so and if it is specific enough to enable a conclusion to be drawn as to the possible effect of that set of circumstances or event on the prices of financial instruments or related derivative financial instruments.

Both Ireland and the United Kingdom have implemented this provision verbatim into their implementing legislation.

The first part of the definition deals with the content of the information involved. It provides that the information should refer to circumstances or events which exist or have occurred, that is, matters of fact, or alternatively, those which might "reasonably be expected" to exist or to occur in the future. This reference to a reasonable expectation is consistent with CESR advice that the information should be based on firm and objective evidence, which can be communicated accurately as opposed to rumors (CESR/02-089d, para. 20). The advent of the information age and the emphasis on the Internet as a means of instantaneous communication has made it extremely easy to start and spread rumors. Clearly, it is neither desirable nor workable that all such rumors would give rise to restrictions on dealing under the directive. Thus a rumor concerning a merger, even one which may be at an advanced stage of negotiation between the parties, will not be included unless there is firm and objective evidence to substantiate it. CESR has advised recently that in considering what may reasonably be expected to come into existence, the key issue is whether it is reasonable to draw this conclusion based on the ex ante information available at the time (CESR/06-562b, para. 1.5). What is not clear is whether the reasonable expectation is that of the potential insider (in the context of his own knowledge and experience) or whether it is again a completely objective standard based perhaps on the reasonable investor.

If the information derives from a stage process, "every fact to do with the process, as well as the totality of the process itself," is precise information and therefore could constitute inside information (CESR/02-089d, para. 20). Thus, a factual statement that the parties have had a first meeting to discuss even the possibility of a merger may be inside information even if those discussions subsequently prove fruitless. Furthermore, the information may constitute inside information even if it lacks significant precision about certain elements of the event. Thus, inside information concerning a takeover bid need not refer, for example, to the terms of the bid or the timing of the bid. Finally, CESR has advised that a piece of information could be considered precise even if it refers to matters or events that could be alternatives. For instance, if the information was that a bidder proposed to acquire one of two companies, this would be considered precise information constituting inside information which could be used by an investor buying shares in the two companies.

The second part of the definition of "precise" information relates more to the effect of the information. Wymeersch has made the point that rumors harden into inside information when market traders take the information into account when valuing the underlying securities (1991, 114). The same view emerges in Directive 2003/124/EC. In order to be considered precise under MAD, the information must be "specific enough to allow a conclusion to be drawn about its impact on prices." CESR has acknowledged that the "precise" condition is very much linked to the "likely to have a significant effect on the price of the financial instrument" condition (discussed below) and that the characteristics of each condition may play an intensifying role on the occurrence of the other (CESR/02-089d, para. 18). The reference to a conclusion on its impact on prices may thus be viewed as a reflection of this relationship. CESR has stated that information would meet this particular requirement either when "it would enable a reasonable investor to take an investment decision without (or at very low) risk" or "when it is likely to be exploited immediately on the market" (CESR/02-089d, para. 20). Again, an example of this would be the identity of a takeover target despite the fact that many details are unavailable.

By way of contrast, it is interesting to consider the Australian Corporations Act 2001, which does not have a requirement of specificity or precision. The insider trading prohibition applies there to any information that is materially price-sensitive. Thus section 1042A of the Australian Corporations Act 2001 states that inside information includes:

> matters of supposition and other matters that are insufficiently definite to warrant being made known to the public and matters relating to the intentions, or likely intentions, of a person.

The Corporations and Markets Advisory Committee in its Report on Insider Dealing in 2003 advised against altering this position (Recommendation 22). It acknowledged the arguments made during a public consultation exercise that the inclusion of such a requirement would make it harder to prove insider dealing. In particular, concern was expressed that in many cases the prosecution might not be in a position to identify the precise information the defendant possessed and that the prosecution would have to rely on evidence of the defendant's access to information and inferences from the defendant's subsequent conduct. The committee concluded that to introduce a requirement of precision or specificity could "unduly narrow the application of the legislation and create artificial distinctions between what does and what does not constitute information" (para. 2.60). It advised that the need to establish the price sensitivity of the information was sufficient on its own and a requirement of precision or specificity was unnecessary. While undoubtedly the removal of the requirement at a European level would make it easier to prove insider dealing, it could also be argued that it would restrict a significant number of dealings by persons who happened to be aware of vague and imprecise rumors of dubious accuracy.

7.3.2 Information Relating to One or Several Issuers of Transferable Securities or to One or Several Transferable Securities

Moloney (2002, 753) has noted that the term "relating to" implies a rather elastic control on the type of information caught by the directive. Clearly, the definition of insider information refers to information which pertains to the issuer or its securities both directly and indirectly. This is deliberate and acknowledges the fact that the abusive potential of insider dealing is not dependent on whether it has a direct or indirect effect on the issuer or whether it is located inside the issuer's sphere or outside the issuer's sphere (CESR/02-089d, paras. 30, 31). Either way, insider dealing can afford the insider an unfair advantage in the marketplace.

Information which directly concerns the issuer might include changes in control, changes in management and supervisory boards, changes in auditors or any other information related to the auditors activity, new licenses, patents, registered trademarks; and decisions to increase or decrease the share capital (CESR/02-089d, para. 47). In addition, any kind of information that is relevant to the market position of an issuer can be regarded as relating to that issuer. CESR explained that this could be "information on events that impact the issuer's assets and liabilities, the financial position, general business operations or organisation and personnel matters as well as material market information about that industry or sector, caused by political, economic or even environmental events" (CESR/02-089d, para. 44). Examples of such information would be Central Bank decisions concerning interest rate or governmental decisions concerning taxation, industry regulation, or debt management. Orders to trade the issuer's securities would also be regarded as relevant information (CESR/02-089d, paras. 49, 50). Although CESR set out a list of examples of information relating to issuers or financial instruments, these examples were not incorporated as Level 2 implementing measures. The list is non-exhaustive and merely indicative and a final determination as to whether it constitutes inside information depends on the specific circumstances in each single case. In addition, in determining whether the events constitute inside information, CESR advised that the materiality of the event needs to be considered (CESR/06-562b, para. 1.15). In order to constitute inside information, the information must be sufficiently material. (This is discussed further below.) The above examples should be used therefore merely as guidance.

The distinction between information of direct or indirect concern is extremely important. Where the information is merely of indirect concern to the issuer or the financial instrument, it is treated as inside information only as far as the prohibition to enter into transactions and to communicate inside information is concerned. However, it does not have to be disclosed under Article 6(1) of MAD as that article applies only to information that directly concerns the issuers. That said, it should also be borne in mind that in the case of events which already have an indirect effect on the issuer, the consequences arising from these events may subsequently directly concern the issuer when they become public knowledge and may become notifiable at that stage (CESR/06-562b, para. 1.15).

7.3.3 Information Which Users of Derivative Markets Would Expect to Receive in Accordance with Accepted Market Practices

The definition of inside information which applies in relation to derivatives on commodities relies not on a price-sensitivity test but rather a test based on the expectation of the market. Article 4 of Directive 2004/72/EC provides that for the purposes of applying the MAD definition of inside information in relation to derivatives:

> users of markets on which derivatives on commodities are traded, are deemed to expect to receive information relating, directly or indirectly, to one or more such derivatives which is:
>
> (a) routinely made available to the users of those markets, or
> (b) required to be disclosed in accordance with legal or regulatory provisions, market rules, contracts or customs on the relevant underlying commodity market or commodity derivatives market.

This expectation is based on "accepted market practices." The notion of accepted market practices also arises as a defense to a charge of certain forms of market manipulation in the MAD and is more contentious in that respect. That term is defined in Article 1(5) of MAD as:

> practices that are reasonably expected in one or more financial markets and are accepted by the competent authority in accordance with guidelines adopted by the Commission in accordance with the procedure laid down in Article 17(2).

Thus, the practices must both be reasonably expected in the market and accepted by the competent authorities. Directive 2004/72/EC sets out the relevant Level 2 implementing measures in this respect. The comitology process was used in order to allow the commission to take account of new developments in the market and to ensure a uniform application of the directive. Therefore, the guidelines referred to above were adopted by the Commission in the form of implementing measures. The difficulty in agreeing guidelines for all member states is that the markets on which the underlying commodities are traded are not regulated to a uniform standard, or indeed in some cases, at all. Thus the disclosure obligations relating to the commodities vary significantly from one market to the next. Furthermore, disclosure obligations may arise from disparate sources. For example, they may be imposed by stock market listing regulators, electricity regulators, or banking regulators. CESR advised that in considering the appropriate implementing measures, it was necessary to take account of the markets on which the underlying commodities are traded, the characteristics of those commodities, the information relating to them which is expected to be disclosed, the perceived function of commodity derivatives markets of allowing market users to transfer risk safely, the characteristics, structures, and rules of the commodity derivatives markets, and the characteristics of users of those markets (CESR/03-212c, para. 16). As a consequence, the decision to accept any market practice applies only in relation to a specific national market.

Article 2 of Directive 2004/72/EC sets out a nonexhaustive list of factors to be taken into account before deciding whether or not to accept a market practice. These include:

the level of transparency of the relevant practice to the whole market (and Recital 2 notes that the less transparent a practice is, the more likely it is not to be accepted) the need to safeguard the operation of market forces and the proper interplay of the forces of supply and demand

the degree to which the relevant market practice has an impact on market liquidity and efficiency

the degree to which the relevant practice takes into account the trading mechanism of the relevant market and enables market participants to react properly and in a timely manner to the new market situation created by that practice

the risk inherent in the relevant practice for the integrity of, directly or indirectly, related markets, whether regulated or not, in the relevant financial instrument within the whole community

the outcome of any investigation of the relevant market practice by any competent authority or other authority mentioned in Article 12(1) of Directive 2003/6/EC, in particular whether the relevant market practice breached rules or regulations designed to prevent market abuse, or codes of conduct, be it on the market in question or on directly or indirectly related markets within the community

the structural characteristics of the relevant market including whether it is regulated or not, the types of financial instruments traded, and the type of market participants, including the extent of retail investors participation in the relevant market

This definition and the relevant criteria are set out verbatim in Regulation 2(1) of the Irish Regulations and section 130A(3) of the FSMA 2000 and MAR 1 Ann 2G.

The Recitals to Directive 2004/72/EC state that competent authorities, in considering the acceptance of a particular market practice, should consult other competent authorities. However, there might be circumstances in which a market practice can be deemed acceptable on one particular market and unacceptable on another comparable market within the EU. The role of CESR in such a case would be to identify "a solution." A similar issue which should be considered but is not determinative is the prevalence of a practice. Although the use of the word "accepted" might be thought to imply an established practice, Article 2(2) of Directive 2004/72/EC requires member states to ensure that practices, in particular new or emerging market practices, are not assumed to be unacceptable by their competent authorities simply because they have not been previously accepted. This reflects a concern which was expressed during the consultation process. A related concern was that a widespread practice, especially one which developed quickly, might automatically be deemed acceptable. In order to ensure that permanent market developments are considered, Article 2(2) also provides that the assessments undertaken by competent authorities be subject to periodic review. Article 3 of Directive 2004/72/EC sets out the process to be followed by the competent authority in accepting any particular market practice. It imposes various obligations on competent authorities in order to ensure a high degree of consultation and transparency vis-à-vis market participants and end users and public disclosure of their decisions regarding the acceptability of market practices. These decisions are then displayed on the CESR Web site.

7.3.4 Information Which Has Not Been Made Public

The omission of a definition or explanation of the concept of making information "public" caused problems for the implementation of the 1989 Directive and led to the adoption of different approaches among member states. This undoubtedly undermined the directive's objective of ensuring a level playing field among investors throughout the EU. It is surprising thus that MAD too fails to define this concept and no Level 2 measures deal with this point.

A number of uncertainties arise in attempting to understand the concept of making information public. The first issue to be considered is the mode of publication. Although Article 2(1) of Directive 2003/124/EC requires member states to ensure that the inside information is made public by the issuer "in a manner which enables fast access and complete, correct and timely assessment of the information by the public," CESR has clarified that information can be publicly available even if it was not disclosed by the issuer in the manner specified by the competent authority. This applies whether the information became public through an incorrect disclosure by the issuer or through a third party (CESR/06-562b, para. 1.9). That said, it is not clear whether the information must be given to the public at large or whether giving it to a section of the public is sufficient. For example, while providing information on the national broadcasting channel clearly makes it public, the situation is not as clear if the information is published in a provincial newspaper with a small circulation. In such a case, the information is certainly available to the public at large but may have only come to the attention of a section of the public. Hopt (1991, 134) has argued that the information should be available to the investing public by having appeared on the stock exchange ticker or having been reported by public radio. CESR itself has clarified that publicly available information may include information which is made accessible on a commercial basis which would include paid-for wire services such as Bloomberg and Reuters (CESR/06-562b, para. 1.9). In *Kinwat Holdings Ltd v. Platform Ltd.* (1982, QR 370), information was deemed "generally available" because it was pleaded in court proceedings and published in a newspaper. In *Johnson v. Wiggs* (443 F.2d 803), information was deemed to be in the public arena because it had been reported in the newspapers and on a local television station. The case law on the meaning of the term "public" in the context of the regulation of prospectuses may also be useful in understanding this concept. In *Nash v. Lynde* (1929, AC 158), Lord Buckmaster in the House of Lords stated:

> a document is not a prospectus unless it is an invitation to the public, but if it satisfied this condition it is not the less a prospectus because it is issued to a defined class of the public.

Similarly, Lord Sumner said:

> "The Public" in the definition section ... is of course a general word. No particular numbers are prescribed. Anything from two to infinity may serve: perhaps even one, if he is intended to be the first of a series of subscribers but makes further proceedings needless by himself subscribing the whole. The point is that the offer is such as to be open to anyone who brings his money and applies in due form, whether the prospectus was addressed to him on behalf of the company or not. A private communication is thus not open to being deemed to be made to the public.

In the context of the publication of information, it might thus be possible to argue that the information is made public when it is capable of being accessed by the public. This is consistent with the manner of implementing MAD in the United Kingdom. Section 118C of the Financial Services and Markets Act 2000 defines inside information as information which "is not generally available" rather than "not been made public." The FSA Code of Market Conduct (para. 1.2.12E) then lists several factors which are to be taken into account in determining whether or not information is generally available, and are indications that it is. These involve considering:

(1) whether the information has been disclosed to a prescribed market through a regulatory information service (or RIS) or otherwise in accordance with the rules of that market;

(2) whether the information is contained in records which are open to inspection by the public;

(3) whether the information is otherwise generally available, including through the Internet, or some other publication (including if it is only available on payment of a fee), or is derived from information which has been made public;

(4) whether the information can be obtained by observation by members of the public without infringing rights or obligations of privacy, property, or confidentiality (the code gives the example of a passenger on a train passing a burning factory who calls his broker and tells him to sell shares in the factory's owner, and

(5) the extent to which the information can be obtained by analysing or developing other information which is generally available.

In Ireland, in implementing MAD as was the case in implementing the 1989 Directive, no attempt was made to define the term "public."

The movement toward an access test for the determination of the category of insider poses a problem in relation to classifying information derived from published information by sophisticated or professional investors with sufficient time, knowledge, and resources. Arguably, such information may not be available to every person on the street. An example of this would be analyst reports prepared for private clients. The FSA Code of Market Conduct addresses this point directly by providing that in relation to the factors in paragraphs (3), (4), and (5) set out above, "it is not relevant that the observation or analysis is only achievable by a person with above average financial resources, expertise or competence" or that the information is only generally available outside the United Kingdom (MAR 1.2.13.2E, 1.2.13.1E). Unusually, Recital 31 of MAD also refers to this situation. It notes that "research and estimates developed from publicly available data should not be regarded as inside information."

It is submitted that the audience issue discussed above must be seen as interlinked with the issue of the assimilation of the information. This leads to a second problem with understanding the term "public" as it applies in MAD. Like the 1989 Directive before it, MAD does not clarify whether information ceases to constitute inside information as soon as it has been released or whether time must be allowed for the information to be absorbed by investors. For example, is an investor free to deal the instant a profit warning is issued on a Stock Exchange's Announcement Service or must he or she wait for a period of time in order for the market to reflect that information? Commenting on the 1989 Directive, Ashe and Murphy suggested that a period of delay was required in order to allow the information to be absorbed.

> It would be odd to think that it was the intention of the directive to allow insiders to deal at the instant after the news had been released since they would still have the trading advantage which the measure is seeking to strip from them. (p. 47)

Such a delay would thus be necessary in order to achieve the directive's intention of placing investors on an equal footing. In *Fyffes Plc v. DCC* ([2005] IEHC 477), the defendants had argued that to be "generally available," the information had to be "internalized by the market." The Irish High Court interpreted the term "generally available" meant that the share price should fully reflect the fact that the information is in the market. In the context of the hypothetical component of the test, it held that information would have been generally available, if it would have been "accessible by investors." In *SEC v. Texas Gulf Sulphur Co.* (401 F.2d 833 (2nd Cir, 1968)), the U.S. Court of Appeals suggested that there should be a time lapse following the disclosure of information in order to allow the information to be assimilated. This gives rise to the question: if a waiting period is required, how long should that be? The *Texas Gulf Sulphur* was decided in 1968 in advance of the substantial developments in information technology. Information can be communicated more quickly today and it is clear that if a waiting period is required, it should be much shorter than that envisaged 40 years ago. It would seem appropriate that the waiting period would depend on the original audience and the mode of communication of the information. For example, where information is made available to a very restricted circle of people, more time should be allowed for this information to be absorbed than, for example, in the case of a company announcement on the Stock Exchange's Regulatory Announcement Service. It is submitted that this would have been a useful issue to have subjected to CESR technical advice.

7.4 PRICE SENSITIVITY

The final condition to be met in order to categorize information as inside information is that the information be price sensitive. Article 1(1) of MAD provides that to constitute inside information, the information must be such as "would be likely to have a significant effect on the prices of those financial instruments or on the price of related derivative financial instruments." In demonstrating how this condition can be the most difficult to determine and to prove, reliance will be placed on the recent judgments in the case of *Fyffes Plc v. DCC Plc & Others* ([2005] IEHC 477 (High Court) and [2007] IESC 36 (Supreme Court)). This case involves the only civil action to be taken in Ireland on the basis of alleged insider dealing.

In advising on Level 2 implementing measures to apply this concept of price sensitivity, CESR correctly rejected the use of fixed thresholds of price movements or quantitative criteria to determine the significance of a price movement. Even differentiation on the basis of markets, market segments, or financial instruments was rejected. The reason for this is that even within such groups, excessive differences and individualities arise to justify a common rate of price movement. For example, a similar threshold should not be applied to a small company's ill-liquid stocks as to a blue chip company's indexed stocks. Furthermore, as the various markets and market segments within member states are not comparable, determining EU-wide common thresholds would be impossible. Instead, CESR recommended developing common rules or guidelines on the evaluation of the likelihood of a significant effect on prices. This approach was deemed to be consistent with the directive's goal of creating a common framework and enhancing conformity in all member states. Consequently, Article 1(2) of Directive 2003/124/EC notes that for the purposes of the definition of insider dealing in MAD:

> information which, if it were made public, would be likely to have a significant effect on the prices of financial instruments or related derivative financial instruments, shall mean information a reasonable investor would be likely to use as part of the basis of his investment decisions.

This definition was applied verbatim in the Regulation 2(1) of the Irish Market Abuse (Directive 2003/6/EC) Regulations 2005 and in section 118C(6) of the U.K. Financial Services and Markets Act 2000.

7.4.1 Reasonable Investor Test

In relation to the reasonable investor, two interrelated issues arise. The first concerns the type of information which such a person would be expected to take into account and the second concerns the profile of such an investor. Recital 1 of Directive 2003/124/EC expressly states:

> the question whether, in making an investment decision, a reasonable investor would be likely to take into account a particular piece of information should be appraised on the basis of the ex ante available information.

The logical rationale for this is that reasonable investors are deemed to base their investment decisions on ex ante available information, that is, information already available to them. Whether or not the reasonable investor uses particular information in making his or her investment decision clearly depends on the reliability of the source of information and the relevance of the information as regards the main determinants of the financial instrument's price (CESR/02.089d, para. 27; CESR/06-562b, para. 1.13; Recital 1 of 2003/124/EC). The Financial Regulators in Ireland and the United Kingdom, while acknowledging that it is not possible to prescribe how the reasonable investor test would apply in all situations, have listed examples of information which is likely to be considered relevant to a reasonable investor's decision. This involves information which affects: the issuer's assets and liabilities; the performance or expectation of performance of the issuer's business; the issuer's financial condition; the course of the issuer's business; major new developments in the issuer's business; information previously disclosed to the market; and events that may significantly affect the issuer's ability to meet its commitments (Market Abuse Rule 5.3 (Ireland) and FSA Disclosure Rules and Transparency Rules 2.2.5 and 2.2.6 (United Kingdom)). Any market variables such as prices, volatilities, liquidity, volume, and so forth likely to affect the related financial instrument or the derivative financial instrument in the given circumstances should also be considered in the assessment of the information's effect on prices. The Financial Regulators suggest that in conducting the reasonable investor test, one must take into account that the significance of the information in question will vary widely from issuer to issuer, depending on a variety of factors such as the issuer's size, recent developments, and the market sentiment about the issuer and the sector in which it operates. They also require the issuer to assume that a reasonable investor will make investment decisions relating to the relevant financial instruments to maximize his or her economic self-interest. In determining, in the light of the above factors, whether the information is likely to have a significant effect on prices, CESR suggested considering: whether the information is the same type as information, which has, in the past, had a significant effect on prices; whether preexisting analysts' research reports and opinions indicate that the type of information in question is price sensitive; and whether the issuer itself has already treated similar events as inside information (CESR/02.089d, para. 28).

In the *Fyffes* case, the Irish High Court and, on appeal, the Supreme Court considered the reasonable investor test in the context of an action under section 108 of the Irish Companies Act, 1990 which implemented the 1989 Directive. That case was initiated by Fyffes Plc, Europe's leading fresh produce distribution company. Fyffes Plc is listed on both the London and the Irish Stock Exchanges. The defendants in the case included DCC, a listed industrial group, and its chief executive, Mr. Jim Flavin. DCC owned a 10.5 percent stake in Fyffes and Mr. Flavin was a nonexecutive director. In an announcement of its preliminary results for the 1999 financial year on December 14, 1999, Fyffes reported that profit before tax and exceptional items in that year had increased over the previous financial year by 5.1 percent and that, while turnover for the period decreased marginally, the total operating profit was up 3.8 percent. In the "Outlook" section of the announcement, having recorded that the results for the year had maintained the group's record of continuous growth, it was stated that the board believed that 2000 would be "another year of further growth for Fyffes." The legal action in question arose as a consequence of the sale of DCC Group's entire shareholding in three tranches on February 3, February 8, and February 14, 2000, at prices of €3.20, €3.60, and €3.90, respectively, grossing in excess of €106 million. On February 17, 2000, a Fyffes announcement predicted further growth and developments and the share price the following day experienced a high of €3.98. From this time onward, the share price declined and on March 17, 2000, (the last day of trading before the AGM), it fell to €3.16. On March 20, 2000, at its AGM, the company issued a profit warning and the share price closed that day at €2.70. By the end of April 2000, the share price had fallen to €1.85. In January 2002, the plaintiff initiated an action against the defendants claiming that the share sales were unlawful because they were effected by Mr. Flavin who at the time was in possession of price-sensitive information by reason of his directorship of Fyffes. The specific information alleged to constitute inside information was contained in the November and December 1999 Trading Reports ("the Trading Reports") which were made available to the Fyffes board on January 6, 2000, and January 25, 2000, respectively. The figures set out in the trading reports inferred that both Fyffes' own expectations and analysts' expectations for the year would not be met.

The reasonable investor approach was proposed by the defendants using case law from the United States. In the High Court, Justice Laffoy accepted this case law as useful in identifying the proper approach to the application of the section 108(1). In her judgment she referred to the U.S. Supreme Court decision in *TSC Industries Inc. v. Northway Inc.* (426 U.S. 438) dealing with section 14(a) of the Securities Exchange Act of 1934, which prohibited the use of false or misleading proxy statements. In that case, the Court held that an omitted fact is material if there is a substantial likelihood that a reasonable shareholder would consider it important in deciding how to vote. In *Basic Inc. v. Levinson* 485 U.S. 224, the U.S. Supreme Court held that the standard of materiality set forth in the *TSC Industries* case was appropriate in the context of section 10(b) and Rule 10(b)-5 of the Securities Exchange Act of 1934. Justice Blackmun elaborated further on the profile of such an investor noting that:

> The role of the materiality requirement is not to "attribute to investors a child-like simplicity, an inability to grasp the probabilistic significance of negotiations," ... but to filter out essentially useless information that a reasonable investor would not consider significant, even as part of a larger "mix" of factors to consider in making his investment decision.

Reference was also made in the High Court to the adoption of the "reasonable investor" approach by a court of first instance in Singapore in *Public Prosecutor v. Allen Ng Poh Meng* ([1990] 1 M.L.J v) and a Malaysian appeal court in *Public Prosecutor v. Chua Seng Huat* ([1999] 3 M.L.J. 305). In the latter case, the "reasonable investor" was described as:

> an investor who possesses general professional knowledge as opposed to the said daily retailer or a person who has made specific researches.

The profile of the "reasonable investor" was of paramount importance to Justice Laffoy's ultimate findings in the High Court. At the time of Mr. Flavin's share dealings the market was described by expert witnesses as in the throes of "dot com mania." All of the expert witnesses agreed that Fyffes' Internet venture, world-of-fruit.com, was the principal driver of Fyffes' share price on the date of the sale and the cause of the unprecedented share value. However, the plaintiff had argued that the reasonable investor test indicates that the behavior of irrational forces within the market is not relevant to the resolution of the issue of price sensitivity and that "the reasonable investor is not to be found at the extremes of the market." The question the trial judge asked herself was whether the reasonable investor had been "infected by, or immune from, the market's infatuation with internet stocks or stocks with an internet dimension." Arising from a consideration of the aforementioned cases, she determined that the concept of the reasonable investor represents the type of investor who was typically found in the market at the time of the dealing. Justice Laffoy stated that the statutory hypothesis assumes that the information is introduced into "the actual world of stock prices" and so "if that investor, on the evidence, was one who was anxious to own internet stocks or stocks with an internet element, the likely consequences of such predilection are a relevant factor." The need to make this type of decision led Justice Finnegan in the Supreme Court to comment:

> The judge may be well fitted to identify the conduct to be expected of the reasonable man but may not be fitted by knowledge or experience to fulfil the same function in relation to the reasonable investor.... The difficulty in using the reasonable investor as a test, and not just to catergorise the test as objective, is compounded in that it is, to my mind, impossible to profile the reasonable investor. There are innumerable categories of investor from the small private investor who will check the value of his shares but now and then to the institutional investor who is in touch with the market throughout the trading day. Is it the dealer who trades within the account or the trustee whose shareholding dates back decades? One investor may concentrate on return, another on capital gain. An investor may be cautious or adventurous and to a greater or lesser degree.

Justice Denham identified similar problems. She described the test as a method of interpretation which removes the analysis one step from the law as created, creating "a system where the law is being looked at through the eyes of a notional person." Rather than clarifying the issue of price sensitivity, she suggested that it renders the situation opaque.

> there are a myriad of factors and investors in a market and to choose some or either as representative of a reasonable investor appears subjective and arbitrary.

The High Court in the *Fyffes* case accepted evidence to the effect that the strength of the sentiment for Fyffes' wof.com venture at the time, as evidenced by what was happening to the share price, was such that the reasonable investor would have concluded that an adverse share price reaction was not likely. It is submitted that this test is more like identifying the average investor rather than the reasonable investor. It may not be entirely consistent with the test envisaged by CESR in relation to MAD. In that context, CESR defines "a reasonable investor" as a person who thinks and behaves in a rational way (CESR/02.089d, para. 27). While the Supreme Court on appeal unanimously agreed that the reasonable investor test was not an appropriate test in that case (on the basis of a literal interpretation of the 1990 Act which did not refer to the reasonable investor), it very usefully examined the use of the test by the High Court. It rejected what it viewed as the use by the High Court of a modified version of the reasonable investor test. Justice Fennelly stated that Justice Laffoy:

> used the reasonable investor not as a representative of all investors in the market whose response to the information might or might not lead to a material effect on the share price, but rather as a test of opinion as to how the market would respond.

What the High Court judge was doing was ascertaining the reasonable investor's opinion of how the market would respond to the release of the information in question. This approach might be said to be resonant of the argument often made that the trading market represents a form of derived demand. For example, Keynes equated professional investment to a newspaper competition in which the competitors have to pick out the six prettiest faces from a hundred photographs, the prize being awarded to the competitor whose choice most nearly corresponds to the average preferences of the competitors as a whole. As Keynes pointed out:

> each competitor has to pick, not those faces which he himself finds prettiest, but those which he thinks likeliest to catch the fancy of the other competitors, all of whom are looking at the problem from the same point of view. It is not a case of choosing those which, to the best of one's judgment, are really the prettiest, nor even those which average opinion genuinely thinks the prettiest. We have reached the third degree where we devote our intelligences to anticipating what average opinion expects the average opinion to be. And there are some, I believe, who practice the fourth, fifth and higher degrees. (p. 156)

By contrast, the reasonable investor test actually requires one to ascertain the investor's likely reaction to the information. It is likely that an application of the latter formulation would yield a completely different result—arguably the same as that declared by the Supreme Court—that the information in the trading reports constituted bad news.

7.4.2 Likelihood of Effect

In the *Fyffes* case, the High Court was required to interpret the meaning of the expression "likely to materially affect the price of securities" in the context of the Irish Companies Act 1990. The High Court determined that in light of the express inclusion in the Act of the "would be likely" criterion, it would not be appropriate to apply the "substantial likelihood" standard applied by the U.S. Supreme Court in *TSC Industries* (426 U.S. 438). Both parties agreed that the word "likely" imports more than a mere possibility. They agreed with the interpretation of the term "likely" given by Justice Cooke in *Colonial Mutual Life Assurance Limited v. Wilson Neill Limited*, a decision of the New Zealand Court of Appeal ([1994] 2 N.Z.L.R. 152). There the statutory definition of inside information under consideration required that the information "would, or would be likely to" affect materially the price of securities. The New Zealand Court of Appeal held that the term constituted more than "a bare possibility, however remote." Instead, Justice Cooke stated that "a real or substantial risk is required." Although counsel for the plaintiff in the *Fyffes* case argued that "likely" is synonymous with "probably" and counsel for the defendant referred to "proof in the balance of probabilities," the Court viewed the two interpretations as essentially the same. This would appear to be consistent with the advice of CESR in this respect. It suggested that the conclusion that information is "likely to have" a significant effect on prices involves determining "the degree of probability with which at that point in time an effect on the price (due to the information) could reasonably have been expected." It advised that "the mere possibility is not enough, as on the other hand a degree of probability close to certainty is not necessary either" (CESR/02-089d, para. 23). During the consultation process, arguments were made that the term should be interpreted as "clearly probable" or "beyond all reasonable doubt" (CESR/07-402, para. 17). These arguments were rejected by CESR on the grounds that they would involve a change to the meaning of the directive.

7.4.3 Gauging Share Price Effect

As stated above, in determining price sensitivity the emphasis is placed on ex ante objective criteria. The crucial factor is deemed to be the time at which the relevant action by the insider takes place. Thus, CESR has advised that in this context it is irrelevant whether or to what extent the price actually changes when the information eventually becomes publicly known. It opined that a piece of information could be considered as likely to have a significant effect on the prices of financial instruments even though, when that information is published, it does not actually produce any effect (CESR/02-089d, para. 22). It accepted merely that "the actual impact on prices might be relevant as an indicator for the investigation of a possible infraction" (CESR/02-089d, para. 26). Directive 2003/124 itself goes slightly further and Recital 2 acknowledges that ex post information may be used to check the presumption that the ex ante information was price sensitive. (However, it emphasizes that it should not be used to take action against people who drew reasonable conclusions from ex ante information available to them.)

The importance of ex post information was also considered in the *Fyffes* case. The High Court had been asked to consider the relevance of what happened in the market after the date of disclosure on March 20 as a measure or proxy of price sensitivity on the date of dealing. The release of that announcement had an immediate and substantially negative effect on the share price of almost 15 percent. The plaintiff referred the court to the decision of the English High Court in *Chase Manhattan Equities v. Goodman* ([1991] B.C.L.C. 897). In that case, Goodman, a director of Unigroup Plc, sold shares in Unigroup Plc while being aware that he was about to resign as a director and that a substantial company debt was undisclosed in the company's balance sheet. The Stock Exchange suspended dealing before the market became aware of these facts at a share price of £1.72.

After the suspension was lifted, the shares traded at 50p to 55p. Justice Knox determined that knowledge of the resignation and the undisclosed debt if generally available would have been likely materially to affect the price of the company's shares. He noted:

> The proof of that pudding is in the eating in that when the suspension which followed almost immediately was lifted the price of the company's shares was very sharply down.

However, Justice Laffoy in the High Court had rejected this "proof of the pudding approach" and excluded this evidence relying on two cases advanced by the defendants. In *SEC. v. Bausch & Lomb Inc.* (565 F.2d 8, 1977) the approach was referred to by the Court as a "facile inference." In *Elkind v. Liggett & Myers Inc.* (633 F.2d 156, 1980) the Court considering the sensitivity of a tip to an analyst that there was a good possibility that earnings would be down referred to its "serious vulnerabilities." Justice Mansfield stated:

> It rests on the fundamental assumptions (1) that the tipped information is substantially the same as that later disclosed publicly, and (2) that one can determine how the market would have reacted to the public release of the tipped information at an earlier time by its reaction to that information at a later, proximate time. The theory depends on the parity of the "tip" and the "disclosure." When they differ, the basis of the damage calculation evaporates. One could not reasonably estimate how the public would have reacted to the news that the Titanic was near an iceberg from how it reacted to the news that the ship had hit an iceberg and sunk.

Two other U.S. cases referred to in support of the plaintiffs argument to include the evidence *SEC v. Lund* (570 F.Supp. 1397, 1983) and *SEC v. Falbo* (14 Supp. 2d 508, 1998) were distinguished on their facts by the High Court. The High Court accepted the following proposition advanced by the defendants as the two prerequisites to a post-market event being of evidential value in applying the price-sensitivity hypothesis:

(a) the information alleged to be price-sensitive should be substantially the same as the information which gave rise to the share price movement which is proffered as a proxy, and

(b) the market conditions on the date at which the hypothesis is being applied are identical with market conditions on the date on which the supposed proxy event occurred.

By contrast, the Supreme Court found that the March 20 announcement was relevant to the consideration of whether the information in the trading report was price sensitive. It rejected the proposition that evidence of a comparator requires complete parity of information and market viewing this as an "extraordinarily rigid" approach which was not supported by the case law. *Bausch & Lomb* and *Elkrind* were both deemed fact specific and not useful in determining whether, had the information in the trading reports been hypothetically released on the market on particular dates, this would likely have materially affected the share price. In relation to the first of the High Court's propositions, Justice Macken explained that there should be "functional equivalence" between the two sets of information. She also criticized the interposing by the High Court of the "reasonable investor" in the application of the parity test on the basis that such a test was not supported by the case law cited and that the determination of parity should instead be a fact-finding exercise for the Court. The Supreme Court unanimously found that there were "significant similarities" between the two documents. In relation to the second proposition, Justice Macken stated market conditions could not be expected to be identical:

> it would be next to impossible to find, on any two dates, even those quite close together, absolutely identical market conditions, save in fortuitous or highly exceptional circumstances. The test to be applied therefore cannot be based on a requirement that in all circumstances the market conditions must be identical.

The Supreme Court found that there were no significant differences between the market in February 2000 and March 2000 which would justify excluding the effect of the March 2000 Trading Statement entirely from consideration. Any differences in the state of the market generally and the market for the particular share at the relevant dates would be taken into account in determining the weight to be given to the evidence. Having admitted the evidence of the implications of the March 20 announcement, it clearly altered the balance of the evidence in this case. It illustrated that the release of information very similar to that in the trading reports had a very negative effect on the share price. This the Court found was "a useful pointer" to the likely market effect of the release of substantially similar information on the dates of the share sales in early February. It is submitted that this cautious approach is consistent with Recital 2 of Directive 2003/124.

7.5 CONCLUSION

MAD sought to promote legal certainty by providing as comprehensive and as market-appropriate a definition of "insider dealing" as possible. However, as has been demonstrated, categorizing information as "inside information" remains far from easy. Making a determination that information, if published, would be likely to have a significant effect on the prices of securities involves a complicated series of steps and questions. This will lead in many cases to delays or failures in disclosure and incorrect dealing decisions. It will also make the task of the courts enforcing the implementing legislation more arduous. The *Fyffes* case provides a striking example of this. That action was based on a single set of undisputed facts described by Justice Fennelly in the Supreme Court as "comparatively simple facts ... the sort of facts upon which common-sense judgments and opinions can be formed without the input of an extraordinary degree of expertise." Yet the purported application of the objective test involved lengthy expert testimony which yielded disparate conclusions from distinguished international academics and market experts. The High Court hearing alone lasted 87 days. Although Justice Fennelly opined that the length of the trial may have been "the product of the large amounts of money at stake and the depth of the respective corporate pockets rather than of the complexity of the issues," the same financial incentives and resources are likely to be evident in most insider dealing actions. It may be, however, that there is no way of avoiding completely any uncertainty. The nature of insider dealing is such that an overly prescriptive approach would be neither workable nor desirable. For example, in determining price sensitivity, an assessment on a case-by-case basis of a myriad of different factors is required. It may be necessary thus to sacrifice a degree of certainty in order to ensure a more equitable marketplace. In this context, the Lamfalussey process, with its emphasis on technical measures and the achievement of a coordinated response at market levels, appears to be the appropriate response. In time, through this process, it is hoped that a greater degree of clarity can be achieved in relation to the unresolved issues detailed in this chapter.

REFERENCES

Ashe, M. 1992. The directive on insider dealing. Company Lawyer 13:15–19.
Ashe, M. and Y. Murphy. 1992. *Insider dealing*. Dublin: Roundhall Press.
Australian Corporations and Markets Advisory Committee. 2003. Insider trading report. Australian Government, Sydney (November).
British Institute of International and Comparative Law. 2005. Comparative implementation of EU directives (I): Insider dealing and market abuse. City Research Series 8 (22 December).
Carlton, D., and D. Fischel. 1983. The regulation of insider trading. *Stanford Law Review* 35:857–95.
Conceicao, C. 2006. Tackling cross-border market abuse. *Journal of Financial Regulation and Compliance* 14:29–36.
Enriques, L., and M. Gatti. 2007. Is there a uniform EU securities law after the financial services action plan? SSRN: http://ssrn.com/abstract=982282.
Ferrarini, G. A. 2004. The European market abuse directive. *Common Market Law Review* 41:711–41.
Forum of European Securities Commissions. 2000. A European regime against market abuse (September) (Fesco/00-096l).
Gjersem, C. 2003. Financial market integration in the Euro area. Economics department working paper 368, OEDC, Paris.
Hansen, J. L. 2002. The new proposal for a European Union directive on market abuse. *University of Pennsylvania Journal of International Economic Law* 23:241–68.
Hansen, J. L. 2004. Mad in a hurry. *European Business Law Review* 183–21.
Iemma, P. 2005. The implementation of the market abuse directive in Italy. *Journal of International Banking Law and Regulation* 20:457–60.
Karmel, R. S. 2005. Reform of public company disclosure in Europe. *University of Pennsylvania Journal of International Economic Law* 26:379–408.
Keynes, J. M. 1936. *The general theory of employment, interest and money*.
Mangelsdorf, A. 2005. The EU market abuse directive: Understanding the implications. *Journal of Investment Compliance* 6:30–37.
Manne, H. G. 1966. In defense of insider trading. *Harvard Business Review* 44:113–22.
McVea, H. 1995. What's wrong with insider dealing. *Legal Studies* 15(3):390–414.
Moloney, N. 2002. EC securities regulation. *Cambridge Law Journal* 62:509–21.
Parker, C., C. Scott, N. Lacey, and J. Braithwaite. 2004. *Regulating law*. Oxford: Oxford University Press.
Reynolds, C., and M Rutter. 2004. Market abuse—A pan-European approach. *Journal of Financial Regulation and Compliance* 12:306–14.
Rider, B., and M. Ashe. 1995. *The fiduciary, the insider and the conflict*. Dublin: Brehon Sweet & Maxwell.
Welch, J., M. Pannier, E. Barrachino, J. Bernd, and P. Ledeboer. 2005. Comparative implementation of EU directives (I)—Insider dealing and market abuse. *British Institute of International and Comparative Law* 8:1–88 (mimeograph, London).
Wymeersch, E. 1991. The insider trading prohibition in the EC member states: A comparative overview. In *European insider dealing*, ed. K. Hopt and E. Wymeersch. London: Butterworths.

CHAPTER 8

Insider Trading in Australia

Anna-Athanasia Dervenis

CONTENTS

8.1 INTRODUCTION

Insider trading has long been a pressing issue in the corporate world and Australian share markets are no stranger to its presence and arguably its prevalence. It has been suggested that between 5 to 10 percent of all share trades involve the use of inside information (Chapman and Denniss 2005). A more chilling statistic, however, is that drawn from a recent survey of Australian executives, which found that 52 percent of respondents stated they would be willing to trade on favorable information about their company before that information is released to the market (Chapman and Denniss 2005). This fact is no doubt a problem because trading on information to which very few people are privy has broader implications for financial markets.

In addition to providing an overview of insider trading, this chapter outlines Australia's regulatory regime behind the prohibition and examines some of the high-profile cases that have surfaced in recent years.

8.2 WHAT IS INSIDER TRADING AND WHY IS IT PROHIBITED?

Insider trading is trading financial products based on information that:

is not generally known to the market, and

if such information were generally known, it would have a material effect on the price of the financial products being traded. (Sections 1042A and 1043A Corporations Act 2001)

An example of when such trading may occur is in the context of a corporate takeover. A company is usually the target of a takeover bid if a bidder considers that the target's shares are currently undervalued. Prior to the bidder's disclosure to the market of its intentions to make a bid for the target, an insider with knowledge of the proposed takeover bid might purchase shares in the target. Once the takeover bid becomes public knowledge, it is expected that the value of the target's shares will increase as the market readjusts to account for the perceived undervaluation of these shares. This ultimately means that our insider is able to capitalize on the increase in share value by selling the shares acquired at the higher price.

At first glance, it would appear that such a series of events is harmless to other participants in the market. However, when one considers the nature of the share market and the effects of such trades at macroeconomic levels, the perception of the trades as "victimless" profit-making maneuvers begins to falter (Chapman and Denniss 2005). Going back to our takeover example, though it is true that the insider is likely to gain from the use of the inside information and has therefore not suffered any detriment, the question must be asked: from whom did the insider purchase the shares and what effect did selling the shares have on the seller? The seller of the target's shares has lost the opportunity to capitalize on the subsequent price rise following the announcement of the takeover bid as the seller lacked the level of knowledge that the insider possessed. Had the seller known the bid was pending, the seller may have elected to hold on to his or her shares in the hope that the value of the shares will increase and therefore sell them at a higher price. Investor confidence in the integrity of share markets may be shattered if the public perception is that such inside trades are commonplace, resulting in a withdrawal of capital investment in the market and compromise of the capital-raising ability of firms.

8.2.1 Why Is Insider Trading Prohibited?

Arguments against the prohibition on insider trading have not been met with much favoritism. The leading advocate for the decriminalization of insider trading, Henry Manne, is a proponent of law and economic theory and perceives insider trading as beneficial due to its ability to enhance efficiency in financial markets. According to Manne, efficiency is promoted by the signaling to the market of new information through price movements (Manne 1966; Rubenstein 2002). This view is founded on the efficient markets hypothesis (EMH), which asserts that all available information about securities traded in the principal securities markets is impounded into market prices with such speed that even professional investors cannot systematically and consistently profit from trading on any newly available information. Though this appears to be a logically sound argument, it falls short in terms of practical soundness—the level of trading that is generally undertaken by a single trader is unlikely to have a significant effect on the price of those securities in the market (Baxt, Black, and Hanrahan 2003).

The more commonly accepted view of the effect of insider trading is that it "undermine[s] confidence in the fairness of a market and therefore its broader economic function" (CAMAC 2003). This argument can be justified in one of four ways.

The first is an argument based on the fiduciary theory rationale, which states that if the information holder owes a fiduciary obligation to the company to which the information relates, any trading activity undertaken by the fiduciary based on that information constitutes a breach of that duty. A person who is a fiduciary would be expected to act in a manner that is honest and loyal to the beneficiaries of the relationship who, because of the fiduciary nature of the person's position, have vested their trust in the belief that the fiduciary will not abuse his or her position. Should the fiduciary use price-sensitive information obtained as a result of his or her position to trade on securities, the confidence of stakeholders in the company will surely be compromised.

The primary problem with the fiduciary argument against insider trading is that it has limited scope. Realistically, it is usually only the directors and senior management of a company who would owe any sort of fiduciary obligation and not your average employee, which leads us to the second rationale—the misappropriation of information theory. This theory treats the price-sensitive information of a company as company property, so that any use of this information by persons for their own purposes is analogous to theft of company property. The greater appeal for the misappropriation theory as opposed to the fiduciary argument lies in the fact that it has the ability to capture persons who themselves are not fiduciaries of the company, but who may have received information from fiduciaries and proceeded to trade on such information (*U.S. v O'Hagan* 117 S Ct 2199; 521 US 642, 1997).

Although the scope of the misappropriation theory is wider than the scope of the fiduciary theory, it is still limited by the fact that the information must be disclosed by someone who fits within the narrow definition of fiduciary. The third and fourth rationales, however, have a much broader scope in that their operation does not rely on any form of fiduciary relationship with the company, but rather on the overall impact that trading on nonpublic price-sensitive information may have on financial markets more generally. The market fairness theory asserts that financial markets should be "level playing fields" where all participants have equal opportunities to access and evaluate information relating to trading decisions, whereas the market efficiency theory operates on the view that insider trading should be prohibited in order to prevent any damage to the reputation and integrity of the market through the delay in information disclosure and erosion of public confidence, both of which have the potential to adversely affect the overall liquidity and capital-raising efficiency of the financial market.

So which theory underlies Australia's insider trading laws?

8.3 HOW IS INSIDER TRADING REGULATED IN AUSTRALIA?

Despite the limitations of the fiduciary and misappropriation theories, these are the theories that have been endorsed by the U.S. courts (*U.S. v. Chiarella* 445 US 222, 1980; *Dirks v. SEC* 463 US 646, 1983). Australian courts, on the other hand, are still grappling with which of the four rationales underlies Australia's insider trading provisions. In one instance the court lent its support to all four rationales (*Exicom Limited v. Futuris Corporation Limited*, 1995, 18 ACSR 404; 13 ACLC 1758), whereas in other cases the court has taken the view that the insider trading provisions have "partially and indirectly endorsed the economic-efficiency paradigm as one of the goals of insider trading prohibition" (*R v. Firns*, 2001, 51 NSWLR 548), a view that is supported by the Australian Corporations and Markets Advisory Committee (CAMAC) in its November 2003 Insider Trading Report (CAMAC 2003). Outside of the specific insider trading provisions, however, there are other provisions that punish directors, company officers, or employees of a company who improperly use information they obtain by virtue of their position to gain an advantage for their own benefit or for someone else (section 183, Corporations Act 2001). Such provisions may signify that the fiduciary theory also plays a small part in regulating insider trading in Australia.

Over the past decade, Australia has experienced fundamental changes to its financial services regulations through the passing of the Financial Services Reform Act 2001 (Cth). Among these changes was the replacement of the former insider trading laws with a new regime that harmonized the licensing and regulation of the financial services industry. The current provisions are now found in Part 7.10 Division 3 of the Corporations Act 2001 (Cth) ("the Act").

The new provisions have remained substantially the same since the 1991 reforms resulting from the Griffiths Committee Report. One of the key recommendations of this report was that any insider trading prohibition should not be based on a theory that limits the scope of the prohibition itself: "the basis for regulating insider trading is the need to guarantee investor confidence in the integrity of the securities markets" (CAMAC 2003). It would therefore appear that Parliament aimed for a prohibition supported by an intrinsic desirability of a minimum standard of fairness in the securities market (Baxt, Black, and Hanrahan 2003). This lends support to the market fairness theory on the prohibition against insider trading.

8.3.1 The Primary Prohibition

The primary prohibition is set out in section 1043A of the Act, which states that a person who possesses "inside information" (section 1042A) must not enter into any transaction agreement of any form in relation to "Division 3 financial products" (section 1042A), or procure another person to enter into such an agreement, if the person knows or ought reasonably to know that the information in relation to those financial products is inside information (section 1043A(1)). An insider is also prohibited from communicating inside information to another person if that person is likely to enter into a transaction or procure another person to enter into a transaction, but only if the relevant financial products are able to be traded on a financial market operated in Australia (section 1043A(2)).

Each element that is essential to the prohibition is defined in section 1042A of the Act. Whether it is a prosecution (criminal proceeding) or a civil action that is brought against the insider, these same elements must be proved. The only difference is the standard of proof that is required: in a prosecution, the standard is beyond a reasonable doubt, whereas in a civil action, the standard is on the balance of probabilities which is a lower threshold than the prosecutorial standard (section 1332).

The biggest hurdle in proving any case of insider trading is establishing that the person knew or ought reasonably to know that the information that the person possessed falls within the definition of "inside information" in section 1042A. Inside information is defined as information that is "not generally available," and if it were generally available, a reasonable person would expect it to have a "material effect" on the price or value of the financial products to which it relates.

The question whether the information would have a material effect on price is fairly straightforward. Information's materiality is dependent on its ability to influence a person's investment decision in relation to the financial product (section 1042D). Therefore, if the evidence suggests that a person would (or would be likely to) base his or her decision on whether to buy or sell shares in a company on a certain piece of information, then that information would be held by the court to have a material effect on the price or value of the financial product.

The question whether the information was generally available at the time of the alleged trading, however, is not as straightforward and is discussed in more detail below.

8.3.2 Is the Information "Generally Available"?

"Generally available" is defined in section 1042C as information that consists of readily observable matter, or that has been disseminated into the market after being made known to persons who regularly invest in the relevant financial product, or that consists of deductions, conclusions, or inferences made or drawn from such information. This definition has been the subject of some controversy, particularly in relation to the phrase "readily observable matter" as this has "the potential to cover information that could not, on any reasonable view, be described as being generally available to persons who commonly invest in relevant financial products" (CAMAC 2003). Case Study 1 illustrates when using this definition can become problematic.

8.3.3 Case Study 1: *R v. Firns*

In the case of *R v. Firns* (2001, 51 NSWLR 548), a Papua New Guinean court handed down a decision at 9:30 a.m. on July 28, 1995, that was favorable to a company called Carpenter Pacific Resources NL. News of the decision had reached the defendant Firns in Brisbane, Australia by 10.08 a.m. that same day through a series of telephone calls which can be traced back to the court in Papua New Guinea. Firns had consequently placed an order with his broker to purchase shares in Carpenter Pacific.

The court's decision that Firns was not guilty of the insider trading offense was due to its finding that the information regarding the decision handed down earlier that day in Papua New Guinea was readily observable matter and therefore generally available because the information was understandable and accessible to the public and the number of people who could actually observe the information is irrelevant in determining this question (2001, 51 NSWLR 548: 77).

It appears from the judgment in favor of Firns that the court did not wish to punish "individual initiative and diligence" (2001, 51 NSWLR 548: 57). However, the practicalities of such a decision are questionable in a context where such speedy relay of information through modern telecommunication is not possible.

It is perhaps worthwhile to note here that only positive acts are prohibited by Australia's insider trading laws. This means that should a person who possesses inside information elect not to transact based on that information, or procure another person not to transact, that person would not have contravened section 1043A. This may appear to be an anomaly when one considers the underlying rationale behind the general prohibition, in the sense that an insider will avoid what may be a substantial loss by, for instance, selling shares in a company, based on information that will have a negative impact on that company's share price. Not only does the insider avoid losses, but the person who purchases the shares from the insider will experience a loss, as will all other market participants who do not possess this information prior to its release to the market. Despite there being some basis for punishing inaction by an insider in these circumstances, particularly by proponents of the market fairness theory, the practical effect of adding such a prohibition will result in punishment for those with inside information in whatever they elect to do: if they trade in the financial product, they will breach section 1043A, but if they refrain from trading, they will breach the inaction prohibition. This is indeed an unsatisfactory situation on the grounds of both fairness and common sense, in addition to the fact that it would be nearly impossible to detect inaction by a trader in the market. It is for these reasons that it is only positive acts that are prohibited by the Australian regime.

8.4 PENALTIES AND STATUTORY DEFENSES

Failure to comply with section 1043A constitutes an offense punishable by either a fine of up to 2,000 penalty units, five years imprisonment, or both (section 1311(1A)(db) and Schedule 3), with the penalty being five times the maximum allowable penalty in the case of a corporation found guilty of insider trading (section 1312). Section 1043A is also a civil penalty provision which means that a contravention of the primary prohibition may result in a civil action rather than a criminal prosecution (section 1317E(1)(jf) and (jg)). A pecuniary penalty order may be made whereby a $200,000 fine is made payable by the offender to Australian Securities and Investments Commission (ASIC) as a civil debt (section 1317G) and further compensation may be payable to a person who has suffered a loss as a result of the breach of the financial services civil liability provision (sections 1043L and 1317HA).

8.4.1 Statutory Defenses

Australia's insider trading regime also includes a number of statutory defenses to the prohibition. Some of the main defenses include exceptions for:

a person withdrawing from a registered scheme (section 1043B);

underwriters (section 1043C);

actions undertaken pursuant to a legal requirement (sections 1043D-E);

actions undertaken where Chinese walls or other information barriers exist (sections 1043F-G); and

knowledge of a person's own intentions or a body corporate's intentions (sections 1043H-I).

Case Study 2 below illustrates the Chinese walls defense in the context of one of Australia's most recent high-profile cases between ASIC and the Australian branch of one of the largest global financial services companies in the world, Citigroup.

8.4.2 Case Study 2: *ASIC v. Citigroup Global Markets Australia Pty Limited*

Citigroup's corporate advisory team was engaged by Toll Holdings Limited on August 8, 2005, to advise Toll on its proposed takeover bid for Patrick Corporation Limited. The employees who would work in the advisory team were known as "private side" employees as they were likely to come into contact with confidential information in the course of their business.

On August 19, a proprietary trader, Andrew Manchee, aggressively bought approximately 1 million shares in Patrick Corporation. Manchee was however on the "public side" of Citigroup's investment banking division and not a part of the private side advisory team engaged by Toll. Citigroup, however, like many other large financial institutions, had information barriers more commonly known as Chinese walls in place between the private side and the public side of the business. Such walls are implemented so that confidential and potentially price-sensitive information in relation to other corporations and institutions does not flow from the private side of the wall through to the public side. It is the existence of this wall, Citigroup argued, that meant that Manchee could not have known about Toll's takeover bid and therefore did not trade on inside information.

What had complicated and potentially jeopardized Citigroup's insider trading defense was that, through the course of the day, Manchee's superiors had been informed that someone on the proprietary desk had been trading a large volume of Patrick Corporation shares and were afraid that this could be perceived as a conflict of Citigroup's interest in advising Toll on its takeover bid on the private side, and in trading on its own account in the takeover target's shares on the public side, even if Manchee did not have knowledge of the proposed takeover. At about 3:30 p.m. on August 19, Manchee had a conversation with one of his superiors, Paul Darwell, on the footpath outside Citigroup's offices. Darwell had advised Manchee to stop buying Patrick Corporation shares but did not offer any explanation as to why. The only further trade that Manchee executed that day was the sale of about 200,000 shares in Patrick Corporation.

ASIC alleged that there were two instances of insider trading by Citigroup (*ASIC v. Citigroup Global Markets Australia Pty Limited*, 2007, FCA 963). The first was in relation to the sale of almost 200,000 Patrick Corporation shares in the afternoon of August 19, 2005, which was after Manchee had the conversation with Darwell on the footpath. The success of this argument depended on the ability of ASIC ability to prove that Manchee was an "officer" within the meaning of section 9 of the Act to show that his knowledge was effectively Citigroup's knowledge (section 1042G). The court, however, dismissed this claim stating that Manchee was neither a director of Citigroup nor did he occupy a management role to warrant his characterization as an officer (2007, FCA 963: 479–501).

The second insider trading claim was in relation to the purchase and sale of Patrick Corporation shares by Citigroup while persons from both the public and the private side of the business who were involved in these transactions were aware that Citigroup was advising Toll on its takeover bid, and therefore in possession of inside information. This claim was also dismissed by the court because Citigroup had put in place arrangements in the form of Chinese walls that could reasonably be expected to ensure that the information derived from the private side of the business was not communicated to the public side, making the statutory defense in section 1043F of the Act available to Citigroup (2007, FCA 963: 579–98).

8.5 ARE THE REGULATIONS EFFECTIVE?

In the ten-year period since the 1991 amendments there were only six prosecutions for alleged insider trading and only two of these resulted in convictions (Rubenstein 2002). Within the last two years, there has been one successful prosecution for insider trading (Rene Rivkin in 2003). So why is there such a low prosecution rate?

8.5.1 Detection

The low incidence of prosecuting a case for insider trading may be due to the difficulty in initially detecting the offense. The Australian Stock

Exchange (ASX) is responsible for detecting unusual or irregular trading in the market and reports any irregularities to ASIC for further investigation. But even with detection systems in place, there may be trades based on inside information that go undetected as the trades are not of a significant size in order to have an impact on the price of the financial product being traded, or are quite simply not detected as "irregular" trades. Aside from stock surveillance, however, ASIC may begin an investigation after being tipped off by a whistleblower of a case of insider trading.

Under the current regulations, there is no incentive for whistleblowers to come forward. However, there have been suggestions for the current penalty system to be replaced with a fine mechanism that takes into account an offender's future capacity to pay (Chapman and Denniss 2005). The repayment of the fine will be based on a system similar to that currently in use for Commonwealth debt repayments such as Higher Education Contribution Scheme (HECS). Ultimately, the way it would work is that, depending on the offender's level of income, deductions will automatically be made from the offender's salary in the form of a tax and these deductions would be used to pay the fine. The incentive for whistleblowers is that the whistleblowers themselves will receive a portion of the fine imposed as a reward pending a successful prosecution.

Although such a scheme has it advantages such as the increased certainty of payment, there are issues regarding whether a debt repayment system could actually be transposed into a penalty payment system. A penalty is not a debt and money has time value, which is why debts accrue interest. Imposing some form of indexation on a penalty that is to be paid over a period of time may appear to be unjustified to some as the total dollar value of all the payments made may actually be greater than the dollar value of the fine imposed. There may also be problems regarding the extent of the incentive. Though rewards would only be payable following successful proceedings, there may be an increase in the number of frivolous claims made by those hopeful to achieve an end gain. The risk that an accused may bring a defamation case against a whistleblower in the event that the accused is exonerated may not be enough of a disincentive for the whistleblower when compared to the potential gain obtained through the reward. This aspect of the scheme would thus require further attention, such as the imposition of penalties on whistleblowers for false accusations.

8.5.2 Establishing the Elements of the Offense

Establishing each element that is required to make a case for insider trading is by no means an easy task. Even if the elements can be established, however, the outcome may differ significantly from case to case. Take, for example, the insider trading cases against the late Rene Rivkin and against Steve Vizard.

In 2003, Rene Rivkin, one of Australia's most prominent stockbrokers, was found guilty of insider trading in relation to the trading of shares in QANTAS (*R v. Rivkin*, 2003, NSWSC 447). Having received information from Gerard McGowan, the chief executive officer of Impulse Airlines, of the upcoming merger of QANTAS and Impulse, Rivkin instructed his broker to purchase shares in QANTAS prior to the announcement. The greatest difficulty that the prosecution faced was proving that Rivkin had possession of the inside information beyond a reasonable doubt and generally speaking, unless there is evidence of a "smoking-gun" and the readily available evidence is circumstantial, proof can be difficult (Boulton 2004; Rubenstein 2002). In this case, however, McGowan's evidence of having informed Rivkin of the merger and specifically warning him not to trade in QANTAS shares was the smoking gun that the prosecution needed. Rivkin's claims that he would have traded in QANTAS shares notwithstanding the information on the merger were irrelevant as no causal connection needs to be established between possession of inside information and the relevant offense—proof of possession of the information at the time of the offense is all that is needed for the insider trading provisions to apply (Boulton 2004).

In stark contrast to the *Rivkin* case, the *Vizard* case (*ASIC v. Vizard*, 2005, FCA 1037) was not even tried as a case of insider trading. (ASIC sought an action under section 183, a civil penalty provision by virtue of section 1317E(1)(a), for the misuse of information obtained as a director of a company.) Vizard had obtained confidential information about a merger between two companies in which Telstra had a strategic interest and came into possession of this information by virtue of his position as a nonexecutive director of Telstra. As a result, Vizard purchased shares in one of the merging entities prior to public knowledge of the merger and sold them shortly after the announcement was made at a profit. One would think that this is clearly a case of insider trading, so why was Vizard not criminally prosecuted? The answer probably lies in the inability to prove each of the elements of the insider trading offense. Remember that only the Department of Public Prosecutions (DPP) can prosecute criminal offenses, and if the DPP feels that the evidence is not persuasive enough for criminal trial, the DPP will not commence criminal proceedings. ASIC, however, still had the option of pursuing a civil penalty action under section 1043A, but all this would have done is lower the standard to which it would have to prove all of the elements of the prohibition rather than change the elements themselves. The avenue taken by ASIC in the *Vizard* case dispensed with the need to prove the insider trading elements altogether and was thus a more simplified path to punishment, though what we are left with is two factually similar cases with two strikingly different punishments. It is, however, ultimately up to the regulator to determine how to approach and deal with instances of insider trading.

8.6 CONCLUSION

Given the lack of prosecutions for insider trading, it is arguable that it is the legislation that is to blame for being too difficult and complex in terms of certainty and understanding. In its report on insider trading, CAMAC made thirty-eight recommendations in relation to the insider trading regulations, the vast majority of which have since been agreed upon by the government in March 2007, with the remaining seven reopened for consultation with the market (Commonwealth of Australia Government 2007).

One of these is Recommendation 38, a recommendation that will potentially amend section 1043A to include the concept of inside information being "disclosable information" or "announceable information" (CAMAC 2003). These two new concepts would also be added to section 1042A as separately defined phrases.

Another recommendation that was reopened for consultation was Recommendation 10, amending the definition of "generally available" in section 1042C(1) to dispense with the need for the currently problematic "readily observable matter" test. Removing this part of the definition would simplify the "generally available" test and is perceived by some as "better based in principle, structurally more germane to the way information is, and will be, obtained, and thus likely to be more efficacious" (CAMAC 2003; Jacobs 2005).

The current criticisms and lack of prosecution for insider trading, as well as the lack of consistency in dealing with cases of insider trading, suggest that it is time the legislation is fine-tuned to better suit the behavior of market participants, the types of markets that operate in the financial services industry, and the way in which information is disseminated through each of those markets. It is this notion of harmonizing the law with what happens in the market that both the CAMAC report and the government's March 2007 Consultation Paper are aiming to achieve. Cases should be dealt with in a way that ensures the offender is punished in the most appropriate manner and in such a way that there is consistency between similar-fact cases. At the same time, however, the legislation should not dictate whether ASIC or the DPP, as the case may be, is better off pursuing a civil action or criminal prosecution. ASIC is currently forced to select a process that gives it any kind of result, but this process does not necessarily lead to the best obtainable result. The pluralism of legal redress in insider trading laws as they currently stand must be addressed in order to have a more workable and effective system of punishment, which in turn will give the market more surety in how the insider trading laws in Australia may apply to it.

REFERENCES

Baxt, R., A. Black, and P. Hanrahan. 2003. *Securities and financial services law.* 6th ed. Sydney: LexisNexis Butterworths.

Boulton, J. 2004. The insider. *Law Institute Journal* 78(7):50–53.

Chapman, B., and R. Denniss. 2005. Using financial incentives and income contingent penalties to detect and punish collusion and insider trading. *Australian and New Zealand Journals of Criminology* 38(1):122–40.

Commonwealth of Australia Government. 2007. Insider trading position and consultation paper, Canberra (March).

Corporations and Markets Advisory Committee. 2003. Insider trading report. Australian Government, Sydney (November).

House of Representatives Standing Committee on Legal and Constitutional Affairs (Griffith Committee). 1989. Fair shares for all: Insider trading in Australia. Canberra (October).

Jacobs, A. 2005. Time is money: Insider trading from a globalisation perspective. *Company and Securities Law Journal* 23(4):231–47.

Manne, H. G. 1966. *Insider trading and the stock market.* New York: Free Press.

Rubenstein, S. 2002. The regulation and prosecution of insider trading in Australia: Towards civil penalty sanctions for insider trading. *Company and Securities Law Journal* 20(2):89–113.

CHAPTER 9

The Evolution of Insider Trading Regulations in Japan

Sadakazu Osaki

CONTENTS

9.1 LEGISLATIVE HISTORY

9.1.1 Before the 1988 Amendment

Japan's regulation of insider trading has a relatively short history. Until the 1988 amendment of the Securities and Exchange Law (hereafter SEL), the basic legislation for capital market regulation in post–World War II Japan, the law did not prohibit insider trading explicitly, though it was widely believed that insider trading was not a rarity in the Japanese stock market during this period.

The SEL was originally enacted in 1948, when Japan was occupied by the Allied Powers under the command of General Douglas MacArthur. The SEL enactment was a part of an economic reform program implemented by New Dealers in the GHQ/SCAP (General Headquarters/Supreme Commander for the Allied Powers), and its drafting was strongly influenced by the U.S. Securities Act of 1933 and the Securities Exchange Act of 1934, two of the more prominent achievements of the Roosevelt administration's New Deal programs. Some provisions in the SEL were almost directly translated from the 1933 and 1934 Acts.

It is well known that the 1933 and 1934 U.S. Acts do not have any provisions explicitly prohibiting insider trading. Instead of prohibiting insider trading per se, the Congress tried to control short-swing trading in order to protect "outside" stockholders against short-swing speculation by "insiders" with advance information (Loss and Seligman 2004, 678). The provision was laid out in Section 16 of the Exchange Act.

Japanese legislation faithfully followed the U.S. example by allowing a company to require the surrender of the profits obtained the company's directors or major shareholders when such persons make a profit by selling shares within six months after their purchase, or when purchasing within the same period after their sale (Article 164 of the SEL, originally Article 189). It was widely accepted that this provision was intended to prevent or discourage unfair insider trading, which was not regulated directly.

Nevertheless, this provision did not play any substantial role in the regulation of insider trading in Japan. One of the reasons was that the system for reporting on the shares held by directors or major shareholders of the issuing company (Article 188 of the SEL) was abolished in 1953. It was argued that the system was impractical because directors or major shareholders who were not actively trading their shares would be reluctant to report their holdings to the authorities. At the same time, the authorities themselves did not have any effective means to enforce such a regulation.

As a result of this amendment, listed companies were unable to obtain the information necessary for requiring their directors or major shareholders to surrender unfair profits from short-swing trading, even if such trading had been carried out.

Moreover, Japanese regulators did not dare to utilize an effective tool employed by the U.S. enforcement authorities for regulating insider trading. In the United States, enforcement of Section 10(b) of the Exchange Act and famous Rule 10b-5 issued by the Securities and Exchange Commission (SEC) under the Act enabled the development of sophisticated case law on the regulation of insider trading and other unfair trading practices.

Although Japan's SEL has a similar provision to Section 10(b) in Article 157, this Article has never been used for prosecuting insiders engaged in unfair trading. The Article prohibits any person to "employ, in connection with the purchase or sale, or other transactions of securities, or derivative transactions, unjust means, scheme or contrivance." From the Japanese regulators' and public prosecutors' viewpoints, the wording of Article 157 is too vague to establish facts construed to be a crime (*Tatbestand*), a key concept for convicting criminals under Japanese criminal law.

9.1.2 1988 Amendment

In 1987 a company called Tateho Chemical Industries Co. Ltd., then listed on the First Section of the Osaka Securities Exchange, lost 20 billion yen in the sharp fall of the Japanese bond market. The company was actively trading Japanese Government Bonds (JGBs) and JGB futures. One of the banks that held shares in the company sold its holdings one day before the company announced its losses. The bank was not indicted, but it had acted against informal guidance from the Ministry of Finance and was regarded as "morally responsible" (Oda 1999, 289).

The incident led to widespread criticism of insider trading from the general public and made it difficult to ignore calls for a strengthening of controls over unfair trading practices. In addition, Japan's regulators were under mounting pressure from other countries to impose effective regulations; insider trading was an active topic of regulation in many jurisdictions. Finally, amendments to the SEL in 1988 introduced explicit prohibitions against insider trading.

The amended Article 190-2 (now Article 166) provided that "corporate insiders" who have come to know "material facts relating to the business of a listed company" should not engage in transactions involving securities issued by the company until those material facts have been "made public."

On the other hand, the amended Article 190-3 (now Article 167) provided that "a person who has made a tender offer or associated person thereto" who has come to know of a fact relating to implementation or withdrawal of a tender offer should not engage in transactions involving securities issued by the targeted company until the implementation or withdrawal of the tender offer has been made public.

To be precise, Article 167 applies not only to tender offers but also to similar circumstances, namely, when a person acquires shares in a listed company amounting to 5 percent or more of the total voting rights (Article 31 of the SEL Order). Although in principle the SEL and its subordinate order require the acquisition of more than 5 percent of shares outstanding to be in the form of a tender offer, if the shares are acquired from ten or fewer persons over a sixty-day period and ownership after the acquisition does not exceed one third, a tender offer is not mandatory. Furthermore, purchases on the market operated by stock exchanges are, in principle, exempt from the tender offer requirement (Article 27-2 of the SEL). Accordingly, although it is possible to accumulate greater than a 5 percent stake without going through the process of a tender offer, the impact that this would have on the share price is no different from the impact if it were through a tender offer. Insider trading rules therefore apply to actions that are the equivalent of a tender offer.

Both articles initially provided for criminal penalties of imprisonment for up to six months and/or a fine of up to 500,000 yen for any person in breach of each provision, and the criminal provisions have been amended several times since then. The latest amendment, made effective in July 2006, provides that any person engaged in unfair insider trading and in breach of Articles 166 or 167 of the SEL may be sentenced to up to five years imprisonment and/or imposed a fine of up to 5 million yen, and that profits obtained through illegal trading are subject to forfeiture. This substantial change in the severity of sanctions imposed on insider trading over a period of less than twenty years could be taken as clear evidence of a significant shift in the perception that the general public in Japan has regarding unfair trading practices in the stock market.

The regulation of short-swing trading in Article 164 could have been viewed as redundant after the 1988 amendments, since that regulation was seen as a substitute for the direct regulation of insider trading, as discussed above. However, public pressure for strengthening regulation was so strong that instead of the Article being abolished it was made more effective by reviving the reporting system that had been abolished in 1953 (now Article 163).

Later in 2000, a defendant who was asked to surrender the profits made through short-swing trading in accordance with Article 164 of the SEL challenged the constitutionality of the provision. The defendant argued that the provision was in breach of Article 29 of the Constitution, which protects property rights, since it allowed a company to recover profits made by short-swing trading even if the trading was not based on unpublished material information and did not cause damages to public investors. However, the Japanese Supreme Court rejected this argument in the plenary session by pointing out that the protection of property rights is not unconditional when it conflicts with the public interest and confirmed that Article 164 was based on a justifiable public policy that did not impose undue restrictions on property rights (Supreme Court judgment, February 13, 2002). This interpretation of the provision is basically in line with U.S. case law under Section 16 of the Exchange Act.

Article 164 was further strengthened in 2006, when an amendment to the SEL made clear that the profits made through short-swing trading by an investment fund without legal person status should also be surrendered under certain circumstances (Article 165-2). This amendment was introduced after the Murakami Fund, later indicted under the allegation of insider trading, had argued that its trading would not be subject to Article 164, as the provision only applied to "major shareholders" who owned more than 10 percent of the total voting rights issued by a listed company as a single natural person or legal person, whereas the Murakami Fund was a group of natural persons each holding a part of the Fund's total holdings.

9.1.3 Characteristics of the Legislation

On the surface, the above-mentioned 1988 amendments prohibiting unfair insider trading seem not to be far from similar legislation in other jurisdictions, such as Part V of the Criminal Justice Act 1993, which regulates insider trading in England. Nevertheless, the Japanese provisions have certain characteristics not found in most of the other jurisdictions. In summary, strict and narrow definitions of the relevant terms were used when defining what constitutes insider trading. The reasoning behind this strictness is that general legal principles require that the facts construed to be a crime

(*Tatbestand*) shall always be defined strictly, and that analogical reasoning when interpreting criminal provisions should be strongly discouraged.

First, the terms "corporate insiders" and "a person who has made a tender offer or associated person thereto" are strictly defined in the SEL. This has not been a substantial hindrance to effective enforcement of the regulations, however, because most possible cases are covered by this definition and because persons who directly receive inside information from the insiders described above are also prohibited from making transactions based on such information (Article 166, para. 3; Article 167, para. 3).

Second, the term "material facts relating to the business of a listed company" is strictly defined. For example, when a listed company has published an earnings forecast and its projected sales then were to change by more than 10 percent from the that forecast, the change would be construed as a "material fact" according to a provision in a cabinet order promulgated under Article 166 of the SEL. However, if the new projection was within a 10 percent range from the previous forecast, the change in the estimated earnings would not be construed as a "material fact." These highly technical definitions of "material facts" are listed in first three subparagraphs of Article 166, para. 2.

However, legislators were well aware of the fact that the use of such detailed definitions of "material facts" could lead to the undesirable consequence of having to let dishonest insiders go on the grounds of their not knowing "material facts." They therefore inserted a so-called basket clause in subparagraph 4 of Article 166, para. 2, which provided that "important facts concerning the listed company's operation, business or property that may have a significant impact on an investor's investment decision, other than those facts listed in preceding three subparagraphs" would be regarded as material facts. As discussed later, the Supreme Court used this clause to expand the scope of Japan's insider trading regulations.

Third, what constitutes "public" information is strictly and narrowly defined. Articles 166 and 167 say that corporate insiders, and so forth can trade only after material facts are made "public." According to Article 166, para. 4 and its subordinate Order (Article 30 of the SEL Order), this is defined as either public disclosure of the information through the current report specified under the SEL has been made, or the information has been distributed to at least two news agencies or other information vendors, and twelve hours have elapsed since the distribution.

This highly technical definition (in particular, the latter concerning new agencies) was based on the assumption that when material facts were conveyed to reporters through a press conference, it would take time before the information was actually reported in the media.

However, as the use of the Internet expanded, this definition became problematic. Many listed companies set up their own Web sites, and uploaded their press releases on those sites as soon as they held the press conference. News agencies also started to report the news on their Web sites as soon as it had been conveyed to them, not waiting for the following day's morning papers to be printed and distributed. As a result, if the definition of what is "public" were to be strictly applied to investors, it could lead to some ridiculous consequences.

When an investor sees relevant news on the "Nikkei Net" Web site operated by newspaper publisher Nikkei, which normally reports the news within a few hours after the press conference, the investor can make transactions based on the information, since he or she did not receive the material facts directly from corporate insiders, even though the information had not been made "public" yet. Nevertheless, if he saw the same information on the Web site of the listed company concerned within twelve hours after the press conference, he would be regarded as a direct recipient of the material facts, and transactions based on the information would therefore be considered unlawful insider trading.

The SEL Order was amended in May 2003 to avoid such an unreasonable outcome, even though there had been no prosecutions on the basis of making transaction using material facts obtained on a listed company's corporate Web site. The amendment made it clear that if the material facts were uploaded on certain Web sites operated by stock exchanges or securities dealers associations, the information would be regarded as "public" as soon as it was uploaded. To facilitate this amendment, Tokyo Stock Exchange (TSE) started to post information reported by the listed companies through TD-Net system, the TSE timely disclosure reporting system, immediately on its official Web site. Nevertheless, the so-called twelve hours rule is still in force, and if a listed company ignores its obligation to report through TD-Net but still uploads material facts on its own Web site, the unreasonable outcome noted above would still be possible, at least in theory.

9.2 DEVELOPMENT OF CASE LAW

Since the 1988 amendments to the SEL went into force, a number of cases of alleged breaches of insider trading prohibitions were brought to court. Through several judgments on controversial cases, Japanese courts substantially widened the scope of Articles 166 and 167, despite the strict and narrow definitions in these provisions of what constitutes insider trading. It could be argued that it was the common sense of judges that made it possible to avoid the unreasonable outcomes that inflexibly worded laws may have otherwise produced.

9.2.1 Nippon Shoji Case

The first important insider trading case was Nippon Shoji. In September 1993 a company named Nippon Shoji Kaisha Ltd. (now Alfresa Corporation), then listed on the Osaka Securities Exchange, started to market a new medicine for herpes zoster. Although the company had some experience in the wholesale marketing of pharmaceuticals, this medicine marked the first attempt by the company to market its own product. The medicine proved to be effective and well received by most practitioners, and the share price of the company rose sharply as a result. A month later, however, it had become apparent to the company that using the medicine together with certain other medicines produced serious side effects, to the extent that several patients had actually died from the combination. A doctor heard about the side effects and patient deaths from a company employee before those facts had been made public and sold the company shares short in an attempt to profit from expected fall in share price.

The doctor was prosecuted, but the Osaka Appellate Court decided that it was not apparent from the evidence in court whether news of the side effect was serious enough to be regarded as a material fact defined in Article 166 of the SEL. Under Article 166, para. 2, subpara. 4 "damage related to a disaster or business of the company" is treated as a material fact only if the expected damage exceeds a certain percentage of the company sales or profit. As the production of pharmaceuticals was not the main business of Nippon Shoji and made only a small contribution to its sales and profit, the court could not definitively ascertain that the information was a material fact under the circumstances. The prosecution appealed to the Supreme Court.

The Supreme Court handling the appeal decided that even if the information received by the doctor was related in some respects with a disaster or business of the company, it could be regarded as a material fact under the basket clause, which provides that "important facts concerning the listed company's operation, business or property that may have a significant impact on an investor's investment decision, other than facts listed in preceding three subparagraphs" would be regarded as material facts (Supreme Court judgment, February 16, 1999). The Supreme Court ordered the lower courts to reconsider the possibility of the information falling in the category of a material fact under the basket clause.

The Supreme Court's interpretation seems to be reasonable enough. Although the loss of the medicine's sales in this case would not have much effect on the company's overall sales and profits, it would have seriously harmed its share price, because investors would have been quite disappointed to learn that the company's new business was in serious trouble.

9.2.2 Nippon Orimono Kako Case

The next notable case was Nippon Orimono Kako. In 1995, the business of a listed company named Nippon Orimono Kako Co., Ltd (now ORIKA Capital Co., Ltd) was in duress. In order to continue operating, the company reached an agreement with another company on an M&A (mergers and acquisitions) deal whereby the latter would underwrite the issuance of new shares by Nippon Orimono Kako. The accused, an auditor and advising attorney of the acquiring company, bought a number of shares in Nippon Orimono Kako, expecting a sharp rise in the share price when the deal was made public.

The accused argued that he was not in breach of insider trading rules since the company's "governing entity" mentioned in Article 166 had not made any "decision" at the time of his purchase. The accused argued that at that time it was still uncertain whether the merger deal would be successfully completed. Nevertheless the Supreme Court interpreted the "governing entity of a company" pursuant to the SEL broadly, ruling that it "is not limited only to entities with decision-making authority prescribed by the Commercial Law, but can include entities able to make decisions seen as effectively equivalent to corporate decisions." The court was also flexible in its interpretation of "decisions" by such entities, ruling that such decisions "must have been made with the intention of realizing the issuance of shares, but there does not need to be an expectation that issuance of said shares is certain" (Supreme Court judgment, June 10, 1999).

The decision ensured that improper use of crucial inside information could not be tolerated even if formal decisions were not made by the formal governing entity of the company, such as the board of directors. It was an important step toward the more effective enforcement of insider trading rules.

9.2.3 Murakami Fund Case

The latest controversial insider trading case is the Murakami Fund, which is ongoing. In June 2006, Mr. Yoshiaki Murakami, head of the Murakami Fund, which over time had acquired large stock positions in a variety of listed companies and was aggressively asserting its shareholder rights, was arrested on the allegation of making illegal transactions based on the nonpublic information that Livedoor Co., Ltd was going to acquire a large stake in Nippon Broadcasting Co., Ltd (Osaki 2006). Although it is still premature to discuss the implications the case may have on Japanese insider trading regulation, since not even the court of first instance, let alone the Supreme Court, has given its judgment, the case is worth mentioning.

Mr. Murakami claimed in a press conference held just before the arrest that he learned from Livedoor executives in November 2004 and January 2005 of that company's intention to acquire at least a 5 percent stake in Nippon Broadcasting. Because the fact that Livedoor was planning to gain control over Nippon Broadcasting's stock was not generally known at the time, it is possible that Mr. Murakami could be considered a recipient of information related to actions equivalent to a tender offer under Article 167 of the SEL. Moreover, prosecutors pointed out that Mr. Murakami might have received such information as early as September 2004.

Open to debate is whether the Livedoor intentions communicated to Mr. Murakami constitute "facts related to the initiation of a tender offer or equivalent action" as prescribed in the SEL. Although Mr. Murakami himself admitted at the press conference the possibility that his own actions broke the law, this is clearly not the same as admitting guilt in court. Even if in fact there was a variety of information exchanged between Murakami and Livedoor, no illegal insider trading took place unless there was an actual purchase of Nippon Broadcasting shares following communication of "facts related to the initiation of a tender offer or equivalent action."

According to news reports, Mr. Murakami denied all the allegations made by the prosecution and pled not guilty in court. The Tokyo District Court is expected to hand down a judgment in July 2007. Whatever the decision, the case will surely continue until it reaches the Supreme Court. It will be interesting to see whether the Supreme Court maintains its flexible interpretation of Articles 166 and 167, or possibly even further expands the scope of insider trading rules.

9.3 INTRODUCTION OF CIVIL FINES

9.3.1 How Civil Fines Work

Another important development that has made Japan's regulation of insider trading more effective is the introduction of civil fines. This came with the amendments to the SEL that went into effect in April 2005, following the example of the civil fines used under the Anti-Monopoly Law.

Unlike criminal penalties, civil fines are an administrative measure aimed at preventing violations by making violators of the law pay a monetary penalty. Under the civil fine system, persons who violate the SEL provisions prohibiting such unfair trading practices as fraudulent transactions, market manipulation, and insider trading or the falsifying of annual reports and other disclosure documents are ordered to pay a civil fine based on the procedures outlined below (Osaki 2007).

When illegal activity is suspected, the Securities and Exchange Surveillance Commission (SESC) conducts an investigation, and if the investigation confirms there has been a violation, the SESC recommends to the Commissioner of the Financial Services Agency (FSA) that a civil fine payment order be issued. The recommendation includes a specific monetary amount for the civil fine to be paid. Upon receiving the recommendation, the FSA commissioner makes the decision to initiate the judgment process. The process is carried out by administrative judges in an administrative tribunal established under the FSA. The judges draft a proposal for payment order after following the semijudicial process, and then submit this proposal to the commissioner.

Although the procedures are similar to those in a court of law, in every case to date, the person subject to the civil fine payment order (the defendant) has filed defense documents indicating no objection, and the civil fine payment orders have been issued exactly as recommended by the SESC. The amount of the fine is determined by a detailed calculation method that differs depending on the offense (pursuant to Article 172 of the SEL).

9.3.2 Positive Effect of Civil Fines

In nearly every insider trading case that has been subject to civil fines thus far, the profits earned by the illegal trades have been in the neighborhood of 20,000 yen to several hundred thousand yen. This may suggest that the maliciousness of the violations has been fairly benign compared with past cases in which criminal charges have been filed. For example, in the Nippon Orimono Kako case discussed above, the forfeiture of profits imposed on the accused amounted to 26 million yen.

Furthermore, the simple trading methods used in these cases suggest that trades were made on the spur of the moment and under the expectation that keeping the trades small would make getting caught unlikely. Although making a show of punishing the more serious cases to serve as a warning to others is not a bad approach, sending a strong message to the market that even the small transgressions will not go undetected is probably a more effective way to maintain a fair and orderly market.

Following introduction of civil fine system, the SESC has continued to also pursue criminal charges in the more serious cases, as it has done in the past. The SESC asked public prosecutors to bring criminal charges in six cases, three of which were related to the infamous Livedoor, in 2006. This number and frequency of cases is not much different from the typical year.

By steadily pursuing the traditional criminal cases in parallel with getting the civil fine system on track and starting to discover those milder infractions that would have gone undetected in the past, the SESC has been working to more thoroughly eliminate unfair activities in securities markets, and it should be commended for that.

9.3.3 Remaining Issues

The civil fine system is thus gradually starting to produce results, although at this point its scope is considerably more limited than the equivalent mechanisms in the United States. A key challenge, therefore, is to expand the system's coverage. The report on enhancing and strengthening the public accounting and auditing system issued by the Financial System Council's Subcommittee on the Certified Public Accountant System proposes implementing a system for levying civil fines on auditing firms, as well, in view of the problems that auditing firms have caused their customers (the audited firms). These problems include the recent rash of fraudulent certification and other auditor misconduct as well as the business suspension order applied to Chuo Aoyama PwC. In March 2007 a bill for amending the Certified Public

Accountant Law in accordance with the proposal included in the report was presented to the Diet.

Another problem with the civil fine system is its lack of flexibility. The amount of fines to be paid is calculated mechanically in accordance with the formula laid out in the relevant provisions of the SEL. Thus, in March 2007 Komatsu Ltd, one of the world's largest manufacturers of construction equipment, was ordered by the FSA commissioner to pay a 43 million yen civil fine.

Komatsu was alleged to be in breach of insider trading regulations when it bought back its own shares in July 2005. At that time the company dissolved its subsidiary in the Netherlands. The subsidiary used to manage a fund of several hundred million yen and had been effectively dead when it was formally dissolved. The executive officer of Komatsu in charge of the finance department started the share buy back before the dissolution was made public and continued to do so after the announcement.

The stock market totally ignored the announcement since the subsidiary in question was insignificant to Komatsu, a large corporate group. Nobody accused Komatsu of stealing money from the investing public by capitalizing on its knowledge about a virtually dead subsidiary. However, Article 166 of the SEL requires that any dissolution of a subsidiary by a listed company be regarded as a material fact, without exception. It thus appears that Komatsu was technically in breach of the law.

However the trivial the breach was, Komatsu did in fact fail to observe an important provision of law concerning its business activity. It can therefore at least be argued that the company failed to maintain its internal control system in perfect order, though not many experts would feel that such negligence deserved a fine of 43 million yen. Even SESC and FSA officials probably felt that the punishment was excessive in view of the harm done by the company. Nevertheless, they had to impose a fine of that amount since they do not have the discretion under the SEL to raise or lower the fine depending on the circumstances or on the maliciousness of the offender.

Problems arising from the lack of flexibility in the wording of laws have been remedied by the Supreme Court, as discussed above, but the lack of flexibility in setting civil fines needs to be addressed by a legislative action. It is still unclear whether legislators are prepared to do so any time soon.

9.4 CONCLUDING REMARKS

There are two different ways to enforce securities market regulations: private enforcement through civil actions brought by investors; and public enforcement actions by governmental authorities. As far as insider trading is concerned, effective private enforcement is not easy because of the lack of clarity regarding who has been harmed by the illegal trades, especially in jurisdictions such as Japan where there is no system of class actions. If class action were to be allowed, it might be possible to bring a civil fraud case against the insider claiming the damages incurred by all the investors who traded that stock during the relevant period.

Effective public enforcement is critical to the regulation of insider trading. Over the past twenty years Japan has evolved a fairly effective system for regulating insider trading, as discussed in this chapter. Although the lack of flexibility that characterizes Japan's legal system has hindered this evolution, the achievements over these two decades should not be undervalued. With the Japanese people having begun to put more money into riskier investments, protecting investors and ensuring a fair and orderly market have become all the more important. We are certain to see further important developments in the regulation of insider trading in the coming years.

REFERENCES

Loss, L., and J. Seligman. 2004. *Fundamentals of securities regulation.* 5th ed. New York: Aspen Law & Business.

Oda, H. 1999. *Japanese law.* 2nd ed. Oxford: Oxford University Press.

Osaki, S. 2006. The Murakami fund incident and future fund regulation. *Nomura Capital Market Review* 9(3):14–27.

Osaki, S. 2007. Civil fines under the Securities and Exchange Law take hold. *Nomura Capital Market Review* 10(1):14–19.

CHAPTER 10

Insider Trading in China

Zhihui Liu and Margaret Wang

CONTENTS

10.1 INTRODUCTION

In the West, the collapse of Enron and WorldCom in the United States and Parmalat in Italy in the early 2000s and the ensuing criminal prosecutions of senior executives have led to an increased focus on the duties and liabilities of corporate officers, particularly where personal profits have been made. In these circumstances, it is not surprising that senior executives have become increasingly aware of their responsibilities and more cautious in discharging their duties toward their companies.

The experience in the West is a sharp contrast to the ancient Middle Kingdom (China), which hosts a new, yet robust, capital market, and a foreign, yet all too frequently used, term: corporate governance.

China's capital market, through the reestablished Shanghai Stock Exchange, along with the Shenzhen Stock Exchange, has been in existence only since the early 1990s. Further, it was not until 2005–6 when the share structure reform took place that up to two thirds of the shares of Chinese listed companies were nontradable shares.

As the name suggests, the nontradable shares cannot be traded on a stock exchange and may only be transferred by private sale—which requires government approval in most cases. These nontradable shares can be classified into state-owned shares (held by public authorities, state-owned asset management companies, or by state-owned enterprises, or SOEs) and legal person shares (held by domestic institutions, such as industrial enterprises, securities companies, research institutes, and investment companies).

In September 2005, the Chinese government, through China Securities Regulatory Commission (CSRC), embarked upon a share structure reform program that required all Chinese-listed companies to convert nontradable shares into tradable ones. A tight deadline of December 2006 was imposed. Companies that failed to convert their share structure by the deadline were barred from raising further funds; for example, the Shenzhen Development Bank (which was controlled by TPG Newbridge Capital) was imposed with a 5 percent trading ban and was prevented from raising new capital, which resulted in its capital adequacy ratio to fall to only 3.71 percent—way below the statutory requirement of 8 percent (Asia Private Equity Review, February 2007).

The short implementation period requiring all nontradable shares to be converted to tradable ones resulted in a sudden influx of tradable shares in the market that may be held in private hands—shares which were traditionally nontradable before the reform and were effectively controlled by the state.

The share structure reform, coupled with the trend for management buyout (MBO), led to a sudden influx of shares that can now be held in private hands and can be freely traded. This sudden influx of privately owned shares is believed to have increased the incidences of insider trading and market manipulation, along with other types of white collar crimes in China.

In September 2007, China's securities watchdog, the CSRC, demanded a crackdown on insider trading and price rigging following the stock of a Shandong-based pharmaceutical company, Jintai, which kept rising to the daily limit of 5 percent for 42 consecutive days, even though the company had been in the red for three years. This increase was reported to be driven by the company's false announcement of additional funding into the company (*Xinhua News* 2007c).

This chapter first explores the issue of whether severe penalties imposed on fraudulent executives act as a deterrent to prevent corporate officers from making a personal profit. It is followed by a discussion of the legal regime in place in China which regulates insider trading. Then it brings a practical focus to bear on that regime by examining case studies of recent insider trading in China.

10.2 ARE SEVERE PENALTIES A DETERRENT FOR FRAUDULENT EXECUTIVES?

Insider trading is a criminal offense punishable by imprisonment in the West, including the United States, the United Kingdom, and Australia. This is in contrast to China, where only monetary penalties, and a possible ban from holding positions as senior executives, are imposed. There is a view that imprisonment, as the punishment that is most severe in constraining one's liberty, may constitute a deterrent for fraudulent executives. However, the following factors have been identified as undermining the potential deterrence of criminalization.

10.2.1 No Prior Convictions

Generally, persons who commit white collar crimes, including insider trading, are not career criminals. They are thus unlikely to have prior convictions and are usually older than other types of criminals, such as street offenders. A "clean" criminal record, when combined with other factors (such as their provision of economic support for their families, their well-regarded position in the community, and their lack of preparation for the emotional and physical trauma of imprisonment), means that it is unlikely that a white collar criminal will be imprisoned (Kahan, and Posner 1999; Szockyj 1999).

10.2.2 Leniency of Punishments

Although white collar crimes have the potential to cause more economic damage than other crimes (such as street crimes), the persons who commit white collar crimes are rarely the subject of criminal prosecutions (Recine 2002). This is because of the lack of resources on the prosecution side and the fact that often the complexity of white collar cases may overwhelm the justice system. Further, of those prosecuted, only a small percentage is convicted and, even where convicted, few white collar criminals are actually given custodial sentences (Berg 2003). Additionally, in contrast to persons who commit street crimes, white collar criminals are usually imprisoned in minimum security prisons, which are disparagingly referred to as "Club Fed" or "Country Clubs" (Szockyj 1999).

10.2.3 Background of White Collar Criminals

One characteristic that almost all white collar criminals have in common is that they have business background or expertise. Accordingly, they tend to be more calculating when it comes to the price they may pay (that is, criminal, and also civil liability and the loss of current income) as compared to the potential gains from criminal conduct. When the price of punishment is low (not only the likely lenience of any criminal penalty but also the relatively low prospects of detection), it is possible that an intending offender would view committing a white collar crime as a great low-risk "investment opportunity" with a potentially high return (Dutcher 2005).

Given that the West has difficulties discouraging white collar crimes, including insider trading offenses, even where prison sentences are imposed on the offenders, it is no surprise that China may have even greater difficulties in enforcing its insider trading laws in the absence of such harsh penalty. This chapter outlines the laws and regulations in China, followed by case studies that examine the effectiveness of these laws and regulations in practice.

10.3 INSIDER TRADING LAWS AND REGULATIONS

Although there is no single piece of legislation in China that prohibits insider trading, there are a number of laws and regulation that cover the field. The most important and powerful piece of legislation is the Securities Law of the PRC ("Securities Law"), passed by the Standing Committee of the National People's Congress on December 29, 1998, with the latest amendments made on October 27, 2005. Other regulations which prohibit or restrict insider trading include, initially, the Provisional Measures for the Management of Securities Companies, issued by the People's Bank of China in 1990. This was followed by two measures that were issued in 1993: Provisional Regulations for the Issuance and Exchange of Shares and Provisional Measures on the Prevention of Securities Fraud, issued by the then China Securities Commission (now known as China Securities Regulatory Commission, CSRC). Further, Provisional Rules on the Prohibition from Entry into the Securities Market were issued in 1997.

Additionally, the CSRC has recently toughened up its enforcement efforts with the aim of cracking down on insider trading activities in China (*Xinhua News* 2007b), along with the issuance of a number of new rules and regulations, including Rules for Trading Equities in State Firms and Rules Targeting Insider Trading at State Firms (Qiao 2007).

The Securities Law specifically prohibits people who possess inside information, and those who have illegally obtained inside information, from using the information to engage in securities trading activities (article 73, Securities Law). Article 76 further provides that, before the information becomes public, people with inside information must not trade securities or disclose such information or suggest that another trade.

Article 74 stipulates people who would possess inside information on securities trading to include:

Director, supervisor, and senior managerial staff of the issuer;

Shareholders holding more than 5 percent of the company's shares and its directors, supervisors, senior managerial staff, as well as those having actual control of the company and its directors, supervisors, and senior managerial staff;

The company which the issuer has a controlling shareholding as well as its directors, supervisors, and senior managerial staff;

People who have access to inside information about a company through their employment with the company;

Staff at securities supervisory and regulatory organizations and those whose roles are related to the issuance and exchange of securities;

Staff at custodians, underwriters, stock exchanges, and securities registration organizations and other related organizations; and

Other people designated by the securities supervisory and regulatory organizations of the State Council.

The Securities Law also provides that inside information is defined as the information that is not yet available to the public but relates to a company's management or finance or has substantial influence over the market price of the company's securities (article 75). Further, article 75 stipulates that the following information is deemed as inside information to a company:

Information on the company's plan to declare a dividend or to increase its capital;

Information regarding major changes to the company's share structure;

Information relating to major changes to the company's debt guarantee arrangements;

Information pertaining to where more than 30 percent of the company's operational assets have been used as security, or sold, or disposed of, in one transaction;

Information on the activities of the company director, supervisor, or senior managerial staff which may result in company liability to pay damages and/or compensation in accordance with the law;

Information on the proposed merger/acquisition of a listed company;

Any other important information which the State Council's securities supervisory and regulatory organizations deem to have material influence over stock prices.

Further, the Securities Law provides people who engage in insider trading will be responsible for compensating the loss suffered by other investors as a result of such activities (article 76). Additionally, severe penalties will be applied to people who engage in insider trading activities, or disclose inside information, or suggest to another to trade such securities. Profits made will be confiscated. Further, a hefty fine of a minimum of 100 percent and a maximum of 500 percent of the profit made will be imposed (article 202). Where the profit made is less than RMB30,000, then the penalty imposed shall be between RMB30,000 and RMB600,000 (article 202). Article 202 also stipulates that a particularly hefty penalty shall be imposed on the employees of securities supervisory and regulatory organizations who engage in insider trading activities—though it did not prescribe the extent of this so-called hefty penalty.

It is also important to note that, during the period of investigation for insider trading, the securities supervisory and regulatory organizations of the State Council have the power to restrict the person being investigated from trading securities for no longer than fifteen trading days. However, if the relevant transaction is complicated, then this period may be extended by another fifteen trading days (article 180).

Other regulations, rules, and measures are in place to supplement the main piece of legislation enacted by the central government, that is, the Securities Law—see the above discussions. These regulations include Provisional Rules on the Prohibition from Entry into the Securities Market (issued by the CSRC in March 1997) and Provisional Rules on the Prohibition of Securities Fraud (issued by the CSRC in September 1993). Additionally, both Shanghai and Shenzhen Stock Exchanges have their own listing and trading rules with which all listed companies on these exchanges must comply. Further, the CSRC is in the process of devising new rules that target insider trading at state firms.

The insider trading cases presented in the section below will provide examples on how other rules and regulations operate in practice. Nevertheless, it is worth mentioning here that new rules currently being devised by the CSRC and expected to be issued in the near future are aimed at restricting the sale of previously state shares by requiring, first, large shareholders at the state-held public firms to offload only up to 50 percent of their stakes in the companies within three years of acquiring these shares; and second, top company officials not to transfer more than one quarter of their total shareholding in their firm in each year of their tenure (*China Daily* 2007; *Shanghai Daily* 2007).

10.4 SELECTED INSIDER TRADING CASES IN CHINA

The experience of China is particularly instructive in relation to assessing whether severe penalties imposed by the law have created an effective deterrent against insider trading activities by corporate officers. The following two cases demonstrate the relatively lenient punishment upon executives who commit insider trading offences in China.

10.4.1 Xinjiang Tianshan Co.

In June 2004, Tunhe Co., a substantial shareholder of Xinjiang Tianshan Co., entered into a confidential share transfer agreement with China Non-Metals Materials Company, in which Tunhe was to acquire 51 million shares to become a 29.42 percent major shareholder of Tianshan. This information was not disclosed to the public until June 29, 2004. However, Mr. Chen Jian Liang, the deputy general manager of Tianshan, was aware of this inside information. Mr. Chen used a fake account which he controlled, by the name of "Liming," and bought 1.65 million Tianshan shares and sold approximately 195,000 shares in the company between the period of June 21 and June 29, 2004—which represented the period immediately before the share transfer agreement was entered into to the day when this information became public.

Mr. Chen was investigated by the CSRC and was found guilty of insider trading. He was fined RMB200,000 and was banned from entering into the securities market for five years pursuant to Provisional Rules on the Prohibition from Entry into the Securities Market. According to article 4(3) and article 5 of the rules, Mr. Chen was prohibited from holding the office of a senior executive at any listed company in China as well as any organization that deals with securities as part of its business.

10.4.2 China International Fund Management Co.

China International Fund Management Company (CIFMC) is a company in which JP Morgan Asset Management (UK) Limited holds a 49 percent stake. This case involves a fund manager, Jian Tang, who was once awarded number one analyst by "New Fortune" in 2003. Mr. Tang, while working for CIFMC, traded shares that CIFMC managed, using the account of both his father and a third party in 2006. Mr. Tang used his father's account to purchase 60,000 shares and the third party's to buy 200,000 shares. Together he had made a total profit of RMB1.2 million from trading these shares using the other people's accounts.

After JP Morgan discovered Mr. Tang's conduct, it permitted Mr. Tang's resignation as a fund manager and other associated roles at CIFMC. However, this matter was never investigated by any supervisory body.

10.5 CONCLUSION

In the West, insider trading is a criminal offense punishable by imprisonment. This is in contrast to China where, at most, only monetary penalties are imposed. The experience in the West suggests even imprisonment, though it is the most severe punishment in constraining one's liberty, does not serve as an effective deterrent to white collar crimes for such reasons as no prior convictions, leniency of punishment (even where imprisonment is carried out!), and background of persons committing white collar crimes. As imprisonment does not deter white collar criminals, including those engaging in insider trading activities, in the West, it is hardly surprising to find people being discouraged in China from carrying out such activities when only monetary penalties are imposed.

Nevertheless, China has come a long way since the reestablishment of its capital market only a little more than a decade ago. It has established a legal framework to regulate its securities market which is supported by a supervisory body under the State Council, namely, the China Securities Regulatory Commission. Bearing in mind that up to two thirds of shares in Chinese listed companies were effectively owned by the state, and not by private individuals, until China embarked on its share structure reform program in 2005, China has made tremendous improvements to its capital market operation in a relatively short period of time. Perhaps it is only a matter of time before China will be able to catch up with the West in regulating the more sophisticated white collar crimes, such as insider trading.

REFERENCES

Asia Private Equity Review. 2007. A 5% trading band imposed to Shenzhen Development Bank. February.

Beijing Morning News. 2007. Morgan's removal of its funds manager, Tang Jian, for insider trading. May 17.

Berg, P. F. S. 2003. Unfit to serve: Permanently barring people from serving as officers and directors of publicly traded companies after the Sarbanes-Oxley Act. *Vanderbilt Law Review* 56:1871–1906.

China Daily. 2007. Rules target insider trading at state firms. February 6.

China Securities Net. 2007. Morgan's response: Removing funds manager, Tang Jian, from all titles. May 16.

China Securities Regulatory Commission. 2007a. Decision paper on punishment to Mr Chen Jian Liang. April 28.

China Securities Regulatory Commission. 2007b. Decision paper on prohibiting Mr. Chen Jian Liang from entry into the securities market. April 28.

Dutcher, J. S. 2005. From the boardroom to the cellblock: The justifications for harsher punishment of white-collar and corporate crime. *Arizona State Law Journal* 37:1295–1319.

Jinrongjie (Finance Industry) News. CSRC's decision paper on administrative penalty. May 23.

Kahan, D. M., and E. A. Posner. 1999. Shaming white-collar criminals: A proposal for reform of the federal sentencing guidelines. *Journal of Law and Economics* 42:365–88.

Qiao, X. 2007. Insider trading targeted by new regulations. *Caijing Magazine*, September 19.

Recine, J. S. 2002. Examination of the white collar crime penalty enhancements in the Sarbanes-Oxley Act. *American Criminal Law Review* 39:1535–70.

Shanghai Daily. 2007. New rules for trading equities in state firms. May 29.

Shanghai Securities News. 2007. Morgan's response to the incident involving Tang Jian. May 17.

Sina Finance News. 2007. CSRC's administrative penalty on Mr Chen Jian Liang. May 23.

Sina News. 2007. Decision relating to Tang Jian by Morgan Funds Management Company. May 16.

Szockyj, E. 1999. Imprisoning white-collar criminals. *Southern Illinois University Law Journal* 23:485–503.

Xinhua News. 2007a. Manager fined for insider trading. May 23.

Xinhua News. 2007b. CSRC determined to combat insider trading. September 5.

Xinhua News. 2007c. China ready to crackdown on insider trading. September 6.

CHAPTER 11

Hedge Fund Fraud

Greg N. Gregoriou and William Kelting

CONTENTS

11.1 INTRODUCTION

The rapid growth in the number of hedge funds and the value of the investment assets they control (Schneeweis, Kazemi, and Martin 2001) has raised concerns about the probity of the investment techniques employed by hedge funds. Among these concerns are insider trading and the potential for these "secretive" hedge funds to engage in other types of securities fraud. Because hedge funds are loosely regulated by the Securities Exchange Commission (SEC) and are not usually subject to the various securities laws that the SEC administers, they have largely managed to keep their internal affairs and trading activities private.

This lack of outside scrutiny—together with performance-based remuneration structures of hedge funds and the fact that the hedge fund manager will often have a sizeable portion of his or her own personal wealth tied up in the fund—creates an environment in which the hedge fund may resort to insider trading and other forms of market abuse or legal but questionable trading tactics to boost fund returns. Insider trading is a criminal offense in the major financial markets in which the majority of hedge fund trading takes place. A conviction for insider trading or even the mere publicity surrounding an indictment for insider trading may have disastrous financial consequences for the hedge fund and its investors. This chapter addresses the issue of what investors in hedge funds can do, in the absence of more stringent regulation, to safeguard their interests and thus protect the value of their investments against fraudulent conduct on the part of the hedge fund manager.

Today, nearly 48 percent of hedge fund managers are domiciled in the United States and most are not required to register with the SEC. However, the number of offshore hedge funds is growing faster in anticipation of tighter and stricter future SEC restrictions, along the lines of the recent, short-lived SEC registration requirement. Will this eliminate hedge fund fraud? Obviously not, especially if the barriers to entry are low and if investors are looking to hedge funds for added returns.

Several hedge fund fraud cases during the last few years have captured the attention of the SEC. Because of the number of investors, such as pension funds and endowments which pour billions into this industry, it has warranted a second and closer look by the SEC. A current push is under way by the SEC to force hedge funds to provide greater transparency and make sure that their internal control structure performs soundly. This may unfortunately serve as an incentive to locate offshore in an attempt to hide unwanted news or even conceal the hedge fund manager's background. This is easily done by simply shifting the fund's assets and registering the fund in an offshore jurisdiction, free of SEC regulation.

Having an unregulated environment invites hedge fund managers, of which many are unwilling to play by the rules, to also enter the industry. Additionally, many well-known money managers have also crossed over to the hedge fund industry, some of whom were under the close scrutiny of the SEC and decided to open hedge funds to escape the regulated environment. Information concerning references, qualifications, investment process, performance of the fund, ethics, and ability to have a financially solid and solvent position can be fabricated.

Therefore, data made available by the hedge fund itself or by database vendors in the hedge fund industry can often be misleading. Hedge fund database vendors, such Hedge Fund Research and TASS, simply receive returns net of all fees from hedge funds. In some cases, stellar performance may be a sign of impending doom or may warrant another look at the hedge fund manager's practices. Furthermore, seeing a five-year track record without a single negative month may be a sign that the manger may be "cooking the books."

11.2 THE MEMORANDUM, LACK OF TRANSPARENCY, AND INFLATED RETURNS

An assessment of the offering memorandum of hedge funds may be an unsatisfactory document, especially if the memorandum is not drafted by attorneys specializing in the investment areas. Furthermore, one must make sure the investment objectives of the hedge fund are properly defined and that the document does not consist of vague paragraphs. In many instances, investors blindly hand over money to hedge fund managers without any obvious understanding of how the hedge fund is invested, the amount of leverage used, and the turnover of the fund.

Forcing hedge funds to become more transparent may identify more fraudulent hedge funds. However, the solution also lies in making sure proper administrators, auditors, and prime brokers reduce the risk to investors of being defrauded. The creation of a nonprofit hedge fund organization to carefully monitor hedge funds may also detect fraudulent hedge funds before losses occur.

Because most hedge funds are unregulated, they operate in secrecy as the proprietary investment strategies of the manager cannot be revealed because of the added value they bring to the fund. This is often a good starting point for scams to occur. Why continue hustling for commissions, selling penny stocks, when for the price of having a lawyer draft a believable private placement memorandum, you can call yourself a hedge fund manager? Leave behind any disciplinary problems you may have had as a broker regulated by the National Association of Securities Dealers and, as a money manager unregulated by the SEC, charge clients a hefty "performance fee" far greater than what any conventional money manager would earn.

11.3 BACKGROUND MANAGER SEARCHES

Many private investigators on Wall Street offer specialized services to investigate and make background checks on hedge fund startups and their managers. Picerno (2001) has reported that more than 150 funds of hedge fund managers have hired private investigators to examine the hedge funds they are interested in investing in. Many investors today fail to examine criminal records or prior fraudulent activities in related financial industries, or determine if any disciplinary action has been taken against the hedge fund managers within the securities industry.

Not only does the hedge fund manager's background need to be examined but also that of the support staff. A check of a manager's impressive Wall Street client references revealed that the individuals named at each firm were brokers who had simply executed trades on behalf of the fund. Wall Street firms had never actually invested in the hedge fund (Picerno 2001).

Investors are also advised to verify if the hedge fund is registered as an advisor. If not, then the assets of the hedge fund may not even be held at a financial institution. Unaudited monthly statements with over-hyped returns have to be further examined to see if indeed some accounting firm is auditing the fund. In many cases, if investors cannot obtain supporting documents, such as its detailed strategy and monthly returns, it is best to avoid the fund. Furthermore, not obtaining net returns on time or even months later could in fact result from cooking the books.

Investors regularly make the error of assuming that retaining a large accounting or law firm, by a hedge fund, in some way guarantees that everything is legitimate. Usually investors also assume that accounting or law firms are discerning when choosing to represent clients. Unfortunately, if clients pay them exaggerated fees, many law and accounting firms will appreciatively take on the obligation. Indeed, the most thriving scammers frequently outwit the outside experts and may use the abilities and status of these experts to legitimize the fraud. The existence of a well-known and respected accounting or law firm is a factor one should consider but unwarranted dependence should not be placed upon its participation.

Many of the fraudulent hedge funds found in Table 11.1 simply took investors' money and used the funds assets for personal and business expenses, such as purchasing mansions, expensive sports cars, sailboats, and even memberships in exclusive country clubs. Many of these hedge funds constantly lied to investors, by falsifying reports and putting these reports on the company letterhead of respected accounting firms. Many

TABLE 11.1. Fraudulent hedge funds

Name of hedge fund	Name of person	Amount of fraud
Manhattan Fund	Michael Berger	$400 million
Ashbury Capital Partners	Mark Yagalla	$25 million
Apex Investments	Mark Yagalla	$25 million
Maricopa Investment Corp.	David Mobley	$300 million
Pinn Fund USA	Michael Fanghella	$107 million
Iris Programme	Burton G. Friedlander	$7.5 million
Jupiter Fund	Burton G. Friedlander	$7.5 million
Cambridge Partners II LP	John C. Natale	$40 million
Friedlander International Ltd	Burton Friedlander	$2.4 million
Strategic Income Fund LLC	E. Thomas Jung	$21 million
Total		*$935.4 million*

used the technique of inflating returns as well as overstating the manager's management experience, performance fees, and even in some cases manipulating the price of warrants on certain stocks.

11.4 WHERE WERE THE AUDITORS?

The answer to that question is, "it depends." Because most hedge funds are not required to register under any of the SEC Acts (namely, the Securities Act of 1933, the Securities Exchange Act of 1934, the Investment Company Act of 1940, and the Investment Advisers Act of 1940), there is no federally imposed audit requirement. In contrast, registered investment companies are prohibited from offering or selling their securities to the public without filing independently audited financial statements with the SEC. However, it seems that most hedge funds do engage independent audit firms (SEC 2003). Investors, of course, should be highly skeptical of a fund that does not engage an independent auditor. There are potential problems even in those cases where financial information is audited. These problems have to do with the auditor selection process, the qualifications of the selected auditor, and the nature of the audit procedures that the external auditor may choose to apply.

The auditors of registered investment companies must comply with the requirements of the Investment Company Act of 1940 and also, in the case of investment companies whose securities are listed for trading on a U.S. exchange, the Sarbanes–Oxley Act of 2002. The Investment Company Act requires that the auditor be selected by a "majority of directors who are not interested persons" (sec 80-a-31). The Act also requires shareholder ratification of the selected auditor, unless approved by an audit committee of the board (Investment Company Act Rule 32a-4). The audit committee must consist solely of independent directors. The Sarbanes–Oxley Act requires auditors of listed investment companies to meet more stringent independence tests, including restricting the provision of nonaudit services to the company by the auditor and mandating auditor rotation (secs 201–203). The Mutual Funds Integrity and Fee Transparency Act of 2003 has extended the auditor requirements of the Sarbanes–Oxley Act to all registered investment companies (sec 104). Because most hedge funds are not registered, and in many, if not most, cases are organized as limited partnerships, there is no similar requirement for auditor selection. Potential investors should be aware of the auditor selection process used by the hedge fund. In many cases, this may simply be a decision made by the investment advisor.

Auditors of registered hedge funds must adhere to stricter requirements than those of unregistered funds. Auditors of registered funds must follow the rules and guidelines established by the SEC. Auditors of registered funds must also follow some of the rules set forth in the recent Sarbanes–Oxley Act. Auditors of unregistered funds are not required to register with the Public Company Accounting Oversight Board (PCAOB) unless the audit firm happens to audit a public company. The PCAOB has the authority to scrutinize the work done by the audit firm; in some cases, hedge fund auditors would not be subject to the same scrutiny.

There may also be significant differences in the scope of work done by auditors of unregistered funds. Auditors of both registered and unregistered funds should follow the guidelines established by the accounting profession such as those established by the American Institute of Certified Public Accountants (AICPA 2003). However, auditors of registered funds are required to do more extensive procedures in some important areas. The SEC requires auditors of registered funds to report on the company's internal control. Auditors of unregistered funds do not have the same responsibility. They will consider internal control in the course of conducting the audit but to a lesser extent. Because internal controls are of the utmost importance in minimizing opportunities for fraudulent behavior, investors need to be wary. Two important areas where the amount of work done by the auditor differs have to do with custody of securities and valuation of securities. Auditors of registered funds must verify custody of all securities. In the case of unregistered funds, the amount of verification is a function of the auditor's discretion. The same difference applies to checking the valuation of the portfolio. Auditors of registered funds must verify all portfolio valuations as of the date of the financial statements. Auditors of unregistered funds will determine the extent of valuation testing judgmentally. This is a very sensitive area as many hedge fund frauds have involved fraudulent valuation of investments. In fact, in many cases, hedge fund valuations are provided by the investment adviser. Because of the typical fee structure for hedge fund advisers, this provides a tremendous incentive for fraudulent activity.

11.5 NATURE OF THE BILLION DOLLAR FRAUDS

The frauds listed in Table 11.1 are fairly typical as to the approach used by the fraudsters to perpetrate the fraud and in the attempted cover-ups. Consistent with the findings of the SEC, these frauds consisted primarily of schemes designed to misappropriate assets or funds provided by investors (SEC 2003). Investors were persuaded to invest based on promises of extremely high returns. Funds provided by investors were then diverted, in most cases, to finance the lavish lifestyles of the perpetrators in massive Ponzi schemes. When returns did not materialize, the fraudsters provided the investors with false statements showing overstated positions as well as overvalued securities and overstated rates of return.

In the case of the Manhattan Fund fraud, auditors were also supplied with bogus statements. Perpetrators of the fraud involving the Pinn Fund supplied altered financial statements as well as forged audit reports. Friedlander provided investors with financial statements accompanied by a compilation report prepared on the KPMG letterhead. A compilation does not provide the user with any explicit assurance; however, the association of a prestigious firm such as KPMG does provide some level of comfort. Unfortunately, the report was fraudulently placed on KPMG letterhead. In the case of the Maricopa fraud, Mobley (a hedge fund manager) refused to provide investors or potential investors with audited financial reports. He stated that "audits would risk divulging his secret and highly profitable trading strategies" (SEC 2000).

Fraud literature suggests that a fraud has three elements: incentive, opportunity, and rationalization (Albrecht 2003; Wells 2003). In the hedge fund fraud cases cited, it appears that the incentive in each case was greed and the desire to finance exorbitant lifestyles. It is difficult to assess opportunity but it is evident that necessary internal controls were not in place to prevent or detect the fraudulent activity. The strong concentration of power in the hands of the adviser and the lack of a formal governance structure, particularly in the case of partnerships, provided ample opportunity for these white collar criminals to carry out their activities. Rationalization often depends on the ethics of the parties involved. In some cases, the fraudsters may have rationalized that "everything would work out in the long run."

11.6 DUE DILIGENCE QUESTIONNAIRE

Funds of hedge fund managers, as well as other institutional investors, must exercise proper care, making sure a manager adheres to the legal requirement to manage the funds entrusted to it prudently, as enshrined in the Uniform Prudent Investor Act of 1994 (Hambrecht, Spitz, and Scherago 2002). The principle of prudence stated in that Act has been incorporated into the law of the majority of the U.S. states. In essence, what the prudent investor rule says is that those investing the money of others will, as fiduciaries, be required to exercise reasonable care, skill, and caution in selecting investments. Therefore, the manager is under a legal duty when investing a client's money to ensure that the investments selected are consistent with the risk and return objectives of the client when that investment assessed in terms of its impact on the client's portfolio.

Table 11.2 is a suggested checklist which can be employed when selecting hedge funds.

11.7 CONCLUSION

Investors like hedge funds because of their low correlations to stock and bond markets, but they are often mesmerized by their above average returns, even in down markets. When examining hedge fund activity many magazines and commentators have projected that nearly 20 percent of all hedge fund managers participate in deceptive activities.

To minimize one's risk of becoming a victim of fraud, we offer the following suggestions: put hedge fund managers under close scrutiny, perform due diligence fraud tests, examine the manager's track record, perform background checks, meet with current clients of the fund, find out whether other criminal proceedings or prior convictions exist, and make sure financial statements are properly audited.

TABLE 11.2. Checklist for hedge fund investors

Was the hedge fund manager's background verified?
Has the hedge fund manager had prior disciplinary action taken against him or her?
Has a criminal and bankruptcy verification been conducted?
How long has the hedge fund manager been in business?
Has the manager previously managed a poorly performing fund which was eventually shut down?
Are the fund returns provided by the manager accurate?
Is the hedge fund manager listed in any hedge fund databases?
Does the manager have well-known clients?
Are the clients satisfied?
Does the hedge fund manager return phone calls and answer all questions?
How frequently are client reports issued? How transparent is the fund?
How much leverage does the fund employ?
Does the manager provide a detailed risk management philosophy?
Does the fund maintain proper legal counsel and administrators, as well as accountants?
Does the fund of funds (FOF) manager geographically diversify his or her hedge fund managers?
How does the FOF manager control risk?
Does the FOF manager report correctly address and meet all requirements to educate investors on a monthly basis?
Does the fund comply with the Sarbanes–Oxley Act?
Is the fund registered with the Commodity Futures and Trading Organization (CFTR)?
Is the fund registered with the SEC?
If unregistered, are financing statements avoided?
If audited, who are the auditors?
Is the audit firm required to register with the PCAOB?
If audited, what type of opinion was rendered?
Do the promised returns make sense?
Do the monthly statements reflect a reasonable rate of return?

REFERENCES

Albrecht, W. S. 2003. *Fraud examination*. Mason, OH: Thomson Southwestern Publishing.

American Institute of Certified Public Accountants (AICPA). 2003. *Audits of investment companies*. New York: AICPA Audit and Accounting Guide.

Anson, M. 2002. *The handbook of alternative assets*. New York: John Wiley & Sons.

Hambrecht, G. A., W. T. Spitz, and N. F. Scherago. 2002. *Selecting a hedge fund manager*. Charlottesville, VA AIMR Publications.

Picerno, J. 2001. The happy world of hedge funds. *Bloomberg Wealth Manager* November: 64–72.

Schneeweis, T., H. Kazemi, and G. Martin. 2001. Understanding hedge fund performance: Research results and rules of thumb of the institutional investors. Working paper, University of Massachusetts, Isenberg School of Management, CISDM.

Securities and Exchange Commission (SEC). 2000. Litigation Release 16446. February 22.

Securities and Exchange Commission (SEC). 2003. Implications of the growth of hedge funds. Staff report to the Securities and Exchange Commission.

Wells, J. T. 1997. *Occupational fraud and abuse.* Austin, TX: Obsidian.

CHAPTER 12

Extraterritorial Reach of the Insider Trading Regimes in Australia and the United States

Yee Ben Chaung

CONTENTS

12.1 INTRODUCTION

The increasing interconnectedness of world securities markets ensures a need for corporate regulators to remain vigilant and open minded in policing insider trading. The recent explosion in private equity–driven takeover activity has resulted in the escalation of opportunities for individuals to trade on inside information. Certainly, this has not escaped unnoticed. In its Financial Stability Review, the Reserve Bank of Australia specifically identified private equity transactions as a potential catalyst for insider trading (Reserve Bank of Australia 2007). It is contended that private equity transactions uniquely involve a multitude of parties, each with an assortment of advisors, and thus, given the significant number of people with knowledge of the relevant transaction, there exists an elevated risk that someone will trade on market sensitive information relating to that transaction. This is more than a theoretical proposition. In data compiled by Bloomberg in May 2007, in each of the three days preceding the disclosure of each of the seventeen largest takeovers in the past year, options trading activity had surged by an average of 221 percent above the daily average over the previous fifty days (Scheer 2007; see also Scannell 2007).

Highlighting the potential for truly international insider trading offenses is the continuing prosecution of Ajaz Rahim by the Securities and Exchange Commission (SEC) in the United States (see Securities and Exchange Commission 2007). In this case, it is alleged by the commission that Rahim, an investment banker based in Pakistan, received information from Hafiz Naseem, a junior investment banker in the United States, in relation to the impending but as yet unannounced takeover of TXU Corporation by a consortium of private equity firms consisting of Kohlberg Kravis Roberts, Texas Pacific Group, and Goldman, Sachs & Company. The firm for which Naseem was an employee had served as a financial advisor in the TXU deal. The commission further alleged that on the basis of this information Rahim acquired a mixture of call option contracts and shares which ultimately resulted in a trading profit of over U.S.$5 million. At time of writing, this case has yet to reach its conclusion.

As illustrated by the case against Rahim, the extraterritorial ambit of the various insider trading prohibitions is an exceedingly pertinent issue. Notably, should the insider trading regimes of each country only apply domestically, they will become increasingly ineffective in policing the securities markets.

This chapter is intended to provide a glimpse at some of the issues concerning the extraterritoriality of insider trading regimes. It will do so by considering in some detail the Australian and U.S. systems. The key features of each are outlined in order to determine whether, in the Australian and American contexts, there exist any major gaps in jurisdictional coverage. Note that in the following sections entitled "Australian Law" and "U.S. Law," unless otherwise stipulated, all references to legislation and courts will be references to those of the topic jurisdiction, Australian or American as the case may be.

Figure 12.1 and Figure 12.2 broadly illustrate the analytical framework for determining whether the Australian or the U.S. regimes will apply in any specific case. However, the usefulness of these diagrams is restricted to providing guidance as to how the various concepts discussed in this chapter fit together; to that end, they are brief and simplistic and are certainly not intended as a comprehensive summary.

12.2 INSIDER TRADING REGULATION IN AUSTRALIA

12.2.1 Outline

The provisions prohibiting insider trading are found in Part 7.10 Division 3 of the Corporations Act 2001 (Cth). Generally, section 1043A provides that persons who possess inside information must neither trade nor procure other persons to trade in the financial product to which the inside information relates. These persons are also prohibited from communicating the information to persons who are likely to engage in the trading of the relevant financial product. Consequently, in considering the territorial

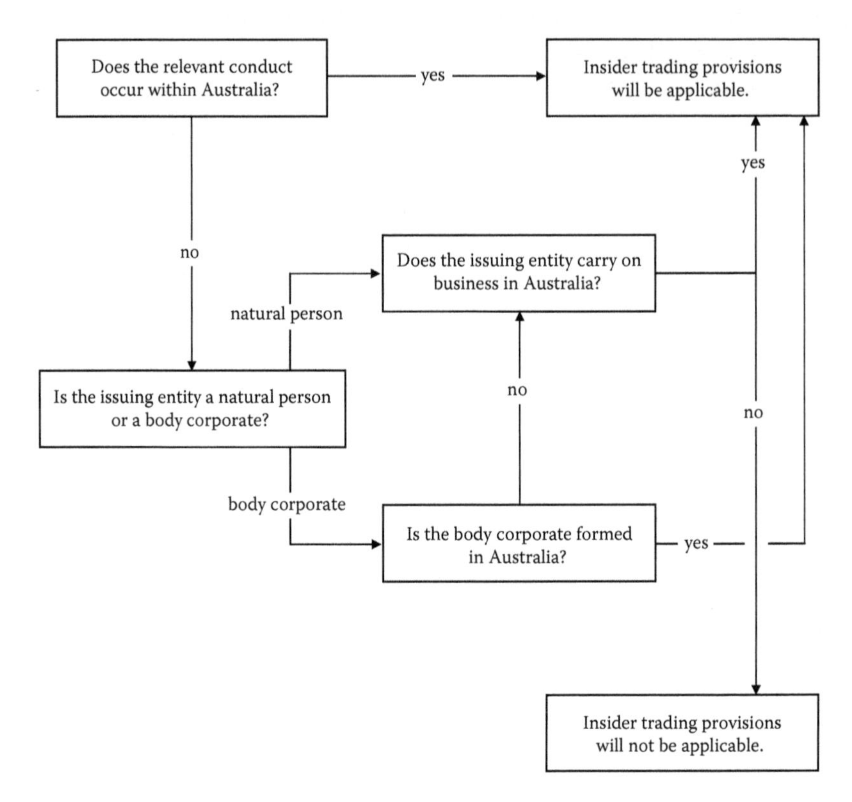

FIGURE 12.1. Australian extraterritorial analysis.

ambit of the insider trading prohibitions, the precise definition of the term "person" must be ascertained.

The insider trading provisions provide further qualification as to their extraterritorial application. Section 1042B states that the insider trading provisions apply to:

> acts and omissions within this jurisdiction in relation to Division 3 financial products (regardless of where the issuer of the products is formed, resides or located [sic] and of where the issuer carries on business); and
>
> acts and omissions outside this jurisdiction (and whether in Australia or not) in relation to Division 3 financial products issued by:
>
> a person who carries on business in this jurisdiction; or
>
> a body corporate that is *formed in this jurisdiction.* [emphasis added]

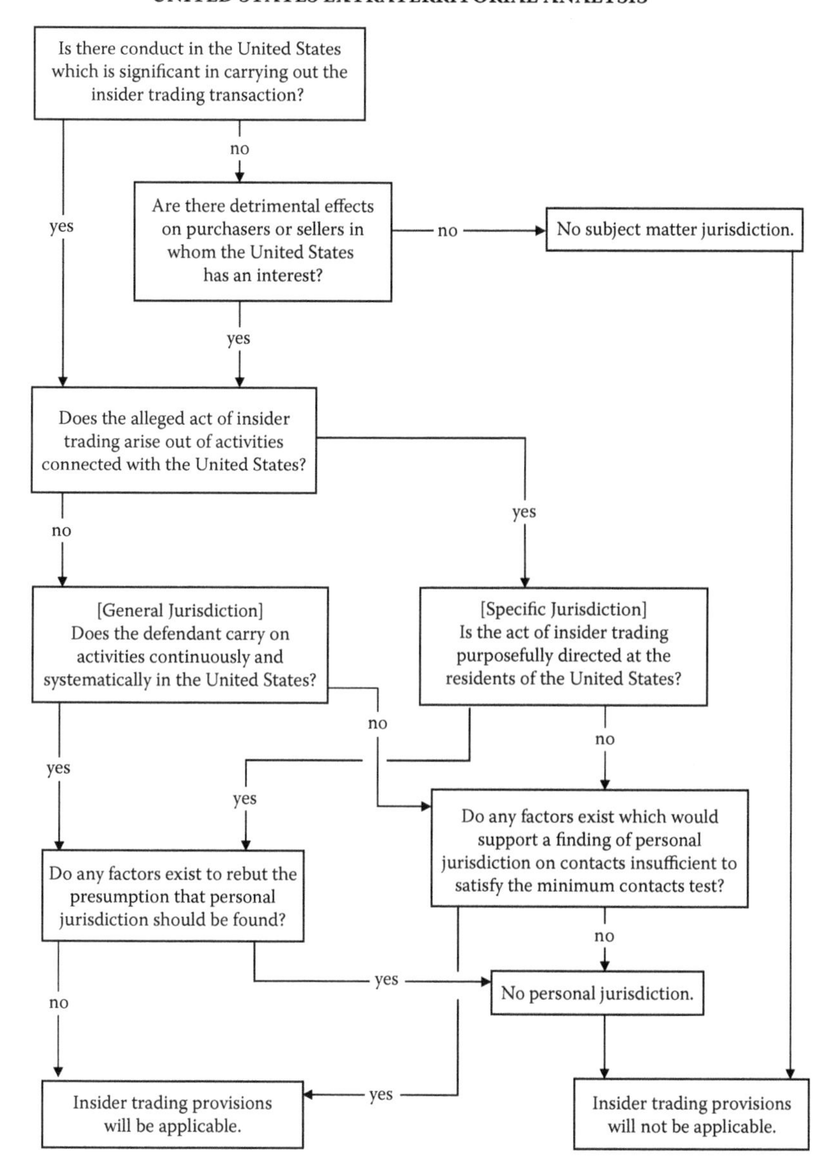

FIGURE 12.2 U.S. extraterritorial analysis.

From this provision it is evident that there is a two-stage analysis regarding whether the insider trading prohibition is applicable. First, where the act or omission in question occurs must be determined. Those acts and omissions which occur within "this jurisdiction," defined by section 5(2) as consisting of the whole of Australia since all the states are referring states, will be subject to the insider trading provisions. Those acts and omissions which occur outside this jurisdiction will only be caught if the issuer of the financial product to which the inside information relates is either a person who carries on business in this jurisdiction or a body corporate that is formed in this jurisdiction.

Thus, there are several issues that warrant further consideration. The definition of the term person, particularly in light of the section 5 prescription as to the territorial application of the Corporations Act, needs to be analyzed. Also, the relevant tests arising from section 1042B need to be more precisely defined. These are principally in regard to the determination of where the relevant conduct occurs and the ascertainment of whether the issuer of the financial product carries on business in this jurisdiction.

There is one further issue which, although important, will not be considered in detail here. The prohibition in section 1043A only relates to Division 3 financial products. The question of whether the definition of a Division 3 financial product is sufficiently broad is certainly a pertinent one in the sophisticated modern economy. Without doubt, the combination of section 1042A and Part 7.1 Division 3 produces an exceedingly extensive definition. However, this definition needs to be examined in further detail before any conclusion regarding its adequacy can be made.

12.2.2 Definition of "Person"

The term person is defined in the Corporations Act only to the extent that it includes a superannuation fund when referred to in Division 2 of Part 2D.2. Logically, this is of no utility in defining the term in the context of the insider trading provisions. However, section 5C provides that the Acts Interpretation Act 1901 (Cth) is applicable to the Corporations Act. Section 22 of the Acts Interpretation Act provides that:

> (1) In any Act, unless the contrary intention appears:
>
> (a) expressions used to denote persons generally (such as "person", "party", "someone", "anyone", "no-one", "one", "another" and "whoever"), include a body politic or corporate as well as an individual; [and]
>
> (aa) individual means a natural person.

Usefully, since this definition of the term person also encompasses bodies corporate, the second stage test in section 1042B in regard to the issuer of the financial product, where the issuer is a body corporate, will be satisfied if the issuer is either carrying on business or formed in Australia. That is, the issuer need not be a natural person for the carrying on business test to be applicable.

No restrictions arise in relation to the residency of persons in regard to the applicability of the insider trading provisions. Subsection 5(7) provides that each provision of the Corporations Act applies according to its tenor to:

> natural persons whether:
> resident in this jurisdiction or not; and
> resident in Australia or not; and
> Australian citizens or not; and
>
> all bodies corporate and unincorporated bodies whether:
> formed or carrying on business in this jurisdiction or not; and
> formed or carrying on business in Australia or not.

Consequently, subject to the satisfaction of the test in section 1042B, the insider trading provisions will be applicable notwithstanding the residency of the relevant person.

12.2.3 Location of Conduct

It is the alleged contravening acts of trading, procuring, or communicating that are relevant for the purposes of the section 1042B jurisdictional analysis. However, the fact that these acts may be found in many species of conduct ensures that problems may be encountered in properly ascertaining where they occur. The lack of any comprehensive judicial authority on this issue further compounds such problems.

In the seemingly simple trading offense there is a real question about where the trade occurred. For instance, if a person is in a foreign country and directs a broker in Australia to purchase shares listed on the Australian Stock Exchange, is the trade taken to have occurred in the foreign country where the directive was given? Alternatively, is the place of occurrence in Australia, the location of the market on which the shares are listed? The limited commentary on this issue suggests that the answer is the former.

One such commentator, Dr. Walker, implies that such an approach was taken by the court in *R v. Kruse*, December 2, 1999, DCt (NSW). In that case, the defendant was the general manager of Carpenter Pacific

Resources NL (Carpenter), an Australian company whose shares were listed on the Australian Stock Exchange. A subsidiary of Carpenter was involved in litigation in Papua New Guinea. On the day of judgment, the defendant was present at the court in Port Moresby and very shortly after the favorable judgment was handed down, the defendant instructed his broker by telephone to purchase Carpenter shares. The judge held that this did not constitute insider trading as the information was a "readily observable matter" and thus was not inside information. It has been suggested by Dr. Walker that the test applied by O'Reilly DCJ in ascertaining whether the information was generally available is tied to the location of the conduct (Walker 2000). Thus, the finding that the information constituted a readily observable matter in Papua New Guinea would necessarily be predicated on an initial assumption that the conduct occurred in Papua New Guinea.

The New South Wales Court of Appeal in *R v. Firns* (2001, 51 NSWLR 548) held that the approach suggested by Dr. Walker to determining whether a matter was readily observable was without basis and thus incorrect. However, no specific comments were made regarding the location of the conduct and therefore the underlying assumption that the act of trading occurred in Port Moresby, where the instruction to trade was given, remains reasonably plausible.

This approach accords with that suggested in *Ford's Principles of Corporations Law* (Austin and Ramsay 2007) in which it is asserted that where trading instructions are sent to a broker in a foreign location, the client's act of trading will be taken as having occurred at the place where the instruction was sent. This *prima facie* seems a suitable approach given the prohibition against the mere application for a trade as it identifies the instruction itself as the impinging act. If applied strictly, this would theoretically result in liability regardless of whether any trade was actually carried out. However, although this may provide flexibility in the finding of a breach of the prohibition, it does create difficulties in regard to pinpointing the location of the conduct. Indeed, it is recognized that such an approach would necessarily mean that should the client communicate trading instructions from outside Australia, the applicability of the insider trading provisions will be subject to the satisfaction of the tests relating to the issuer of the relevant financial product. Although this may in fact be good law, it appears to be somewhat nonsensical.

A more flexible approach to ascertaining where conduct occurs was taken by Justice Merkel in *Bray v. F. Hoffman-La Roche Ltd* (2002, 190

ALR 1). In that case, the judge found that communications from an overseas parent company to an Australian subsidiary, which "for the most part … have been initiated [from] outside Australia [and] were directed to and were expected to be, and were, received in Australia … [can] be regarded as taking place in Australia." Although it is noteworthy that the judge restricted his findings to the context of the applicability analysis under section 45 of the Trade Practices Act 1974 (Cth), there does not logically seem to be any reason against its application in an insider trading context.

Indeed, where a person instructs a broker in Australia to buy or sell a security listed on the Australian Stock Exchange, it would seem reasonable that the trade be taken to have occurred in Australia notwithstanding the location of the person at the time the instruction was given. By comparison, the distinction based on the actual location of the person conducting the trade appears rather arbitrary and can lead to haphazard consequences (Austin and Ramsay 2007).

A slightly more sophisticated option may be to adopt a test which is dependent on the characterization of the alleged impugning conduct. For instance, the trading prohibition in section 1043A forbids persons in possession of inside information from applying for, acquiring, or disposing of the financial product to which the information relates, or from entering into an agreement to that effect. It would appear more sensible if the conduct were specifically identified first, that is, whether there is an application for, as opposed to an acquisition or disposal of, the financial product. If there is an application without an actual trade taking place, then the place where the application was made, or indeed, where the trading instruction was sent, would seem a logical location for that action. Where the trade does take place, it would accord with general notions of common sense that the location ascribed to the trade be the location of the market. Understandably, like any general rule, the potential for complexities and uncertainties exist; an off market transaction provides one such uncertainty. However, as a basis for analysis, this approach seems as reasonable as any of those proffered by commentators. Given the lack of judicial authority on this issue, its future acceptance is eminently possible.

12.2.4 "Carry on Business"

Part 1.2 Division 3 of the *Corporations Act* provides guidance as to which entities carry on business in Australia. Section 18 specifically states that the carrying on of business includes the carrying on of business of a kind that is otherwise than for profit. Section 20 affirms that business may be carried on alone or with other persons.

Section 21 provides that:

(1) A *body corporate that has a place of business* in Australia, or in a State or Territory, *carries on business* in Australia, or in that State or Territory, as the case may be [emphasis added].

(2) A reference to a body corporate carrying on business in Australia, or in a State or Territory, includes a reference to the body:

(a) establishing or using a share transfer office or share registration office in Australia, or in the State or Territory, as the case may be; or
(b) administering, managing, or otherwise dealing with, property situated in Australia, or in the State or Territory, as the case may be, as an agent, legal personal representative or trustee, whether by employees or agents or otherwise.

(3) Despite subsection (2), a body corporate does not carry on business in Australia, or in a State or Territory, merely because, in Australia, or in the State or Territory, as the case may be, the body:

(a) is or becomes a party to a proceeding or effects settlement of a proceeding or of a claim or dispute; or
(b) holds meetings of its directors or shareholders or carries on other activities concerning its internal affairs; or
(c) maintains a bank account; or
(d) effects a sale through an independent contractor; or
(e) solicits or procures an order that becomes a binding contract only if the order is accepted outside Australia, or the State or Territory, as the case may be; or
(f) creates evidence of a debt, or creates a charge on property; or
(g) secures or collects any of its debts or enforces its rights in regard to any securities relating to such debts; or
(h) conducts an isolated transaction that is completed within a period of 31 days, not being one of a number of similar transactions repeated from time to time; or
(j) invests any of its funds or holds any property.

In relation to financial services businesses, section 761A provides that the section 911D definition is applicable. This is to the effect that "a financial services business is taken to be carried on in this jurisdiction by a person if, in the course of the person carrying on the business, the person engages in conduct that is intended to induce people in this jurisdiction to use the financial services the person provides or [if] it is likely to have that effect."

The provision that a person does not carry on business in a jurisdiction solely by virtue of having invested in that jurisdiction is subject to the nature of the investment. Where an investment is merely of a passive nature, then without anything further, the person will not be carrying on business. However, where the investment is such that it requires some degree of administration or management, subsection 21(2)(b) will be applicable and the person will be found to be carrying on business.

It is to be noted that the Corporations Act provides only a partial definition and does not proffer any general rule for determining whether an entity is carrying on business. Consequently, an analysis of case law is necessary.

Application of Campbell, Re Gebo Investments (Labuan) Ltd v. Signatory Investments Pty Ltd (2005, 190 FLR 209) is the only case to date which examines section 21 of the Corporations Act in any great detail. In that case Justice Barrett observed that the use of the word "includes" in subsection 21(2) provides "scope for the operation and application of territorially based concepts of carrying on business derived from the general law." It is according to these concepts that Justice Barrett suggests that the "carrying on of business generally involves conducting some form of commercial enterprise, systematically and regularly with a view to profit." Naturally, however, any profit requirement is rendered redundant by the operation of the aforementioned section 18.

The judge continued to find that subsection 21(3) operates in such a way that none of the factors subsequently mentioned in that subsection could individually be sufficient to lead to a conclusion that an entity carries on business within a particular jurisdiction. It leaves open the possibility that where one or more of the listed activities are engaged in simultaneously, such a positive conclusion is entirely open.

At common law, the concept of "carrying on" generally connotes continuity and repetition. In *Smith (on behalf of National Parks and Wildlife Service) v. Capewell* (1979, 142 CLR 509), Justice Gibbs finds that the carrying on of business "signifies a course of conduct involving the performance of a succession of acts, and not simply the effecting of one solitary transaction." However, the judge goes on to find that a single transaction can amount to the carrying on of a business where the relevant person holds an intention to carry on a business and where that transaction was "undertaken in pursuance of that intention."

An exception to this rule may exist in relation to isolated transactions of sufficient scale. Justice Dawson in *United Dominions Corporation Ltd v. Brian Proprietary Limited* (1985, 157 CLR 1) held that although continuity and repetition are generally indicative of the carrying on of business, these elements are not necessary requirements to such a finding. In that case, the judge found that the construction of a hotel and shopping center, in spite of the fact that it constituted only one transaction, was of sufficient scale to find a carrying on of business.

In regard to the term "business," Justice Mason in *Hope v. Bathurst City Council* (1980, 144 CLR 1) found that it was most accurately described as a "commercial enterprise as a going concern." In that case, the judge found that grazing activities constituted a business where they were of a substantial scale and exhibited a sufficient degree of commerciality. The fact that the relevant activity had a permanent character, involved advertising and the maintenance of financial records, and that the land on which the grazing took place was "put to its best potential use" were highlighted as factors particularly supportive of such a conclusion.

It is noteworthy that the court in *Luckins v. Highway Motel (Carnarvon) Pty Ltd* (1975, 133 CLR 164) held by majority that a permanent place of business is not a necessary prerequisite to finding that an entity carries on business "in" a particular jurisdiction. It was held that a company can carry on business in a state notwithstanding that the central management and control of the business resides elsewhere. On the facts it was found that a Victorian tour company which conducted tours to Western Australia conducted business in Western Australia by virtue of its entering into commercial transactions with persons in that state from time to time over an extended period. This finding was notwithstanding that the company had neither a place of business nor any property in Western Australia.

12.3 INSIDER TRADING REGULATION IN THE UNITED STATES

12.3.1 Outline

Unlike Australia, the United States has not adopted a regulatory regime to specifically prohibit insider trading. Instead, the prohibition exists in the broad judicial interpretations of the antifraud provision of the Securities Exchange Act of 1934; section 10(b). As a result, the regulation of insider trading in the United States bears little resemblance to the somewhat formulaic Australian statutory system. Indeed, since the seminal decision of the U.S. Supreme Court in *Chiarella v. United States,* 445 US 222 (1980), insider trading jurisprudence has ostensibly become an adjunct to general fiduciary obligations.

Where the Australian Corporations Act has specific guidance regarding the extraterritorial reach of its provisions, the U.S. legislation is silent on such issues (Kramer and Murray 2002; *In Re Royal Ahold NV Securities & ERISA Litigation,* 351 F.Supp 2d 334 (D Md, 2004) (*Royal Ahold*)). Instead, the United States has developed through its courts a series of elaborate tests to determine whether a particular case falls within its jurisdiction.

These tests are neatly encapsulated within the umbrella headings of "subject matter jurisdiction" and "personal jurisdiction." The inquiry into whether subject matter jurisdiction exists is in effect asking whether the factual circumstances of the particular case are sufficiently relevant to the United States. The question of personal jurisdiction is an examination into whether the person accused of insider trading holds a sufficient connection to the United States. These two concepts are examined in further detail.

However, there exist a number of other interpretative means for determining the foreign reach of the Securities Exchange Act that are not considered here. For example, the Third Restatement of Foreign Relations Law of the United States is often cited as being informative on issues of jurisdiction. However, it is suggested that its authoritative value is limited (Testy 1994). Also, many commentators and courts have found the SEC regulations regarding the extraterritorial reach of the disclosure requirements under the Securities Exchange Act to be a convenient guide to determining the scope of the antifraud provisions. However, because the courts have recognized that the ambit of the antifraud provisions is necessarily broader than that of the mandatory disclosure provisions, there appears to be minimal utility in considering these regulations (for example, see *Europe and Overseas Commodity Traders, S.A. v. Banque Paribas London,* 147 F.3d 118 (2nd Cir, 1998) (*EOC* case)).

12.3.2 Subject Matter Jurisdiction

There is a long-standing principle that the U.S. legislature has the power to legislate with respect to foreign persons and conduct and that whether such a wide-reaching effect exists is a matter of statutory construction (*Blackmer v. United States,* 284 US 421, 1932). However, there is a general presumption that unless a contrary intention appears, legislation will only be applicable within the United States (*Equal Employment Opportunity Commission v. Arabian American Oil Co.,* 499 US 244 (1991) (*Aramco*); *Microsoft Corporation, Petitioner v. AT&T Corporation,* 127 S. Ct. 1746 (2007) (*Petitioner v. AT&T Corporation*)). As previously mentioned, the Securities Exchange Act is silent on this issue. Therefore, it would *prima facie* appear that the United States' insider trading rules, founded on the general antifraud provisions of the *Securities Exchange Act,* are inapplicable outside the United States.

However, courts have devised two exceptions to this general principle, upon which subject matter jurisdiction may be established. These exceptions have been crafted under a somewhat expansive and judicially activist approach to the ascertainment of legislative intention. It is said that an inquiry into whether the exceptions are applicable is an inquiry into whether "Congress would have wished the precious resources of the United States courts and law enforcement agencies to be devoted to them rather than [to] leave the problem to foreign countries" (*Bersch v. Drexel Firestone, Inc.,* 519 F.2d 974 (2nd Cir, 1975), at 985) (*Bersch*).

The two exceptions are the "conduct" and the "effects" tests. The satisfaction of either one will provide an avenue for the application of the insider trading regime and thus, by virtue of 15 USC § 78aa, bring the case within the exclusive jurisdiction of the various district courts of the United States (*Royal Ahold*). Thus, it is these tests that require further consideration.

12.3.3 Conduct

To date, no Supreme Court authority exists on the operation of the conduct test. As a result, federal courts have been able to exercise significant discretion in crafting the test, often employing "policy considerations" along with the courts' best judgment in determining its ambit (*Kauthar SDN BHD v. Sternberg,* 149 F.3d 659 (7th Cir, 1998), at 664) (*Kauthar*). This has resulted in a continuum of different approaches emerging. To illustrate the varying approaches, the more restrictive interpretations of the Second, Fifth, and District of Columbia Circuits are considered and compared to the somewhat more lenient versions from the Third, Fourth, Seventh, Eighth, and Ninth Circuits.

The relative court groupings have been identified differently by the various courts. For instance, the court in *Royal Ahold* recognizes the Third Circuit as having the most expansive approach, the Second, Fifth, and District of Columbia circuits as having the most restrictive interpretations, and the Seventh, Eighth, and Ninth circuits as occupying the "middle ground." The groupings in the following sections are merely intended to reflect the different poles of jurisprudence and the degrees of variation in between. It is noteworthy that at times one or more of the approaches may be virtually indistinguishable.

12.3.4 Second, Fifth, and District of Columbia Circuits

The approach of the Second Circuit was initially articulated in the case of *Leasco Data Processing Equipment Corporation v. Maxwell,* 468 F.2d 1326 (2nd Cir, 1972) (*Leasco*). In this case, the plaintiff alleged that it was deceived by the defendants into purchasing stock at artificially inflated prices in a British corporation controlled by one of the defendants, a British citizen. It was alleged that the series of misrepresentations made by the defendants in relation to the company amounted to violations of the antifraud provision of the Securities Exchange Act. These misrepresentations took place in both Britain and the United States. The court held that where there is significant conduct within a territory, "a statute cannot properly be held inapplicable simply on the ground that, absent the clearest language, Congress will not be assumed to have meant to go beyond the limits recognized by foreign relations law." To that end, because some of the misrepresentations occurred within the United States, misrepresentations which were an "essential link" in inducing the plaintiff to make the stock purchases, the case fell within the subject matter jurisdiction of the U.S. courts. It was of no consequence that the securities in question were issued by foreign entities.

These principles were further elaborated upon by the court in *Bersch.* Here, the court explained the justification for the conduct test exception, noting that "Congress did not mean the United States to be used as a base for fraudulent securities schemes even when the victims are foreigners." However, the ambit of the test was somewhat narrowed when the court held that the test would not be satisfied in "cases where the United States activities are merely preparatory or take the form of culpable nonfeasance and are relatively small in comparison to those abroad." From this it appears that the satisfaction of the conduct test requires that the conduct be sufficiently significant so as to be classified as an "essential link" in the overall scheme.

The Fifth Circuit Court of Appeals in the case of *Robinson v. TCI/US West Communications Incorporated,* 117 F.3d 900 (5th Cir, 1997), expressly adopted the Second Circuit approach. The court made specific mention of the fact that in order to satisfy the conduct test, the domestic conduct must have "material importance" to or have "directly caused" the harm suffered by the plaintiffs.

The District of Columbia Circuit has perhaps the most restrictive interpretation of the conduct test. The Court of Appeals in *Zoelsch v. Arthur Andersen*, 824 F.2d 27 (DC Cir, 1987) (*Zoelsch*), explicitly adopted the approach of the Second Circuit. However, the court's view of the Second Circuit approach appears to be narrower than that emanating from the aforementioned Second Circuit cases. The court noted that "the Second Circuit's rule seems to be that jurisdiction will lie in American courts where the domestic conduct comprises all the elements of a defendant's conduct necessary to establish a violation of section 10(b) and Rule 10b-5: the fraudulent statements or misrepresentations must originate in the United States, must be made with scienter and in connection with the sale or purchase of securities, and must cause the harm to those who claim to be defrauded, even though the actual reliance and damages may occur elsewhere." This narrow test reflects the opinion of the court that because issues of extraterritorial application are principally concerned with policy considerations, the expansion of the ambit of statute by courts usurps the role of Congress. Nowhere is this more evident than in the passage where the Court of Appeals states that "were it not for the Second Circuit's pre-eminence in the field of securities law, and [their] desire to avoid a multiplicity of jurisdictional tests, [they] might be inclined to doubt that an American court should ever assert jurisdiction over domestic conduct that causes loss to foreign investors."

The facts of *Zoelsch* are as follows. The American defendant had prepared a financial statement which was briefly referred to in the prospectus of a German company as the basis of some of the data contained within the prospectus. The plaintiff alleged that this constituted a misrepresentation, and as it was prepared in the United States, it gave subject matter jurisdiction to the courts of the United States. However, the Court of Appeals, in adopting a restrictive conduct test, affirmed the decision of the district court to the effect that the conduct in the United States was so "relatively insignificant when compared with the nature and breadth of the allegedly fraudulent conduct abroad" that it was "merely preparatory" and thus conferred no subject matter jurisdiction.

12.3.5 Third, Fourth, Seventh, Eighth, and Ninth Circuits

A rather more lenient approach was adopted by the Third Circuit Court of Appeals in *SEC v. Kasser,* 548 F.2d 109 (3rd Cir, 1977). There it was held that "the federal securities laws do grant jurisdiction in transnational securities cases where at least some activity designed to further a fraudulent scheme occurs within this country." In applying this test, the court held on the facts that because there was conduct within the United States that was "crucial to the consummation of the fraud" the conduct test was satisfied. It is unclear whether the court, in finding that the conduct was "crucial," implied that such a degree of significance was required or whether it merely reflected the court's view of the facts; one which necessarily satisfied the test. In the most recent consideration of this issue in the Third Circuit, the District Court in *Markus Blechner v. Daimler-Benz AG,* 410 F.Supp.2d 366 (3rd Cir, 2006), found that where the conduct alleged to comprise the fraud occurs "predominantly outside the United States" it will be insufficient to satisfy the conduct test. This suggests that the domestic conduct must bear some degree of significance by reference to the entirety of the alleged fraud.

The Eighth Circuit approves this approach. In *Continental Grain (Australia) Pty Ltd v. Pacific Oilseeds Incorporated,* 592 F.2d 409 (8th Cir, 1979) (*Continental Grain*), the Court of Appeals held that where "conduct in the United States is in furtherance of a fraudulent scheme and is significant with respect to its accomplishment," subject matter jurisdiction will exist. This test was expressly adopted by the Ninth Circuit Court of Appeals in *Grunenthal GmbH v. Hotz,* 712 F.2d 421 (9th Cir, 1983), and the Seventh Circuit in *Kauthar.* In *Kauthar* the court noted that while the "conduct must be more than merely preparatory in nature" it need not "itself satisfy the elements of a securities violation."

Most recently, the Fourth Circuit District Court in the case of *Royal Ahold* adopted the conduct and effects tests and cited and followed the approach of the court in *Kauthar.* In *Re Ahold,* it was held that because reasonable investors would ordinarily rely on the information contained within SEC filings, misleading and deceptive statements made in such filings could form the basis of a finding of subject matter jurisdiction. These statements were held to be material in the overall fraud and a contributing cause to the loss suffered by the plaintiffs.

12.3.6 Effects Test

Often described as the "Mother Court of securities law," the Second Circuit is the principal purveyor of jurisprudence on the effects test. This is exemplified by the fact that the genesis of the effects test is found in the Second Circuit case of *United States v. Aluminum Co. of America*, 148 F.2d 416 (2nd Cir, 1945), where it was adopted in an antitrust context. Consequently, the decisions of the Second Circuit necessarily form the central focus of any discussion of the effects test.

The first application of the effects test to the Securities Exchange Act occurred in 1968 in *Schoenbaum v. Firstbrook*, 405 F.2d 200 (2nd Cir, 1968) (*Schoenbaum*). There, the Court of Appeals held that a "District Court has subject matter jurisdiction over violations of the Securities Exchange Act although the transactions which are alleged to violate the Act take place outside the United States, at least when the transactions involve stock registered and listed on a national securities exchange, and are detrimental to the interests of American investors." It is this fundamental concept, that the transactions give rise to detrimental effects domestically, that forms the basis of the effects test.

Most relevantly in the insider trading context, *Schoenbaum* unequivocally endorses the finding of subject matter jurisdiction in cases where the relevant security is listed on an American market. The Court held that "Congress intended the Exchange Act to have extraterritorial application in order to … protect the domestic securities markets from the effects of improper foreign transactions in American securities." This authority is now well entrenched; for example, see *MCG, Inc. v. Great Western Energy Corp.*, 896 F.2d 170 (5th Cir, 1990).

However, issues arise in relation to the application of the Securities Exchange Act to transactions involving foreign securities. There appear to be two key considerations in these circumstances: the nature and significance of the American interest in the foreign securities, and the characteristics of the American investors.

The first matter was highlighted by the Second Circuit Court of Appeals in *Bersch* where it was stated that there must be "injury to purchasers or sellers … in whom the United States has an interest" and that the alleged harm suffered cannot simply be an adverse effect on either American investors or the economy generally. Although this test appears restrictive on its face, courts have been quite willing to ascribe a broad interpretation. In *Consolidated Gold Fields PLC v. Minorco, S.A.*, 871 F.2d 252 (2nd Cir, 1989) (*Minorco*), the issue arose as to whether subject matter jurisdiction existed in an antitrust context with respect to the takeover of Gold Fields, a British corporation, by Minorco, a Luxembourg corporation. There, the court held that because American residents represented some 2.5 percent of Gold Fields' shareholders, both directly and through nominees, the takeover necessarily had a sufficient effect in the United States to found subject matter jurisdiction. This scenario illustrates the two key issues regarding the securities interests. First, there is a question of what proportion of the securities holders need to be American before the test is to be satisfied. This has never definitively been determined. Still, the *Bersch* case illustrates that the requirement may not be particularly onerous; there, subject matter jurisdiction was held to be sustainable on the fact that twenty-two American residents held just over 1 percent of the shares. The second issue is in regard to the quality of the interest. In the *Minorco* case, the Court noted that if the interests were only held indirectly, for example, through a trust as was the case in *IIT v. Vencap, Ltd,* 519 F.2d 1001 (2nd Cir, 1975) (*Vencap*), subject matter jurisdiction would not be established.

The other key consideration concerns the American investors themselves. In *Bersch* the court observed that "Congress surely did not mean the securities laws to protect the many thousands of Americans residing in foreign countries against securities frauds by foreigners acting there." Underpinning this concept is principle that the conduct of foreigners must be directed toward U.S. purchasers in order for subject matter jurisdiction to exist. Thus, if an American investor obtains securities in a foreign country while resident in that country, U.S. courts will not provide an avenue for redress. The fundamental question here is whether a particular foreign security is intended to be open to investment by U.S. citizens. Courts have analyzed this issue through adopting inferences regarding the intentions of the issuers. For instance, in *Bertsch*, the court found that because American citizens had actually invested in the foreign security, the relevant promotional information disclosure materials must have been disseminated in America. Whether there was an actual intention that the information be sent to America was held to be irrelevant to the inquiry at hand. Logically, the mere fact that securities are held by domestic residents gives rise to an inference that at the very least, the issuers are not adverse to the idea of having American investors. For the courts, this has been sufficient to provide a foundation for subject matter jurisdiction.

Prima facie, the effects test already places an extensive range of cases within the jurisdiction of American courts. However, it has been forwarded that the test will become only more expansive with time. As the world markets are becoming increasingly interconnected, the United States and its residents will be progressively more exposed to the actions of foreign persons in foreign exchange markets (Testy 1994). Logically, this suggests that in at least as far as the effects test is concerned, any existing gaps in the jurisdictional coverage of the U.S. insider trading regime will gradually diminish.

12.3.7 Judicial Discretion

There is significant flexibility in the application of the conduct and effects tests. As mentioned earlier, courts have repeatedly stated that the satisfaction of either test will be sufficient to establish jurisdiction and that therefore, meeting the requirements of both is not necessary (see, for example, *Continental Grain*, 592 F.2d at 417). However, this appears to be a general proposition, the application of which depends on the facts of the case and the persuasion of the particular court.

The Second Circuit Court of Appeals in *Itoba Ltd. v. LEP Group PLC,* 54 F.3d 118 (2nd Cir, 1995) (*Itoba*), held that there is no strict requirement that the tests be applied "separately and distinctly from each other." The court continued to add that "an admixture or combination of the two often gives a better picture of whether there is sufficient United States involvement to justify the exercise of jurisdiction by an American court." As a finding of fact in *Itoba*, because the fraud at the center of that case had occurred on an American exchange and harmed thousands of U.S. shareholders, jurisdiction was found as a matter of course. It is likely that this case merely avails the possibility for both the conduct and effects tests to be applied concurrently; if it is interpreted as mandating the application of both in every case, it will be impossible to reconcile with preceding authority.

Rather perplexingly, some courts suggest that the satisfaction of either the conduct or the effects tests will not automatically lead to a finding of subject matter jurisdiction in exceptional cases. Most interesting, however, is the fact that this somewhat divergent view comes from the Second Circuit; the origin of much of the authority on the conduct and effects tests and indeed the principal source of the long-standing authority that the tests are individually sufficient to determine jurisdiction.

The *EOC* case illustrates this approach. The facts of the case are fairly straightforward and are as follows. Europe and Overseas Commodity Traders (EOC) is a Panamanian company which is managed and wholly owned by a Canadian citizen named Carr. EOC held an account with the defendant Banque Paribas in London. During a visit to England, Carr was recommended an investment opportunity by Arida, an account manager with Banque Paribas in London. Carr expressed interest in the proposal but soon after left for Florida. While in Florida, discussions commenced via telephone between Carr and Arida regarding the investment. These discussions resulted in Carr executing a large purchase of securities with the capital of EOC. It is alleged by Carr that these purchases were made on the basis of misleading information tendered by Arida in the course of their discussions.

The Court of Appeals held that although the facts necessarily satisfied the conduct test, no subject jurisdiction existed. While the court recognized the efficacy of the conduct and effects test in analyzing issues of jurisdiction, the ultimate aim of ascertaining congressional intention was held to be paramount. Excluding the occurrence of the telephonic discussions, the court noted that the case had no connection with the United States and highlighted that on the facts, there existed "no U.S. entity that Congress could have wished to protect from the machinations of swindlers." Consequently, it was concluded that notwithstanding the satisfaction of the conduct test, it would be unreasonable to find subject matter jurisdiction as there is "no U.S. party to protect or punish." This result, and indeed approach, appears especially commonsensical in light of the fervent reaffirmation of the presumption against extraterritoriality by the Supreme Court, albeit in the context of Title VII legislation, in *Aramco.*

However, it is difficult to reconcile this finding with the abundance of judicial authority that champions both the conduct and effects tests as being individually sufficient to determine subject matter jurisdiction. The policy imperative which underpins the conduct test is most clearly stated by the Second Circuit Court of Appeals in *Vencap* when it remarked that it did "not think Congress intended to allow the United States to be used as a base for manufacturing fraudulent security devices for export, even when these are peddled only to foreigners."

Two subsequent cases from the District Court of the Southern District of New York have scrutinized this issue. In 1998, the court observed that "a simple mechanical application of the jurisdictional tests is insufficient" and that "a proper analysis should focus on the policy considerations that led to the extraterritorial application of [the] laws in the first place—protecting or punishing U.S. parties and markets" (*Interbrew SA v. Edperbrascan Corp.*, 23 F.Supp. 2d 425 (SDNY, 1998), at 429). Logically, this perspective conflicts with those cases that found subject matter jurisdiction as a matter of course upon the fulfillment of either the conduct or effects test. Oddly, the same court, albeit with a different judge presiding, took a slightly different view in 2004. In the case of *In Re Vivendi Universal* (2004 US Dist. LEXIS 21230), the court saw the decision in the *EOC Case* as one founded on a novel factual circumstance and one which did not actually constrict the applicability of the conduct and effects tests. However, the court did note that jurisdictional issues tend to revolve around the question of whether the alleged conduct implicates an interest in the United States. This tends to suggest that overarching policy considerations can impinge on the analytical sphere occupied by the conduct and effects tests.

The operation of the conduct and effects tests is certainly not straightforward. Generally, it appears that the fulfillment of the requirements under the tests will be sufficient to establish subject matter jurisdiction. In circumstances where only the conduct test is satisfied and neither party is a United States citizen, the result will be less clear. Given the explicit recognition by courts that jurisdictional analyzes necessitate policy considerations, there exists the possibility that the tests for extraterritoriality to be applied in the future will become less systematic and less transparent. This may provide an opportunity for plaintiffs to assert the existence of jurisdiction based on the totality of the factual circumstances, even where neither the conduct nor the effects tests are strictly met.

12.3.8 Personal Jurisdiction

In cases where the defendant is a foreign citizen, notwithstanding the existence of subject matter jurisdiction, liability cannot be imposed if personal jurisdiction is not established. The concept of personal jurisdiction arises out of the restriction in the ambit of the Securities Exchange Act by the requirement of due process in the Fifth Amendment of the Constitution (*Leasco*, 468 F.2d at 1339). This doctrine is premised on the notion that persons should have "fair warning" as to where liabilities may arise so as to give "a degree of predictability to the legal system that allows potential defendants to structure their primary conduct with some minimum assurance as to where that conduct will and will not render them liable to suit" (*World-Wide Volkswagen Corp. v. Woodson,* 444 US 286 (1980), at 297) (*Volkswagen*).

The case of *Burger King Corp. v. Rudzewicz,* 471 US 462 (1985) (*Burger King*), is an exceedingly comprehensive summary of the analytical framework for issues of personal jurisdiction. There, the Supreme Court synthesizes the main aspects of preceding authority into a single, albeit complicated doctrine. This doctrine appears to take the form of a multistage inquiry. First, there is an initial question as to whether the defendant holds the basic "minimum contacts" with the particular jurisdiction. These "contacts" are essentially factual circumstances which connect the defendant with the jurisdiction. The existence or absence of the requisite contacts then becomes an input for the subsequent reasonableness evaluation. Here, the court essentially weighs up countervailing factors to determine whether, in the circumstances, a finding of jurisdiction would be reasonable. Unsurprisingly, this confers a tremendous degree of discretion on the courts.

12.3.9 Minimum Contacts

The minimum contacts requirement was first devised by the Supreme Court in *International Shoe Co. v. Washington,* 326 US 310 (1945), at 317 (*International Shoe*), where it held that for personal jurisdiction to exist, a defendant needs to hold certain minimum contacts with the jurisdiction which are sufficient to ensure that the maintenance of the particular suit does not "offend traditional notions of fair play and substantial justice." This principle was expanded upon in the later case of *Hanson v. Denckla,* 357 US 235 (1958), with the finding that "it is essential in each case that there be some act by which the defendant purposefully avails itself of the privilege of conducting activities within the forum State, thus invoking the benefits and protections of its laws." The enjoyment of such benefits and protections ensures that as a corollary, the defendant must "submit to the burdens of litigation in that forum as well" (*Burger King*, 471 US at 476).

In *Burger King*, the Supreme Court noted that there is no requirement that the defendant physically enter the jurisdiction. Consequently, while such a physical presence may enhance the quality of the requisite contacts, its absence will not be fatal to finding jurisdiction where the defendant's efforts are "purposefully directed" toward the residents of the particular jurisdiction.

These are broad observations regarding the minimum contacts inquiry. In each case, however, it will be necessary to identify whether the issues concern general or specific personal jurisdiction. A finding of general jurisdiction will enable a court to hear a case arising out of activities completely unrelated to the forum jurisdiction itself. As set out by the Supreme Court in *International Shoe*, general jurisdiction will be found in instances where "the continuous corporate operations [of a person] within a state … [are] so substantial and [are] of such a nature [so] as to justify suit against it on causes of action arising from dealings entirely distinct from those activities." Specific jurisdiction is premised on the cause of action arising out of the activities of the defendant which are related to the forum state. Thus, in each case whether general or specific jurisdiction is required will depend on whether a nexus exists between the activities of the defendant which give rise to the suit and the particular jurisdiction (*Helicopteros Nacionales de Colombia, SA v. Hall,* 466 US 408 (1984), at 414).

12.3.10 General Jurisdiction

In what remains the most authoritative guidance on general jurisdiction, the Supreme Court in *Perkins v. Benguet Consolidated Mining Co.,* 342 US 437 (1952), held that there is no universal requirement that the cause of action in any case arise out of the contacts of the defendant with the particular territory. Where the activities of a defendant in a territory are sufficiently substantial, conduct entirely unrelated to the territory will be reviewable by American courts. In applying these principles, the court held that because the foreign defendant had carried on continuous and systematic activities in the territory, a finding of jurisdiction was entirely open. In that case, the defendant company had conducted activities consisting of "directors' meetings, business correspondence, banking, stock transfers, payment of salaries [and the] purchase of machinery." This finding can be contrasted to that in *Helicopteros.* There, the Supreme Court held that the making of purchases in a territory, albeit at regular intervals, would not be sufficient to establish general personal jurisdiction.

Conceptually, the principle of general jurisdiction appears to be an extension of the ancient notion that American courts have jurisdiction over "nonresidents who are physically present in the state" (*Burnham v. Superior Court of California,* 495 US 604, 1990). The analysis can usefully be viewed as the ascertainment of the strength of the connection between the defendant and the forum state, a sufficient nexus resulting in the defendant effectively being deemed constructively present in that state.

12.3.11 Specific Jurisdiction

Principles regarding specific jurisdiction are possibly of greater utility in insider trading cases as specific jurisdiction may be easier to establish than general jurisdiction. Indeed, it has been suggested that the apparent abandonment of the general jurisdiction analysis by the lower courts is due to the perceived lower standard required to find specific jurisdiction (Condlin 2004). While this may be true of the quantitative requirements under the minimum contacts test, an analysis for specific jurisdiction imposes onerous qualitative hurdles.

In *Burger King* the Supreme Court held that mere foreseeability of causing injury is insufficient. It was found that the activities of the defendant must be purposefully directed at the residents of the particular jurisdiction. This is to ensure that jurisdiction will not arise as a result of "random, fortuitous or attenuated contacts." Expressly acknowledged, however, is the fact that even a single act can support jurisdiction where it creates a "substantial connection" with the forum either as a result of its significant scale or its creation of "continuing obligations" with the residents of that forum. Oddly, however, these principles appear to have been misinterpreted and watered down by lower courts in insider trading cases.

For instance, in *SEC v. Unifund Sal*, 910 F.2d 1028 (2nd Cir, 1990) (*Unifund*), the Court of Appeals observed that while not every transaction involving the shares of an American corporation will satisfy the effect requirement, where the securities of the corporation are exclusively traded on an American exchange, insider trading can be reasonably expected to harm U.S. shareholders. It was said that this will suffice to establish personal jurisdiction over a foreign defendant. The facts of the case involved a Lebanese company which had purchased shares on an American exchange through a foreign office of an American brokerage firm. It was held that because the shares were exclusively traded in the United States this activity or contact was sufficient to satisfy the requirements to establish personal jurisdiction. However, the language of the court suggests that the requirements involve a test of foreseeability instead of the "purposeful direction" test set out by the Supreme Court in *Burger King*.

Somewhat comically, even this mistaken test of foreseeability is subject to confusion. The District Court in *SEC v. Softpoint Inc.* (2001, US Dist

LEXIS 286) (*Softpoint*) construed the judgment in *Unifund* to hold that "a district court may exercise its personal jurisdiction in a securities action so long as the defendant's activities [have] an unmistakably foreseeable affect [sic] within the United States." A careful reading of *Unifund* does not disclose any such finding; it was merely stated that the plaintiff had alleged that the effect of the defendant's conduct in the United States was "unmistakably foreseeable."

This accumulating mass of misinterpretations was seemingly followed in *SEC v. Alexander*, 160 F.Supp 2d 642 (SDNY, 2001) (*Alexander*). There, one of the defendants successfully brought a motion to dismiss the insider trading suit for lack of personal jurisdiction. The defendant Toffoli was an Italian resident who had traded a small amount of shares on an American exchange through her bank in Italy. It appears that Toffoli was unaware that the shares were traded on the New York Stock Exchange and she argued that this fact, coupled with her old age and the fact that she did not speak English, meant that it would be unfair to find jurisdiction. In granting the motion to dismiss, the court held that the circumstances made it "unlikely that Toffoli's acts presented unmistakably foreseeable effects within the United States that could reasonably be expected to be visited upon United States shareholders." Here, to support its approach, the court quoted passages from *Softpoint* verbatim. Although the court has mistakenly applied the wrong test, it is notable that a strict application of the "purposeful direction" test from *Burger King* would most likely result in the same finding in this case.

12.3.12 Reasonableness

The relevance of considerations arising under the "reasonableness" analysis has been explained by the Supreme Court in *Burger King* as follows.

> These considerations sometimes serve to establish the reasonableness of jurisdiction upon a lesser showing of minimum contacts than would otherwise be required.... On the other hand, where a defendant who purposefully has directed his activities at forum residents seeks to defeat jurisdiction, he must present a compelling case that the presence of some other considerations render jurisdiction unreasonable.

The precise mechanics of the reasonableness test described in this passage are unclear. There appear to be two alternatives. The first is that the minimum contacts test serves to determine which of two presumptions should operate. Should the minimum contacts be satisfied, there will be a presumption that jurisdiction will exist unless it is shown to be unreasonable by the defendant. Conversely, should the minimum contacts not exist on the facts, it will be up to the plaintiff to show cause as to why jurisdiction should be upheld. The second alternative is that the degree to which the relevant contacts are demonstrated is merely an input for the reasonableness test, a positive finding under the minimum contacts test only serving to tip the balance of the analysis toward a finding of jurisdiction, and vice versa. In any case, this difference may only be academic.

The variety of considerations relevant in evaluating reasonableness was highlighted in *Volkswagen*. There, the Supreme Court noted that courts may evaluate "the burden on the defendant," "the forum state's interest in adjudicating the dispute," "the plaintiff's interest in obtaining convenient and effective relief," "the interstate judicial system's interest in obtaining the most efficient resolution of controversies," and "the shared interest of the several states in furthering fundamental social policies." Additionally, in *Burger King*, it was held that jurisdiction will not be found where "litigation is so gravely difficult and inconvenient that a party is at a severe disadvantage in comparison with his opponent."

There is a policy imperative that is frequently imported into the reasonableness inquiry: the fact that the "United States has a substantial interest in the integrity of its securities markets" (*SEC v. Euro Security Fund*, 1999, US Dist LEXIS 1537). This arises out of the perceived unlikelihood of foreign nations policing the American markets. Thus, in cases involving American securities exchanges, courts have been willing to broadly construe personal jurisdiction even in circumstances where a strict interpretation of the minimum contacts test would not be satisfied.

12.4 COMMENTS ON JURISDICTIONAL COVERAGE

To evaluate the overseas reach of either the Australian or the U.S. insider trading provisions, specific scenarios must be borne in mind as to what conceivably ought to be covered. Logically, there are four circumstances involving both foreign and domestic elements for which coverage would be desirable. These are where:

1. a domestic resident while in the domestic country engages in the insider trading of securities listed on a foreign exchange (scenario one);

2. a domestic resident while in a foreign country engages in the insider trading of securities listed on a foreign exchange (scenario two);
3. a foreign resident while in a foreign country engages in the insider trading of securities listed on a domestic exchange (scenario three); and where
4. a foreign resident while in the domestic country engages in the insider trading of securities listed on a domestic exchange (scenario four).

Although somewhat simplistic, these four scenarios provide a convenient vehicle to assess the reach of the insider trading regimes. The comments that follow are intended as very general observations on the application of extraterritoriality principles; more complex factual scenarios will certainly alter the analysis if not also the ultimate conclusion.

12.4.1 Scenario One

Australia

Prima facie, where an Australian resident initiates the trade of a security listed on a foreign exchange while in Australia, the act of trading will be taken to have occurred in Australia and, thus, the insider trading provisions of the Corporations Act will be applicable.

United States

Similarly, an American resident who trades a foreign security while in possession of inside information will be subject to the U.S. insider trading regime. Subject matter jurisdiction will exist as the conduct test will be satisfied. The *EOC* case exception is unlikely to apply because, in this case, there will be an American resident in whom the United States will have an interest in punishing for securities-related fraud.

Naturally, personal jurisdiction will exist as a result of the person's residence and physical presence in the United States.

12.4.2 Scenario Two

Australia

Given that the act of trading probably occurs overseas, the applicability of the insider trading provisions is predicated on the issuer of the securities either being formed or carrying on business in Australia. As a general observation, the fact that the vast majority of foreign listed securities would be issued by entities without any connection to Australia ensures that in most cases which fall under "scenario two" umbrella, the Australian insider trading provisions would not be applicable.

United States

It is difficult to see this particular scenario satisfying the effects test. Given that the relevant conduct occurs outside the United States, satisfaction of the effects test is logically the only available option to establish subject matter jurisdiction. Consequently, cases which match the scenario two description are likely to fall outside the American insider trading provisions for lack of subject matter jurisdiction.

12.4.3 Scenario Three

Australia

In relation to those equity securities that are listed on the Australian Stock Exchange, the issuing entities will necessarily use "a share transfer office or share registration office in Australia," thus satisfying the carrying on business test under section 21 of the Corporations Act. Consequently, notwithstanding that the act of trading occurs overseas, the foreign resident will be subject to the insider trading provisions.

United States

The effects test is likely to be satisfied since the security being traded is listed on an American exchange; one which consequently will have a significant number of American investors. Consequently, subject matter jurisdiction is likely to be established.

Personal jurisdiction will depend on a number of factors such as the scale of the transaction and whether the security is listed on any other exchange in the world. Although it is likely that personal jurisdiction will be found, there may be some cases where it is abjectly unreasonable to find jurisdiction for a number of factors; the facts of the *Alexander* case are illustrative of this point.

12.4.4 Scenario Four

Australia

The result here will be the same as that in scenario three for essentially the same reasons.

United States

As in scenario three, subject matter jurisdiction will be established on the satisfaction of the effects test as the security is listed on an American exchange. However, the case is strengthened here by the fact that the act of trading occurs in the United States, and hence, the conduct test is also likely to be satisfied.

Personal jurisdiction will be found as a matter of course based on the long-standing principle that nonresidents physically present in the United States are subject to its laws.

12.5 CONCLUDING THOUGHTS

Examining the various authorities regarding the extraterritorial ambit of the Australian and U.S. insider trading regimes, it quickly becomes evident that Australian jurisprudence is relatively primitive. Not only is there enormous disparity in the amount of material available for consideration, but also the operation of some of the relevant tests regarding extraterritoriality under the Corporations Act appears entirely nonsensical in light of their apparent objectives. For example, as has been noted in *Ford's Principles of Corporations Law,* "it is hard to see why the place of formation or the place of business of the [issuing entity] should be a material consideration at all."

That being said, despite the massive amount of material relating to the extraterritorial application of the U.S. Securities Exchange Act, the precise scope of the various tests is not definitive and fluctuates from circuit to circuit. However, it appears in cases of insider trading that there is ample flexibility in the analytical framework to enable courts to find the requisite subject matter jurisdiction and personal jurisdiction in the majority of cases involving an American securities exchange. It is certainly well recognized by American courts that there is a strong policy incentive to find jurisdiction in order to protect the domestic markets.

Notwithstanding the apparent difference in sophistication between the Australian and American approaches, the actual effect of each, certainly insofar as this simplistic analysis is concerned, is broadly the same. However, as mentioned earlier, more complex factual circumstances may reveal the subtle nuances of each system and in so doing highlight genuine discrepancies in extraterritoriality.

REFERENCES

Austin, R. P., and I. M. Ramsay. 2007. *Ford's principles of corporations law.* 13th ed. Sydney: LexisNexis Butterworths.

Kramer, D., and M. Murray. 2002. The extraterritorial application of United States securities laws to punish insider trading. *S&P's The Review of Securities & Commodities Regulation* 35(7):65–71.

Reserve Bank of Australia. 2007. Financial stability review. March.

Scannell, K. 2007. Insider trading: It's back with a vengeance—SEC dials up enforcement as merger boom fuels misdeeds. *Wall Street Journal.* May 5.

Scheer, D. 2007. Insider trading runs rampant on Wall Street; options trades surge prior to buyouts. *Bloomberg News.* May 11.

Securities and Exchange Commission. 2007. Third amended complaint from *SEC v. Rahim et al.*

Testy, K. 1994. Comity and cooperation: securities regulation in a global marketplace. *Alabama Law Review* 45:927–68.

Walker, G. R. 2000. Insider trading in Australia: When is information generally available? *Company and Securities Law Journal* 18:213–17.

CHAPTER 13

An Investigation of the Whistleblower–Insider Trading Connection

Evaluation and Recommendations

Edward J. Lusk and Michael Halperin*

CONTENTS

* We thank Professor James J. Coffey, Esq. of the Department of Accounting: SUNY Plattsburgh, Plattsburgh, NY for his reading and comments and Ms. Ellen Slack of the Lippincott Library of the Wharton School for her excellent editorial assistance. Regarding the citations made from the Web site of Morrison and Foerster, LLP, we acknowledge the helpful comments of Alison Cleaver, acleaver@mofo.com, and the kind permission of Morrison and Foerster, LLP to use these citations under the condition set forth following: Rains, D. P., and S. R. Kulkarni. (2003) Legal Updates and News: Executives! Start Using Rule 10b5-1 Trading Plans. (December). http://www.mofo.com/news/updates/bulletins/bulletin02013.html. Because of the generality of this chapter, the information quoted may not be applicable in all situations and should not be acted upon without specific legal advice based on particular situations.

13.1 INTRODUCTION

The securities-trading markets are driven by many forces. In the 1970s we believed that a simple regression could tell us all we needed to know about "The Market." Today, we better understand just how little we really know about this serpentine time series called "The Market." This is relevant in trying to understand insider trading (ITr) simply because the one thing that we know for sure, and really have always known, is that one can profit from information asymmetries. If my distant early warning (DEW) system is better than yours, in every scenario I win. In this way one actuates the trader's golden rule: "Do unto others, before they do unto you." This is not cynicism, but rather "The Street." This primordial dictum is the reason that there needs to be regulation of the trading markets and, more importantly, of those who trade. The question, and the point of departure for our inquiry, is what sort of ITr regulation there should be and how it might be best effected. We are interested in the positive and negative aspects of ITr—as an activity that sometimes needs to be controlled, but that under some circumstances functions as a DEW of possible problems and so should be permitted. To this end we will (a) examine an intriguing model of control offered by Macey (2007) that conditions the legality ITr on the whistleblower's rationale, (b) enlarge our inquiry to incorporate the stream of legislation that aims to control the dark side of ITr, and (c) make recommendations for dealing with this elusive and contentious issue of ITr.

13.2 MACEY: THE DEW IS THE WHISTLE

In his 2007 article, Macey draws a distinction between the permissible ITr that society should condone—legitimize and promote—and ITr that should be actionable based on the violation of property rights that accrue to the organization for the information it creates. This argument goes back to Locke's concept of the entitlement which accrues to *honest-industry*. In

Macey's view, if certain information is produced in the legitimate execution of an organization's mission, then making use of such information to gain a market advantage is illegal; and penalties, criminal and civil, should accrue to those who have misappropriated this information and turned it into ill-gotten gain. However, if the honest-industry criterion is not satisfied then there are two options that should be *legally* available: whistleblowing (WB) and ITr.

Macey deftly crafts a model that demonstrates the social desirability, fairness, and effectiveness of ITr compared to WB as a means to identify fraud. He shows that when circumstances justify WB, they also legally and logically justify ITr. Macey further demonstrates that whistleblowers and inside traders are indistinguishable on material issues, in particular, because: "(a) they are informational intermediaries; (b) they have information not widely known or not already reflected in share prices; and (c) they are in a pre-existing contractual or quasi-contractual relationship with the source of the information" (Macey 2007, 1912).

He then argues that between the two legal options, WB and ITr, that ITr is more desirable in almost every important dimension. Specifically, Macey indicates that "Given the complexity of whistleblowers' motives, their inability to make a credible commitment about the veracity of their information, and the necessity for bureaucratic investigation of the information being disclosed, it is not surprising that whistleblowing is often unsuccessful" (2007, 1917). Here, "unsuccessful" is used in the context of revealing fraud. He continues that, because ITr is itself a risk-taking strategy compared to WB, it has certain advantages. Specifically, ITr possesses more credibility as an information signal, it is not subject to interpretation, and it is immediately perceivable as an economic event. Additionally, ITr tends to distribute the risk-bearing for insiders; while outside shareholders can diversify their portfolios, the insiders can only trade to minimize the high risk they bear as option holders and as invested employees in the organization. Finally, Macey points out that ITr works in the right direction to lower the share price and so mitigates against the larger loss that would occur if ITr were not permitted. These reasons all argue for ITr as an efficient, effective, and so desirable means of fraud discovery.

Giving these arguments additional credibility, recently Altucher wrote in the *Financial Times* (2007):

> Here are the benefits to making insider trading legal: (1) The more information in a market, any market, the more efficient prices

> become. If informed investors start buying or selling based on privileged information, asset prices will rise [here he means adjust] to their "correct" level. (2) Fraud will be exposed earlier. Enron is an example where tens of thousands of investors got burned because they were piling into the stocks during the later stages of its fraud. If insiders were selling we would've seen a much swifter move down, and probable fraud exposed. (3) Companies will either become more transparent, to keep the retail investor happy, or will themselves enforce secrecy rather than being complacent with the idea that the law somehow protects their secrets. (4) One concern is that there will be a flight of liquidity because people will be concerned about the legitimacy of our markets. Rather, the opposite will occur. More enforcement dollars will be used to uncover actual frauds such as the next Enron or WorldCom. Arguably, these frauds are a thousand times more dangerous for the retail investor than what is probably a victimless crime such as insider trading. (5) Insider trading is almost impossible to prosecute and the government wastes countless dollars trying.

This sums up the justification, and the logical and legal basis, for permitting WB; and for preferring ITr to WB as a means of detecting and so correcting fraud. However, this is just a part of the story. For it to work there has to be a way to decide if the insider trade was conditioned on the failure of the honest-industry criterion. Only then is ITr legal, so this is a precondition. We return to this important consideration after we examine the other side of ITr—where it is in fact the misappropriation of honest-industry information—in which case it is illegal and should be penalized.

13.3 THE DARK SIDE OF INSIDER TRADING

There is no clear legal definition of ITr. Gorman (2004, 478) points out that nowhere in the Securities and Exchange Commission (SEC) Act of 1934 is the phrase "insider trading" used. Its de facto definition is derived from case law and precedent, making it difficult for traders to know the current SEC, Sarbanes–Oxley Act of 2002, or other legal spins on ITr; see Gasparino (2005, 44) and Kakabadse, Kakabadse, and Kaspurz (2006, 33). However, we offer a discussion that we feel conveys the spirit of ITr by SEC Associate Director Thomas Newkirk and Melissa Robertson, Senior Counsel in the SEC Division of Enforcement (1998, 2):

> It is the trading that takes place when those privileged with confidential information about important events use the special advantage of that knowledge to reap profits or avoid losses on the stock market, to the detriment of the source of the information and to the typical investors who buy or sell their stock without the advantage of "inside" information.

This discussion enlarges the narrow fiduciary context to include the market context, which is currently thought of as consistent with misappropriation theory; also see Gorman (2004, 479). Newkirk and Robertson (1998) continue and point out the relevant sections of the Act of 1934 where aspects of ITr are discussed:

> Section 16(b) prohibits short-swing profits (profits realized in any period less than six months) by corporate insiders in their own corporation's stock, except in very limited circumstance. It applies only to directors or officers of the corporation and those holding greater than 10% of the stock and is designed to prevent insider trading by those most likely to be privy to important corporate information.
>
> It shall be unlawful for any person, directly or indirectly …
>
> (a) to employ any device, scheme, or artifice to defraud,
> (b) to make any untrue statement of a material fact or omit to state a material fact necessary in order to make the statements made, in light of the circumstances under which they were made, not misleading, or
> (c) to engage in any act, practice, or course of business which operates or would operate as a fraud or deceit upon any person, in connection with the purchase or sale of a security.

According to Hamilton and Trautmann (2002, ch. 5), H.R. 3763: Sarbanes–Oxley of 2002 (SOX) amends the 1934 Act's Section 16 to require directors, officers, and 10 percent equity holders to report their purchases and sales of securities more promptly, that is, by the end of the second day following the transaction, or by such other time established by the SEC where the two-day period is not feasible. The purpose of the requirements is to make information about insider transactions available to investors more promptly so that they can make better-informed investment decisions. These transparency provisions are set out by Hamilton and Trautmann (2002, 65):

(1) The two-day statement reporting insider trades must be filed electronically.
(2) The SEC must provide the two-day statement on a publicly accessible Internet site by the end of the business day following the filing.
(3) The company, if it maintains a corporate website, must provide the statement on that website by the end of the business day following the filing. See [SOX] Act Section 403 at para. 1031.

With this information, it is clear that one needs a way to legitimately trade securities, in particular given the prevalence of stock-option plans. This is where Rule 10b-1 Plans (R10P) comes into play. These are plans which, when properly executed, allow insiders to sell their holdings without violating ITr laws and regulations.

13.4 RULE 10B5-1 PLANS (R10P)—THE ESSENTIAL DETAILS

The key issue, which is currently the subject of SEC interest (Searcey and Scannell 2007), is that when the R10P is adopted the seller has no inside information; this is the affirmative-defense justification of these plans. In this case, sellers are protected from ITr liability/prosecution *even if* they come into possession of material, nonpublic information by the time the R10P sales are executed. This can of course be challenged. A challenge is almost always made on the basis of undue influence where the executive has acted to change the normal course of events with an eye to a profit. An excellent example of this is the case of Mr. Tevanian, Apple's CTO, who reported selling his option shares on April 3, 2005, for an average price of $63.31, under an R10P adopted January 31, 2005. Had he waited a few days, he could have increased his take by more than $8.50 a share, or by more than $2.6 million (see Brulliard 2006). The reason for this is that a few days after the R10P contract date of April 3, Apple announced its Mac/Windows XP agreement. If, per chance, the R10P sales date happened to postdate the announcement, the SEC would probably have been very interested in Mr. Tevanian's activities in relation to the timing of the announcement. Let us now consider the details of these important financial plans.

According to Morrison & Foerster, LLP (2003), a law firm with a wealth of experience in designing Rule 10b5-1 plans, an R10P must be in writing *and* the following must be specified:

> Number of shares to be bought or sold. This can be designated as a number of shares, as a percentage of the executive's holdings, or as the number of shares needed to produce a specific dollar amount. Rule 10b5-1 even allows the number of shares to be generated by an algorithm or computer program. Rule 10b5-1 plans can provide for multiple transactions, and so different amounts of shares can be designated for each purchase or sale.
>
> Prices at which the shares will be bought or sold. This can be designated as a specific dollar price, a limit order price, or as the prevailing market price. Again, prices can also be determined by an algorithm or computer model, so multiple transactions at different prices can be ordered.
>
> The timing of the purchases or sales. This can be designated as a specific date or time, or as the time at which a specific event occurs.

13.5 THE FLEXIBILITY OF RULE 10B5-1 PLANS

Morrison & Foerster, LLP (2003) suggest many executives worry that R10Ps might force them to sell stock at inopportune times or at unfavorable prices. But they note that Rule 10b5-1 is sufficiently flexible to accommodate almost all business and personal objectives. A properly executed R10P is almost by definition SEC-compliant and often can provide trading flexibility when there are "blackouts" or other company-designated "trading windows" (see SOX: Title III, Section 306). Morrison & Foerster, LLP (2003) also recommend the following strategies for maximizing the flexibility of an R10P:

> Rule 10b5-1 plans can be of any duration. So one frequent objection to Rule 10b5-1 plans—that they lock into executives' trading strategies that may become outmoded over time—can easily be overcome. We recommend plans as short as six or nine months in duration. That way, if conditions change, the executive's plan and trading strategy can change as well.

> ... Some executives mistakenly believe that Rule 10b5-1 plans must cover all their holdings. Not so. We recommend that an executive makes only a small part of his holdings—perhaps 20 or 25 percent—subject to Rule 10b5-1 instructions. Alternatively, an executive could provide for sales of a small portion of his holdings on a regular schedule, but provide for sales of a larger portion if certain price targets are reached.
>
> ... No one wants to sell stock at a low price. One easy way to prevent this is by including a minimum price floor in every Rule 10b5-1 plan. Multiple price floors, which increase over specified periods of time, can also be used.
>
> ... Executives who want to maximize their return can create a matrix of future price targets. This strategy addresses a common concern of executives—that plan sales will occur at prices that will feel, in hindsight, to have been too low.
>
> ... Some executives base trading decisions on how their company's stock performs relative to various market or industry indices, or relative to certain selected competitors. Rule 10b5-1 is flexible enough to accommodate these types of strategies. A plan could, for example, provide for sales when a particular market indicator rises 10 percent in a two-month span, or when one company's stock outperforms a benchmark index (or a competitor's stock) by 10 percent over a specified period.
>
> ... Executives' stock sales are often driven by a number of personal financial considerations, including home purchases or remodels, college tuition payments, and the like. All of these can be built into custom-tailored Rule 10b5-1 plans. For example, a plan could provide for sales 15 days before a college tuition payment for the executive's child is due, with the number of shares to be sold linked to the average cost of tuition as published by the college.

13.6 RECOMMENDATIONS AND SUMMARY

As is evident, the use of R10Ps will go right to the heart of the ITr temptation, and ITr under the WB context is to be encouraged. In this regard, we offer some recommendations, first for the inside trader under the WB argument, and then for those creating R10Ps.

13.6.1 Regarding the WB Context

Considering the positive side of ITr, that is, where the honest-industry argument is invalid, inside traders must prove that they had reason to believe there was material fraud. According to recent case law, they do not have to be correct; they just have to have believed it. This is the WB protection and we agree with Macey that it must logically extend to ITr in the WB context. Here the inside traders would be wise to consider the following pre-ITr activities to establish, as a defense, the reasonability of their belief in the existence of material fraud. We recommend that those contemplating ITr under the WB protections should:

Inform, in writing by means of a hard copy, management, the auditor of record—and for firms with publicly traded stock, the audit committee—of their concerns. This is consistent with what the auditors must do according to AS 2 of SOX.

Seek out others in the organization who may have occasion to notice the same events that have led to suspicion of fraud or illegal activities. Talking over one's concerns with these individuals would be a good reality check.

Ascertain if the suspected fraud rises to the level of materiality in comparison with other events that fall under honest industry. For example, lapping being done by a junior accounts receivable clerk is hardly a justification for the vice president of strategic planning to do a massive short sell when the real motivation was that late on Friday the vice president found out that the firm's China-partnering venture had just fallen apart, that this would likely reduce productive activity by 37 percent starting in six months, and this devastating news would likely be published in the Financial Times on Monday next.

Inform the proper authorities with reasonable dispatch. This can include the federal, state, or local authorities, in particular, for actions which fall under the False Claims Act (31 USC 3729).

Not sit on the information by quietly creating a series of small short sells over time. This will look as if you are trying to game the system by hoping your "inaction" will allow the stock price to fall even farther than it would have fallen. If you get greedy with your short sells, you place your WB credibility at risk and it is your WB credibility as your defense that is at issue. On the other hand, reacting too quickly is not wise in that you will need to collect "reasonable" justification information to back up your claim that you believed that there was evidence of material misdoings. This will be a balancing act, sometimes without a net.

Be clear that ITr, in the WB context, makes one a whistleblower. There have been, and even après SOX continue to be, problems with the enforcement of WB protections. See the David Welch story ("Accounting Web" 2007) and the Ted Beatty saga (Sapsford and Beckett 2002) for some heart-rending and chilling stories.

Realize that as inside traders try to cover themselves their actions will likely alert various parties of the possibility of fraud. These individuals are de facto "tippees." They may act on this information and either become whistleblowers themselves or even start ITr of their own. These actions can drive the price of the stock farther down or even result in the suspension of trading. This can be a plus or a minus depending on how the ITr is executed.

13.6.2 Regarding Rule 10b5-1 Plans

According to the recommendations of Morrison & Foerster, LLP (2003), there are three steps needed to minimize the risk of adverse publicity from trades made under an R10P:

> Publicly disclose new plans.... The best way to prevent unhappy surprises is to make a public disclosure each time an executive adopts a new Rule 10b5-1 plan. Investors are less likely to react negatively to stock sales if they know in advance about an executive's plan. Some CEOs have issued press releases to disclose the adoption of new Rule 10b5-1 plans, but we prefer using Form 8-Ks....
>
> Delay transactions until after public disclosure. We recommend that Rule 10b5-1 plans have an effective date at least 30 days after the plan is publicly disclosed to shareholders. This minimizes the risk of adverse publicity and also should help combat any "good faith" challenge to the plan by the SEC....

> Report plan sales on Form 4s. Executives must file Form 4s reporting all transactions in their company's stock. We recommend that Form 4s for sales made pursuant to Rule 10b5-1 plans specifically note that the sales were made pursuant to the plan....

In addition to the three Morrison & Foerster's recommendations above, we offer the following:

It is all about transparency—the more the better, as we see in Section 403 of SOX. Although under the current regulations R10Ps are not required to be made public, we recommend that *all* R10P information be made public. This will essentially eliminate suspicion about prior or inappropriate actions after the R10P is effected (Kakabadse, Kakabadse, and Kaspurz 2006).

All company employees as well as those with contracts, including outsourcing, should be required by company policy to file R10Ps that deal with any stock of the organization or direct market competitors. This is a small extension of the Form 4s requirement. If it seems far-fetched, rent the film *Wall Street* and watch the Buddy Fox cleaning services segment again!

The company institutes, as part of its internal control programs, *random* checking on all stock transactions that are executed by employees or those with contracts with the organization. Many organizations have "blackout" or "trading-window" policies and this would be a minor extension of the control and scrutiny permitted under those blackout policies. We hope that at some time this scrutiny could extend to immediate family members.

We like the "related-industry" argument—for example, on the anticipation of bad news, rather than selling short, insider traders "buy long" in their competitor's stock (Chen and Zhao 2005; Scott and Xu 2004). In this regard, we recommend that the SEC or the company require informational filings similar to Form 4s, to deal with such hedge-trading in the competitor market.

13.7 A FINAL RECOMMENDATION DEALING WITH REGULATION

Regulation seems to be too diffuse. Occupational Safety and Health Administration (OSHA), the SEC, and the Department of Justice all play a role. The key word is, of course, *consolidation*, which would be a "surefire" step to save investigation and legal resources. Also we should learn from the other groups that face the same problem: *rule-breaking for gain*. How about the sporting world where, if I take performance-enhancing prohibited substances, I am the winner against those whose panoplies are less well stocked than mine. Track and field competition faces the same problems that we find in the trading markets. They have decided after years of embarrassing failures that random testing, where everyone is fair game, and 100 percent testing of winners, is the way to go. The evidence looks most impressive. Perhaps we should use the same model.

We can go with the SEC 10 percent or even 5 percent radar screen. And depending on the budget, a group of inside traders is selected as the sample to be investigated—that is, tested. Trying to follow the founding fathers' idea of a judiciary that delivers a speedy and public trial, the sample could be processed as follows: the traders are evaluated by an SEC arbitrator who has subpoena rights regarding any electronically available information. The statute of limitations for the arbitration: one month. If the arbitrator finds evidence *for*, then the case moves into the legal system, where the statute of limitations is six months. If the Department of Justice finds for a violation, then penalties are exacted and the appeal clock starts; the defendant would have one year to file for redress. The final finding is due six weeks after the filing. Defendants found not guilty would receive all of the resources consumed in their defense plus interest at five times the thirty-day T-bill composite (tax-free). During this process defendants cannot be fired or otherwise penalized.

13.8 CONCLUSION

With these recommendations, we believe that ITr can be encouraged in the proper context and limited in those cases where there is likely to be misappropriation. Realizing that certain bureaucratic necessities are inherent in our recommendations, on net we expect that societal resources will be conserved. In summary, we opt for transparency, well designed and executed Rule 10b5-1 plans, and encouraging WB–ITr as a designed market-oriented mechanism to deal with problems.

REFERENCES

Accounting Web. 2007. First whistleblower unprotected by SOX. http://www.accountingweb.com/cgi-bin/item.cgi?id=103603. June 8 (accessed July 8, 2007).

Altucher, J. 2007. The case for legalising insider trades. *Financial Times* July 10, p. 12.

Brulliard, N. 2006. Stock-sale plans help, hurt insiders. *Wall Street Journal* (Eastern edition) April 12, p. C12.

Chen, R., and X. Zhao. 2005. The information content of insider call options trading. *Financial Management* Spring:153–72.

Gasparino, C. 2005. Let's make an inside deal. *Newsweek* March 7, p. 44.

Gorman, C. 2004. Are Chinese walls the best solution to the problems of insider trading? *Fordham Journal of Corporate and Financial Law* 9:475–99.

Hamilton, J., and T. Trautmann. 2002. *Sarbanes-Oxley Act: Law and explanation.* Chicago: Commerce Clearing House.

Kakabadse, A., N. Kakabadse, and A. Kaspurz. 2006. Insider out. *Business Strategy Review* Spring:33–35.

Macey, J. (2007) Getting the word out about fraud: A theoretical analysis of whistle-blowing and insider trading. *Michigan Law Review* 105:1899–1940.

Morrison & Foerster, LLP. 2003. Legal updates and news: Executives! Start using Rule 10b5-1 trading plans (December). At http://www.mofo.com/news/updates/bulletins/bulletin02013.html (accessed August 15, 2007; site information by D. P. Rains and S. R. Kulkarni).

Newkirk, T., and M. Robertson. 1998. Insider trading—A US perspective. Speech presented at the 16th International Symposium on Economic Crime, Jesus College, Cambridge, England. http://www.sec.gov/news/speech/speecharchive/1998/spch221.htm. September 19 (accessed August 15, 2007).

Sapsford, J., and P. Beckett. 2002. Informer's odyssey: The complex goals and unseen costs of whistle-blowing. *Wall Street Journal* November 25, p. A1.

Scott, J., and P. Xu. 2004. Some insider sales are positive signals. *Financial Analysts Journal* May/June:44–52.

Searcey, D., and K. Scannell, K. 2007. SEC now takes a hard look at insiders' "regular" sales. *Wall Street Journal* April 4, p. C1.

Part 2

Regulating Insider Trading

B. Legal Insider Trading

CHAPTER 14

A Middle Ground Position in the Insider Trading Debate

Deregulate the Sell Side

Thomas A. Lambert*

CONTENTS

* A longer version of the argument asserted here appears in Lambert (2006).

14.1 INTRODUCTION

For more than four decades now, corporate law scholars have debated whether the government should prohibit insider trading, commonly defined as stock trading on the basis of material, nonpublic information. Participants in this long-running debate have generally assumed that trading that decreases a stock's price should be treated the same as trading that causes the price to rise: either both forms of trading should be regulated, or neither should. This chapter argues for a middle-ground position in which "price-decreasing insider trading" (sales, short sales, and purchases of put options on the basis of negative information) is deregulated, while "price-increasing insider trading" (purchases of stock and call options on the basis of positive information) remains restricted.

The reason for the proposed asymmetric treatment is that price-decreasing insider trading provides significantly more value to investors than price-increasing insider trading. Most notably, price-decreasing insider trading provides an effective means of combating the problem of overvalued equity—that is, a stock price that is so high that it cannot be justified by expected future earnings. Overvalued equity, scholars are finding, leads corporate managers to take a number of value-destroying actions (Jensen 2005). Deregulation of price-decreasing insider trading would create a means by which corporate insiders—those in the best position to know when a stock is overvalued—could signal the market that the stock price is too high, thereby reducing the costs associated with overvalued equity. While deregulation of price-increasing insider trading could similarly remedy *under*valued equity, undervaluation causes fewer problems than overvaluation, and there are numerous other mechanisms for addressing that sort of mispricing. Moreover, the potential investor losses resulting from price-increasing insider trading are higher than those caused by price-decreasing trading.

This chapter first briefly summarizes the long-running policy debate over insider trading. It then describes the problem of overvalued equity and explains why price-decreasing insider trading creates greater investor benefits than does price-increasing insider trading. It next considers the cost side of the balance, explaining why price-decreasing insider trading imposes lower investor costs than does price-increasing insider trading.

The chapter concludes that investors are best off under an asymmetric insider trading regime that generally permits price-decreasing insider trading while restricting price-increasing insider trading.

14.2 THE INSIDER TRADING DEBATE

Ever since Henry Manne published his classic book, *Insider Trading and the Stock Market* (Manne 1966), scholars have debated whether there truly are harms associated with insider trading and, if so, whether they outweigh the harms created by an insider trading ban. Defenders of the ban on insider trading insist that it is fundamentally unfair for some traders to have an informational advantage over others, particularly when the advantaged traders are corporate insiders who are supposed to be acting as agents for those who lack the informational advantage (Schotland 1967, 1439). Ban defenders also contend that insider trading causes efficiency losses by (1) discouraging investment in the apparently rigged stock market, thereby reducing the liquidity of capital markets (Asubel 1990, 1022–23); (2) encouraging insiders to delay disclosures and to make management decisions that increase share price volatility but do not maximize firm value (Haft 1982, 1054–55; Levmore 1982, 149); and (3) increasing the "bid–ask" spread of stock specialists, who systematically lose on trades with insiders (whom they cannot identify *ex ante*) and who will thus tend to "insure" against such losses by charging a small premium on each trade (Copeland and Galai 1983; Glosten and Milgrom 1985). Finally, some defenders of the ban assert that it is justified as a means of protecting the corporation's property rights in valuable information regarding firm prospects (Bainbridge 2002, 598–607).

Proponents of the deregulation of insider trading discount these arguments and assert that insider trading can be beneficial on the whole and ought to be limited, if at all, only by corporations themselves via contract. With respect to the fairness argument, deregulation proponents retort that insider trading cannot be "unfair" to investors if they know in advance that it might occur and nonetheless choose to engage in the purportedly unfair trades (Scott 1980, 807–9). Moreover, deregulation proponents assert, the purported efficiency losses occasioned by insider trading are overblown (Carlton and Fischel 1983). There is little evidence, they say, that insider trading reduces liquidity by discouraging individuals from investing in the stock market, and it might actually increase such liquidity by providing benefits to investors in equities. With respect to the claim that insider trading creates incentives for delayed disclosures and value-reducing management, advocates of deregulation claim that such mismanagement is unlikely for several reasons. First, managers face reputational constraints that will discourage such misbehavior. In addition, managers, who generally work in teams, cannot engage in value-destroying mismanagement without persuading their colleagues to go along with the strategy, and any particular employee's ability to engage in mismanagement will therefore be constrained by his or her colleagues' attempts to maximize firm value or to gain personally by exposing proposed mismanagement. With respect to the argument that insider trading raises the cost of trading securities by increasing the bid–ask spread, proponents of deregulation point to empirical evidence discounting that purported effect (Dolgopolov 2004). Finally, deregulation proponents assert that any "property right" to material nonpublic information need not be a non-transferable interest granted to the corporation; efficiency considerations may call for the right to be transferable and/or initially allocated to a different party (e.g., to insiders) (Macey 1984).

In addition to rebutting the arguments for regulation, proponents of deregulation have offered affirmative arguments for liberalizing insider trading (Carlton and Fischel 1983; Manne 1966). First, they maintain that insider trading should generally be permitted because it increases stock market efficiency (i.e., the degree to which stock prices reflect fundamental value), which helps guarantee efficient resource allocation. Corporate insiders, after all, generally know more about their company's prospects than anyone else. When they purchase or sell their own company's stock, thereby betting their own money that the stock is mispriced, they convey valuable information to the marketplace. Assuming their trades somehow become public, other rational investors will likely follow their lead, which will cause stock prices to reflect more accurately the underlying value of the firm. More efficient stock prices, then, will lead to a more efficient allocation of productive resources throughout the economy.

Deregulation advocates further maintain that corporations ought to be allowed to adopt liberal insider trading policies because permitting insider trading could be an efficient form of managerial compensation. The argument here is that competition in the labor and capital markets will lead corporations to adopt efficient insider trading policies. On the one hand, the market for managerial labor may reward corporations with liberal insider trading policies, for the right to make money through insider trading is valuable to potential managers. On the other hand, capital market pressures will prevent corporations from adopting insider trading policies that are, on balance, harmful to investors. Thus, deregulation advocates maintain, the interaction of the labor and capital markets will assure that firms will adopt insider trading policies that are, on the whole, value maximizing.

Not surprisingly, the affirmative case for liberalizing insider trading has not gone unchallenged. With regard to the argument that insider trading leads to more efficient securities prices, ban proponents retort that trading by insiders conveys information only to the extent it is revealed, and even then the message it conveys is "noisy" or ambiguous, given that insiders may trade for a variety of reasons, many of which are unrelated to their possession of inside information (Gilson and Kraakman 1984). Ban defenders further maintain that insider trading is an inefficient, clumsy, and possibly perverse compensation mechanism (Bainbridge 2002, 591–92).

One of the most striking aspects of the well-worn insider trading debate is its starkness. Assuming that insider trading must be treated as a whole, ban defenders and opponents have argued over liberalization in all-or-nothing terms; they have not considered whether some species of insider trading should be treated differently than others. The remainder of this chapter attempts to demonstrate that price-decreasing insider trading, which consists of trading by insiders on the basis of negative nonpublic information, provides greater net benefits to investors than price-increasing insider trading, which consists of trading by insiders on the basis of positive nonpublic information. Accordingly, the law should treat price-decreasing insider trading less harshly than price-increasing insider trading.

14.3 PRICE-DECREASING INSIDER TRADING CONFERS GREATER BENEFITS ON INVESTORS

For reasons explained below, stock overvaluation is more likely than undervaluation to persist and tends to cause greater harm to investors when it occurs. Accordingly, insider trading that reduces the price of overvalued equity toward fundamental value will provide greater investor benefits than will insider trading that increases stock prices.

14.3.1 Why Overvaluation Is More Likely to Persist

Empirical evidence suggests that the bulk of securities mispricing occurs in the direction of overvaluation rather than undervaluation (Finn et al. 1999). This should not be surprising, for the two groups of individuals most likely to provide the information that would correct stock mispricing—corporate managers and professional stock analysts—are much more likely to do so, and have better tools for doing so, when the mispricing is in the negative direction.

14.3.1.1 Corporate Managers

First, consider insider managers. While scholars have articulated persuasive arguments in favor of the view that corporate managers, seeking to protect their reputations for trustworthiness, will have a tendency toward candor (Easterbrook and Fischel 1984), there are numerous reasons to believe that managers will tend to be systematically optimistic in their portrayals of their corporation's prospects, and will thus be less likely to correct overpricing than underpricing (Langevoort 1997).

First, corporate managers may fail to be forthcoming with stock price-correcting bad news because they face "last period" and "multiple audience" problems. The last period problem exists when the undisclosed news is so bad that it might cause insolvency or some kind of managerial shake-up (Arlen and Carney 1992). If senior managers think the undisclosed bad news will result in company insolvency or in their being fired or demoted, they may rationally decide that the costs to them of misleading disclosures (or omissions) are less than the costs to them of candor. The multiple audience problem results from the fact that corporate managers cannot make targeted disclosures of negative information only to shareholders. When managers make a corporate disclosure, they inform not only shareholders, but also such corporate constituencies as consumers, employees, and suppliers. They may wish to conceal price-decreasing information in order to protect relationships with those constituencies (Langevoort 1997, 116–17). It may be quite rational, then, for corporate managers to conceal price-decreasing information, despite their interest in maintaining a reputation for candor.

Well-documented cognitive biases may also lead managers to overemphasize good news. For example, cognitive psychologists have observed that individuals, such as corporate managers, who must process a large volume of information frequently adopt heuristics, or mental shortcuts, to assist them with that task (Kiesler and Sproul, 1982). Often, those heuristics involve the creation of coherent "stories" into which the individuals attempt to fit the information they receive. Because "story revision" requires the use of scarce cognitive resources, it is disfavored. Accordingly, individuals unconsciously tend to construe information and events in a manner that confirms their prior beliefs, attitudes, and impressions (Lord et al. 1979). For corporate managers, this tendency may result in a "commitment" bias under which the managers strongly resist evidence that previously selected courses of action were ill-chosen (Tetlock et al.

1989). In addition, managers may be falsely optimistic because they officially "control" corporate endeavors. There is substantial empirical support in the psychology literature for the proposition that individuals systematically overrate their own abilities and achievements (Bazerman 1994). Thus, one should expect corporate managers to overestimate the chances of success of the businesses under their control.

Perhaps more significant than these cognitive biases are the dynamics of information flow within the corporation. Much of the information concerning the success of a firm's endeavors—particularly nonquantifiable, "soft" information, such as the degree of consumer enthusiasm for new products, the progress of products through the research and development pipeline, and so forth—is not immediately available to the firm's senior managers. Instead, the agents with the most direct access to this information tend to be nonmanagerial employees and low- to mid-level managers. Senior managers, then, must rely on their underlings to provide them with information regarding crucial aspects of the firm's prospects (Dutton et al. 1997).

The problem with this hierarchical system is that there is a danger at each stage of the information-relay system that material information will be suppressed or exaggerated in some fashion, as each information provider will be tempted to tweak the message to conform to his or her self-interest. Seeking promotion or other rewards, information providers have an incentive to inform their superiors of every bit of value-enhancing information of which they are aware. By contrast, if they know their endeavors are not going as well as expected, they may positively spin that information or keep it to themselves in the hope that things will turn around soon. By the time the price-affecting information reaches the senior managers in charge of corporate disclosure, it is likely to have been "massaged" so as to make underlings look good. In other words, it is likely to be positively biased (Langevoort 1997, 119–25). Unaware of negative information, the senior managers in charge of corporate disclosures can neither directly disclose the bad news nor factor it into their more general forecasts.

Finally, even if corporate managers were as likely to perceive overvaluation as undervaluation and were equally motivated to correct both forms of mispricing, they would be more likely to correct undervaluation than overvaluation because they have more effective means of doing so. Consider a manager confronted with evidence that his or her company is undervalued. The manager might issue a press release explaining why the market was undervaluing the firm, or he or she could initiate a stock repurchase, thereby signaling management's strong belief that the stock is undervalued. Managers finding undervalued equity to be a chronic problem could adopt equity-based compensation schemes for executives (e.g., payment in stock or stock options). A manager confronting overvalued equity, by contrast, is somewhat strapped. As a practical matter, managerial candor is not an option, for a manager who directly announced to the market that the corporation's stock was overpriced probably would not keep his or her job for very long. Nor could the manager correct the mispricing by engaging in a sale transaction that would send the reverse signal of a stock repurchase. Whereas the signal sent by a stock buy-back is relatively unambiguous, a sale transaction designed to signal overvaluation (e.g., an equity offering or a sale of treasury shares the corporation previously purchased) is much noisier. It could easily be interpreted as a means of raising capital for some sort of corporate undertaking. And, of course, equity-based compensation, which helps prevent undervaluation, exacerbates overvaluation by inducing managers to drive the share price higher even when they know the company is overvalued. There is thus an asymmetry in the degree to which managers and market forces are able to correct the different species of mispricing: the primary options available for correcting negative mispricing are not practically available when the mispricing is in the positive direction (Jensen 2005, 14).

14.3.1.2 Stock Analysts

Stock analysts, the other individuals who are well positioned to identify and correct stock mispricing, are also less likely to correct overvaluation than undervaluation. Consider the optimism bias exhibited in the Enron debacle. In the autumn of 2001, just weeks before Enron's December 2, 2001 bankruptcy, each of the fifteen largest Wall Street firms covering Enron's stock had buy recommendations in place (Senate Committee on Governmental Affairs 2002). And as late as October 26, 2001—*after* Enron's chief financial officer had been forced to resign, the Securities and Exchange Commission (SEC) had initiated an Enron investigation, and the *Wall Street Journal* had run several stories about Enron's earnings management problems—ten of the fifteen largest Wall Street firms covering Enron maintained buy recommendations (Kroger 2005, 102), as did fifteen of seventeen top Wall Street analysts surveyed by Thompson Financial/First Call (Craig and Weil 2001). Sadly, Enron was no outlier. The ratio of buy to sell recommendations has recently been as high as 100 to 1 (Coffee 2004, 316–17), and in the period immediately preceding a 60 percent drop in the NASDAQ, only 0.8 percent of analysts' recommendations were sell or strong sell (D'Avolio et al. 2001, 14). Thus, the evidence suggests that analysts, quick to report undervaluation by issuing buy recommendations, are less responsive to mispricing in the positive direction (Dreman and Berry 1995; Stickel 1990).

Like corporate managers, stock analysts face a set of incentives that systematically biases them toward optimism. Because most stock analysts are employed by firms that make the lion's share of their money by providing brokerage and investment banking services, they have an incentive to issue optimistic "buy" recommendations, which may be acted upon by anyone, rather than pessimistic "sell" recommendations, which (absent short-selling) can be acted upon only by incumbent shareholders. More importantly, the more lucrative investment banking side of a brokerage firm's business benefits from optimistic analyst reports. Issuers of securities want to make sure that the analysts employed by their investment bank will drum up investor enthusiasm for the issue, so as to command the highest price possible. They also want to ensure that the analysts continue to support the stock after the offering so that it increases in value. Managers thus carefully consider the optimism and enthusiasm of an investment bank's analysts in determining whom to hire. Indeed, CEOs report that the reputation of the analyst covering the relevant industry is an important determinant of their choice of an underwriter for their company's initial public and seasoned equity offerings (Hong 2004, 2–3).

Empirical evidence suggests that analysts' employers have structured their promotion and compensation schemes accordingly. Harrison Hong and Jeffrey Kubik, for example, analyzed the earnings forecasts and employment histories of 12,000 analysts working for 600 brokerage houses between 1983 and 2000 and found that analysts were systematically rewarded for being optimistic as long as the optimism was within a range of accuracy that maintained the credibility of the analysts (Hong and Kubik 2003, 313–15). Hong and Kubik also found that relatively optimistic analysts were much less likely to be fired or to leave a top brokerage house, were much more likely to be hired by a better house, and were given better assignments than their more pessimistic (realistic?) colleagues. It thus seems that analysts face personal incentives to issue enthusiastic and optimistic recommendations and cannot be counted on to provide investors with the "bad news" necessary to correct instances of overvalued equity.

14.3.2 Why Overvaluation Is More Likely to Cause Investor Harm

Not only is overvaluation more likely than undervaluation to occur and persist, it also tends to cause greater harm to investors when it does occur. Perhaps most importantly, overvaluation creates much larger agency costs than does undervaluation. Agency costs are the costs that arise from individuals' cooperative efforts. They appear whenever any principal hires an agent to act on his or her behalf, for the agent will always have an incentive to act opportunistically or to shirk, and the principal must therefore take steps to prevent or ensure against such behavior. Agency costs may thus be defined as the sum of the contracting, monitoring, and bonding costs incurred to reduce the conflicts of interest between principals and agents, plus the residual loss that occurs because it is generally impossible to perfectly identify the interests of agents and their principals (Jensen and Meckling 1976). In a corporation, agency costs arise because the directors, officers, and other managers charged with running the corporation's business have interests that conflict with those of the corporation's residual claimants, the shareholders. Although capital markets generally operate as a powerful tool for minimizing agency costs (because firms that have developed effective mechanisms for lowering such costs will be most attractive to investors), recent economic developments suggest that, when equity becomes overvalued, securities markets tend to *exacerbate* agency costs.

Before examining why overvaluation creates substantial agency costs, consider why undervaluation does not do so. When a firm's equity is undervalued, the incentives of shareholders and managers are likely to be closely aligned: both groups will usually want to drive the stock price upward toward fundamental value. Shareholders will desire that result because price appreciation adds to their long-term wealth and enhances the corporation's overall health (and thus its value) by making it easier for the firm to raise large sums of money in the capital markets. Managers will typically want that result because a higher stock price enhances their job prestige and frequently their compensation and enables the corporation to be more flexible (because it can use its high-priced stock as currency or raise more money for expansion in the capital markets). Given the overlap in shareholders' and managers' desires, it is unlikely that undervaluation will occasion any managerial behavior that diverges from shareholder interests.

The situation is markedly different when a firm's stock is overvalued. In that situation, the interests of shareholders and managers are likely to diverge substantially. Managers are unlikely to prefer that the stock price fall to fundamental value, for, as noted, they reap a host of benefits from a high stock price. While most managers will realize that overvaluation cannot last forever and that price correction is likely to occur eventually, they will probably refrain from taking steps to reduce price to fundamental value. Their tendencies toward optimism will likely lead them to believe either that they can eventually cause the firm to generate cash flows that will justify the currently inflated price or that they will be able to exit the corporation (by resigning their positions and selling their stock) prior to the inevitable price correction.

On first glance, one might suppose that shareholders would similarly desire for equity overvaluation to persist; after all, the higher the stock price, the greater a shareholder's wealth. Because overvaluation tends to be eventually corrected, however, medium- to long-term shareholders generally cannot capture the transitory wealth increase stemming from overvaluation and thus will not care to extend periods of overvaluation. While short-term shareholders may be able to profit from transitory periods of overvaluation, they can do so only if they sell their stock prior to the inevitable price correction. Such a "bail before correction" strategy is much riskier for shareholders than for managers, for shareholders know little about corporate events that may reveal overvaluation and are thus more likely to delay too long before selling their stock. Moreover, shareholders possess neither actual nor apparent control over the events likely to reveal overvaluation and will thus tend to be less optimistic than managers about their ability to sell their stock before the inevitable price correction. Accordingly, even short-term stockholders will value periods of overvaluation less than managers will. In addition, any upside experienced by shareholders during periods of overvaluation is likely to be counteracted by a significant downside: managers seeking to maintain stock prices at artificially high levels tend to take a series of value-destroying actions.

In order to protect their jobs and reputations, managers of overvalued firms often need to "buy time"—that is, to trick the market into maintaining the high stock price until they can exit the firm (both as shareholders and as managers) or can produce the corporate performance required to justify the stock price (Jensen 2005). Consider, for example, a prominent account of the financial collapse at Enron (McLean and Elkind 2003, 171):

Enron's accounting games were never meant to last forever.... The goal was to maintain the impression that Enron was humming until [CEO Jeff] Skilling's next big idea kicked in and started raking in *real* profits.... In Skilling's mind, though, there was no way he was going to fail. He had always succeeded before, and his successes had transformed the company. Why would it be any different with EES and broadband?

Such continued trickery requires beating analysts' expectations, for the capital markets routinely punish firms that fail to meet such expectations (Skinner and Sloan 2002). The problem is that managers of overvalued firms cannot perpetually meet analysts' expectations by exploiting legitimate value-creating opportunities. Once those options have been exhausted, they will eventually turn to gimmicks that are designed to produce numbers that appease the market but actually reduce long-term firm value. Michael Jensen has identified three such gimmicks that are routinely pursued by managers of overvalued firms (Jensen 2005, 10):

> To appear to be satisfying growth expectations you use your overvalued equity to make long run value destroying acquisitions; you use your access to cheap debt and equity capital to engage in excessive internal spending and risky negative net present value investments that the market thinks will generate value; and eventually you turn to further accounting manipulation and even fraudulent practices to continue the appearance of growth and value creation.

Consider how these three gimmicks work in concert to destroy corporate value.

14.3.2.1 Value-Destroying Acquisitions

Because corporate acquisitions create the appearance of growth (and thus may fool the market for at least a while), corporate managers who have exhausted other growth options may find such acquisitions attractive, even if they are ultimately value reducing. Consider, for example, recent findings by Sara B. Moeller, Frederick P. Schlingemann, and René M. Stulz, who compared how merger announcements affected the stock prices of acquiring firms during the 1998–2001 period, a period of significant equity overvaluation, with the acquiring-firm price effects occasioned by merger announcements in the 1980s (Moeller et al. 2005). The authors discovered that, for the 1998–2001 period, the value of acquiring firms declined by a total of $240 billion in the three-day periods surrounding announcements of acquisitions. During all of the 1980s, by contrast, the loss in value of acquiring firms during the three-day period surrounding merger announcements was only $4.2 billion. Moreover, whereas the acquirers' losses in the 1980s were offset by gains to acquirees for a net synergy gain of $11.6 billion, such an offset did not occur in the 1998–2001 period; rather, the losses to acquirers exceeded acquirees' gains for a net synergy loss of $134 billion.

Equity overvaluation seems to have influenced this value destruction. The authors found that most of the value losses were attributable to eighty-seven "large loss" transactions, in which the loss to each acquiring firm exceeded $1 billion. The bidders in those transactions appear to have been overvalued: they had statistically significantly higher Tobin's q and market-to-book ratios (both proxies for overvaluation) than both the bidders in other deals during the same time period and all bidders in the period from 1980–97. Moreover, a substantially greater proportion of bidders in large loss deals financed their acquisitions using equity: 71.6 percent of the bidders in large loss deals did so, as opposed to 35.2 percent of other bidders during the same time period and 30.3 percent of all bidders in the 1980–97 period. In short, what the authors term "wealth destruction on a massive scale" appears to have occurred because overvalued bidders used their high-priced stock to finance deals that, from an investor's perspective, should not have been pursued.

14.3.2.2 Negative NPV Greenfield Investments and Avoidance of Positive NPV Investments

Equity overvaluation also tends to lead managers to reduce firm value by pursuing certain greenfield investments that have a negative net present value (NPV) and avoiding other investments that have a positive NPV. When equity is overvalued, firm managers effectively have more capital to invest. Most obviously, they may pay for expenses using their firm's inflated stock as currency. In addition, they can raise more cash by issuing new equity at prices reflecting their firm's overvaluation. Empirical data indicate that managers do, in fact, take advantage of periods of overvaluation by issuing equity (Baker and Wurgler 2002; Graham and Harvey 2001). Equity overvaluation thus increases the resources with which managers may pursue firm expansion, creating a version of what Michael Jensen has termed the "agency costs of free cash flow" (Jensen 1986). Those agency costs arise because managers with the resources to do so are likely to pursue firm expansion beyond the point that is optimal for stockholders. Whereas the rational stockholder desires the firm to expand to the point at which its marginal cost of expansion equals the marginal value added to the firm because of such expansion, managers will tend to seek expansion to the point at which their private marginal benefits occasioned by the expansion equal their marginal cost of seeking that level of expansion (including, of course, the cost of any "punishment" they expect to receive because they have pursued expansion excessively). Because managers receive a disproportionate share of the benefits of firm expansion, they tend to pursue a level of investment that is excessive in that it fails to maximize firm value.

In addition to causing active value destruction through imprudent acquisitions and unwise greenfield investments, overvaluation may cause passive value destruction by encouraging managers to forgo positive NPV projects. Because the dominant strategy of managers of overvalued firms is, in the words of Jensen, to "postpone the day of reckoning until [they] are gone or [they] figure out how to resolve the issue" (Jensen 2005, 10), they will look for opportunities to conceal their firm's overvaluation from the market. One way to do so is to delay value-enhancing investment expenditures in order to meet quarterly earnings expectations and avoid the value reassessment that accompanies missing such an expectation. Recent research suggests that this sort of value-sacrificing behavior is widespread. In a recent survey, 80 percent of corporate chief financial officers stated that they would be willing to delay discretionary expenditures on research and development, advertising, and maintenance in order to meet earnings expectations, and more than 55 percent stated that they would "delay starting a new project even if this entails a small sacrifice in value" in order to meet a target (Graham et al. 2005). Overvaluation thus tends to cause passive value destruction as managers attempt to buy time by delaying positive NPV investments.

14.3.2.3 Eventual Fraud

Once managers of overvalued firms have exhausted their opportunities to boost or maintain apparent firm value through acquisitions and greenfield investments, they face a temptation to pursue more direct means of duping the market. They may begin with "earnings management," the well-accepted practice of smoothing earnings by strategically timing the recognition of revenues and expenses in order to meet market projections. In the chief financial officer survey mentioned above, 40 percent of respondents stated that they would "book revenues now rather than next quarter" if their company were in danger of missing an earnings target (Graham et al. 2005, table 6). The problem is that earnings management can evolve rapidly into outright fraud, for managers who recognize revenues early and push recognition of expenses into the future will face more difficult accounting challenges in subsequent quarters and will eventually have no choice but to lie or have their company revealed as overvalued (Jensen 2005, 8).

It should be obvious that accounting manipulation imposes significant costs on a firm. In the likely event that the firm's accounting manipulations are revealed, customers will be less willing to do business with the firm; compliance costs will rise as regulators monitor the firm more closely; potential business partners will be less willing to embark on joint ventures; lenders will be less likely to extend credit on favorable terms; and investors will invest their money elsewhere (or demand a higher return on investment). Accounting manipulations thus make it hard for a company to flourish and, in extreme cases (e.g., Enron), may kill the company altogether. Thus, the agency costs created by accounting manipulation, which overvalued equity encourages as a means of buying time, are potentially huge.

Because overvaluation is more likely to occur and persist than undervaluation and tends to impose greater costs on investors when it does occur, insider trading that pushes an inflated stock price down toward fundamental value provides greater investor benefit than insider trading that increases the price of an undervalued stock. We turn now to compare the costs imposed by the two species of insider trading.

14.4 PRICE-DECREASING INSIDER TRADING IMPOSES LOWER COSTS ON INVESTORS

As Dennis Carlton and Daniel Fischel have argued, it would be difficult for managers, who typically work in teams, to cause deliberate harm to a corporation in order to benefit from insider trading; such deliberate mismanagement could cause great reputational harm, and would probably give rise to a successful investor lawsuit based on breach of fiduciary duty, and some colluding managers would therefore likely defect (Carlton and Fischel 1983, 873–74). The most plausible type of corporate harm occasioned by insider trading, then, is the thwarting of valuable corporate transactions that could otherwise be accomplished (Bainbridge 2002, 600–602).

Price-decreasing insider trading is less likely to cause that sort of harm than is price-increasing insider trading. To see why this is so, consider why price-increasing insider trading might prevent such transactions from occurring and why price-decreasing insider trading generally could not do so.

Price-increasing insider trading may injure a corporation seeking to take advantage of nonpublic information regarding an asset's hidden value. Suppose, for example, that managers are aware that some asset the corporation seeks to acquire is undervalued and, if purchased by the corporation, would enhance corporate value. The law generally permits an asset buyer who has discovered information regarding an asset's hidden value to refrain from disclosing that information, and the corporation will thus want to keep such information a secret in order to prevent the asset's price from rising; see Restatement (Second) of Contracts § 161 cmt. d (1981) ("A buyer of property … is not ordinarily expected to disclose circumstances that make the property more valuable than the seller supposes").

If insiders who are aware of the corporation's forthcoming asset purchase attempt to profit personally by purchasing their corporation's own stock, their trading may cause an increase in the corporation's stock price, and that price activity may cause the current owner of the asset not to sell or to demand a higher price. Price-increasing insider trading, then, would squander an otherwise available corporate opportunity. Consider, for example, the classic *Texas Gulf Sulfur* case, *Securities and Exchange Commission v. Texas Gulf Sulphur,* 401 F.2d 833 (2d Cir. 1968), in which geologists from a mining company (TGS) had discovered a valuable ore deposit. Because tremendous value would accrue to TGS if the company could purchase surrounding land at a favorable price, the TGS president ordered insiders to keep the discovery a secret so as not to tip off neighboring landowners before they sold their property to TGS.

Price-decreasing insider trading, by contrast, is unlikely to thwart value-creating corporate transactions that could otherwise be legally accomplished. The relevant situation would be one in which the corporation had an interest in keeping its stock's price inflated above its true value in order to accomplish some transaction. For example, the corporation might desire to use its overvalued stock as consideration for a purchase, to issue new equity at an inflated price, or to secure credit on favorable terms. It probably could not do so. If insiders were aware of information indicating that the stock was overvalued but refrained from disclosing that information, any stock price–dependent transaction entered into during the period of inflation would likely be voidable by the corporation's counter-party; see Restatement (Second) of Contracts § 164 (1981) (permitting rescission of contract by party who is victim of fraudulent or material misrepresentation); id. § 161(b) (stating circumstances under which failure to disclose negative information may give rise to right to void a contract).

Thus, corporate transactions that would be thwarted by price-decreasing insider trading probably could not be legally accomplished in any event.

There is, in short, an asymmetry in the law regarding precontract disclosures (Kronman 1978), and that asymmetry causes price-increasing insider trading to be more value destructive than price-decreasing insider trading. Because a corporation generally need not disclose information about hidden value before transacting on the basis of that information, it may legitimately keep such information a secret. Price-increasing insider trading may prevent it from doing so, and may thereby thwart value-creating transactions. Information suggesting that the corporation is overvalued, however, must generally be disclosed. Accordingly, price-decreasing insider trading would not reveal any corporate secrets that would not otherwise have to be revealed. It is therefore less likely to squander legitimate corporate opportunities.

It seems, then, that price-decreasing insider trading is less likely than price-increasing insider trading to cause the sort of investor harm insider trading is likely to occasion.

14.5 CONCLUSION: AN ASYMMETRIC INSIDER TRADING POLICY CONSTITUTES THE MAJORITARIAN DEFAULT

This chapter has shown that (1) undervaluation is more likely to be self-correcting than overvaluation; (2) in the long run, undervaluation is unlikely to impose significant costs on investors, while overvaluation is likely to do so; and (3) whereas insider trading that pushes a stock's price upward toward actual value may cause harm to the corporation and its investors, insider trading that pushes an inflated price downward toward value is unlikely to do so. Taken together, these observations suggest that an asymmetric insider trading policy that permits some form of price-decreasing insider trading, while generally banning price-increasing insider trading, is the policy investors and managers would likely bargain for were they able (practically and legally) to do so. In other words, an asymmetric insider trading policy that liberalizes only price-decreasing insider trading likely represents the majoritarian default policy. Accordingly, the law should liberalize price-decreasing insider trading (subject only to contractual restraints imposed by corporations themselves), while continuing to regulate price-increasing insider trading. Of course, the devil is in the details. I have elsewhere attempted to flesh out the contours of a workable asymmetric insider trading policy and to demonstrate that such a policy could be implemented under current law (Lambert 2006).

REFERENCES

Arlen, J. H., and W. J. Carney. 1992. Vicarious liability for fraud on securities markets: Theory and evidence. *University of Illinois Law Review* 691–734.

Asubel, L. M. 1990. Insider trading in a rational expectations economy. *American Economic Review* 80(5):1022–41.

Bainbridge, S. M. 2002. *Corporation law and economics.* New York: Foundation Press.

Baker, M., and J. Wurgler. 2002. Market timing and capital structure. *Journal of Finance* 57(1):1–32.

Bazerman, M. H. 1994. *Judgment in managerial decision making.* 3rd ed. New York: John Wiley & Sons.

Carlton, D., and D. Fischel. 1983. The regulation of insider trading. *Stanford Law Review* 35:857–95.

Coffee, J. C. 2004. Gatekeeper failure and reform: The challenge of fashioning relevant reforms. *Boston University Law Review* 84:301–64.

Copeland, T. E., and D. Galai. 1983. Information effects on the bid-ask spread. *Journal of Finance* 38(5):1457–69.

Craig, S., and J. Weil. 2001. Heard on the street: Most analysts remain plugged in to Enron. *Wall Street Journal* October 26, p. C1.

D'Avolio, G., E. Gildor, and A. Shleifer. 2001. Technology, information production, and market efficiency. Harvard Institute for Economic Research, Discussion Paper 1929. http://ssrn.com/abstract=286597.

Dolgopolov, S. 2004. Insider trading and the bid-ask spread: A critical evaluation of adverse selection in market making. *Capital University Law Review* 33:83–180.

Dreman, D. N., and M. A. Berry. 1995. Analyst forecasting errors and their implications for security analysis. *Financial Analysts Journal* 51(3):30–41.

Dutton, J. E., S. J. Ashford, R. M. O'Neill, E. Hayes, and E. E. Wierba. 1997. Reading the wind: How middle managers assess the context for selling issues to top managers. *Strategic Management Journal* 18(5):407–23.

Easterbrook, F. H., and D. R. Fischel. 1984. Mandatory disclosure and the protection of investors. *Virginia Law Review* 70:669–715.

Finn, M. T., R. J. Fuller, and J. L. Kling. 1999. Equity mispricing: It's mostly on the short side. *Financial Analysts Journal* 55(6):117–26.

Gilson, R. J., and R. H. Kraakman. 1984. The mechanisms of market efficiency. *Virginia Law Review* 70:549–644.

Glosten, L. R., and P. R. Milgrom. 1985. Bid, ask and transaction prices in a specialist market with heterogenously informed traders. *Journal of Financial Economics* 14(1):71–100.

Graham, J. R., and C. R. Harvey. 2001. The theory and practice of corporate finance: Evidence from the field. *Journal of Financial Economics* 60(2–3):187–243.

Graham, J. R., C. R. Harvey, and S. Rajgopal. 2005. The economic implications of corporate financial reporting. National Bureau of Economic Research, Working Paper 10550. http://ssrn.com/abstract=491627.

Haft, R. J. 1982. The effect of insider trading rules on the internal efficiency of the large corporation. *Michigan Law Review* 80:1051–71.

Hong, H. 2004. Seeing through the seers of Wall Street: Analysts' career concerns and biased forecasts. Princeton Working Paper. http://www.princeton.edu/~hhong/seers.pdf.

Hong, H., and J. D. Kubik. 2003. Analyzing the analysts: Career concerns and biased earnings forecasts. *Journal of Finance* 58(1):313–51.

Jensen, M. C. 1986. Agency costs of free cash flow, corporate finance, and takeovers. *American Economic Review* 76(2):323–29.

Jensen, M. C. 2005. Agency costs of overvalued equity. *Financial Management* 34(1):5–19.

Jensen, M. C., and W. H. Meckling. 1976. Theory of the firm: Managerial behavior, agency costs, and ownership structure. *Journal of Financial Economics* 3(4):305–60.

Kiesler, S., and L. Sproull. 1982. Managerial responses to changing environments: Perspectives on problem sensing from social cognition. *Administrative Science Quarterly* 27(4):548–70.

Kroger, J. R. 2005. Enron, fraud and securities reform: An Enron prosecutor's perspective. *University of Colorado Law Review* 76:57–137.

Kronman, A. T. 1978. Mistake, disclosure, information, and the law of contracts. *Journal of Legal Studies* 7:1–34.

Lambert, T. A. 2006. Overvalued equity and the case for an asymmetric insider trading regime. *Wake Forest Law Review* 41:1045–1129.

Langevoort, D. C. 1997. Organized illusions: A behavioral theory of why corporations mislead stock market investors (and cause other social harms). *University of Pennsylvania Law Review* 146:101–72.

Levmore, S. 1982. Securities and secrets: Insider trading and the law of contracts. *Virginia Law Review* 68:117–60.

Lord, C. G., L. Ross, and M. R. Lepper. 1979. Biased assimilation and attitude polarization: The effects of prior theories on subsequently considered evidence. *Journal of Personality and Social Psychology* 37(11):2098–2109.

Macey, J. R. 1984. From fairness to contract: The new direction of the rules against insider trading. *Hofstra Law Review* 13:9–64.

Manne, H. G. 1966. *Insider trading and the stock market*. New York: Free Press.

McLean, B., and P. Elkind. 2003. *The smartest guys in the room*. New York: Penguin.

Moeller, S. B., F. P. Schlingemann, and R. M. Stulz. 2005. Wealth destruction on a massive scale? A study of acquiring-firm returns in the recent merger wave. *Journal of Finance* 60(2):757–82.

Schotland, R. A. 1967. Unsafe at any price: A reply to Manne, insider trading and the stock market. *Virginia Law Review* 53:1425–78.

Scott, K. E. 1980. Insider trading: Rule 10b-5, disclosure and corporate privacy. *Journal of Legal Studies* 9:801–18.

Senate Committee on Government Affairs. 2002. Financial oversight of Enron: The SEC and private sector watchdogs. Washington, DC.

Skinner, D. J., and R. G. Sloan. 2002. Earnings surprises, growth expectations, and stock returns or don't let an earnings torpedo sink your portfolio. *Review of Accounting Studies* 7(2–3):289–312.

Stickel, S. E. 1990. Predicting individual analyst earnings forecasts. *Journal of Accounting Research* 28(2):409–17.

Tetlock, P. E., L. Skitka, and R. Boettger. 1989. Social and cognitive strategies for coping with accountability: Conformity, complexity, and bolstering. *Journal of Personality and Social Psychology* 57(4):632–40.

CHAPTER 15

Positive and Negative Information—Insider Trading Rethought

Kristoffel R. Grechenig*

CONTENTS

* I thank Jim Jalil, Hillary Sale, and Ethan Stone for their comments.

15.1 INTRODUCTION

In the past half century, insider trading has been widely discussed in the literature (e.g., Manne 1966; Treynor 1971; Easterbrook 1981; Carlton and Fischel 1983; Bainbridge 1986; Dalley 1998; Pritchard 1998; Goshen and Parchomovsky 2001; Jalil 2003) and it has been subject to decisions of the U.S. Supreme Court (*Chiarella v. United States*, 445 U.S. 222, 1980; *Dirks v. Securities and Exchange Commission*, 463 U.S. 646, 1983; *United States v. O'Hagan*, 521 U.S. 642, 1997). Yet commentators disagree fundamentally both when interpreting the current law as well as regarding efficiency analysis.

A small number of recently published articles have come up with a new line of argument that emphasizes the distinction between positive and negative information with a focus on the distribution of information rather than on the production of it. After the fall of Enron, Manne (2005) was one of the first to explain how insider trading would have prevented managerial misconduct and other corporate fraud. If insiders were allowed to trade on concealed information, the information would have become public. Even though Manne did not explicitly refer to negative versus positive information, he clearly put the insider trading debate in the light of the recent scandals. Kobayashi and Ribstein (2006) and myself (Grechenig 2006) highlighted the positive effects of insider trading on negative information. Whereas Kobayashi and Ribstein focused on outsiders with private information, I analyzed typical insiders and made the distinction between positive and negative information explicit. Macey (2007) applied these insights to the discussion on whistleblowing, arguing that insider trading on negative information has advantages over conventionally rewarding whistleblowers for uncovering information.

A core element of the new approaches is the *Dirks* decision in which the Supreme Court held that Raymond Dirks did not violate the law by trading on negative inside information. This chapter seeks to consolidate the new readings of *Dirks* and other cases according to economic analysis. It argues that the law distinguishes between insider trading on positive and insider trading on negative information and highlights the efficiencies.

15.2 POSITIVE AND NEGATIVE INFORMATION

15.2.1 The Efficiency Debate

Today scholars widely agree that insider trading has both positive and negative effects. On one hand, insider trading enhances informational efficiency of the capital markets by allowing and incentivizing insiders to sell private information to the market, and thereby incorporating a large amount of information in the price. With more information, especially inside information, prices are likely more accurate and the economic resources are distributed more efficiently. On the other hand, insiders impose a "tax" on market participants because insiders trade with an inherent informational advantage. Since investors will anticipate expected losses when they trade against insiders, they will only buy at a lower price, so that some (otherwise) efficient transactions will not be carried out (Leland 1992). This so-called insider trading tax appears in the form of larger bid–ask spreads that market makers use to compensate for their losses when they trade with insiders (Treynor 1971; Manne 2005). They pass some or all of their losses on to the other outside investors by increasing the difference between the price at which they are willing to buy and the price at which they are willing to sell the stock. As a result of this social cost, many scholars have argued that insider trading is unfair. These and many other arguments brought forth both in favor and against insider trading (Bainbridge 1986) were traditionally considered to hold true for *positive* and *negative* information.

15.2.2 Distinguishing between Positive and Negative Information on the Basis of *Dirks v. SEC*

Recent reinterpretations of *Dirks v. SEC* suggest that insider trading on positive information is not to be treated the same as insider trading on negative information. In *Dirks*, an analyst named Raymond L. Dirks, who specialized in providing investment analysis to institutional investors, was alleged to have violated insider trading laws. Dirks received information from a former officer of the company who argued that the corporation's assets were vastly overstated as a result of corporate fraud. This officer urged Dirks to verify the fraud and disclose it publicly. Dirks tried to publish the news but failed because the topic was considered too precarious. However, he told his clients who started to sell their stock. Eventually, the stock price collapsed and fraud was exposed. The Court held that Dirks did not violate the insider trading laws on various grounds. On the face of the decision, it seemed decisive that Dirks had not received any (direct) personal benefits because neither did he himself trade on the information nor was he paid for passing that information on to someone else. The Court's assumption of Dirks altruistic attitude was heavily criticized. It was clear that Dirks received some benefit from forwarding the information to his clients, be it only a reputational gain (Carlton and Fischel 1983). From today's perspective, there seems to have been a more significant element: the fact that insider trading involved negative information that otherwise would have been concealed for a relatively long period of time. Fraud was made public precisely because Dirks told his clients about the inaccurate financial statements who then traded on the information and caused the stock price to decline. The Court explicitly stated that "the central role that [Dirks] played in uncovering the fraud at Equity Funding … is an important one. Dirks' careful investigation brought to light a massive fraud.… But for Dirks' efforts, the fraud might well have gone undetected longer" (Dirks, 659 n. 18). The Court not only recognized that insider trading had some social benefits but also that these benefits outweighed the social costs. It clearly distinguished between positive and negative information, since fraud can only be brought to public knowledge by trading on negative information. The distinction between positive and negative information is consistent with well-known insider trading cases like TGS (SEC v. Texas Gulf Sulphur Co., 401 F 2d 833, 1968) and O'Hagan, which held insiders liable for trading on positive inside information (Grechenig 2006).

15.2.3 Distinguishing between Positive and Negative Information on Efficiency Grounds

We could define information as negative when disclosure leads to a decrease of the stock price and positive when it leads to an increase. The economic basis for distinguishing between insider trading on positive and insider trading on negative information lies in the effect on the dissemination of information. Typically, insiders are willing to tell that they have worked successfully and that their decisions have proved to be the right ones (*ex ante*). They are very likely to disclose positive information for reputational reasons, career ambitions, special monetary rewards such as bonuses, or simply to boost their self-esteem. (To a small extent, disclosing positive news may be delayed in order to decrease volatility.) For the same reasons, insiders have strong incentives not to disclose negative information. They will receive less pay, for example, because stock options are worth less, their value on the job market decreases, they may be subject to social sanctions for being an unsuccessful manager or worker, and so forth. Consistent with our intuition, the incentive structure suggests that positive information reaches the market sooner than negative information.

Where information is published fairly quickly, insider trading can do little to enhance the information flow. To pay insiders for disseminating positive information, by allowing them to trade, makes little sense. They would disclose the information anyway. The opposite is true for negative information. If insiders would otherwise withhold the information for a relatively long period of time, it may well make sense to pay them for disclosing it. (Paying insiders does not mean that insiders receive a higher overall remuneration because trading profits will typically influence the regular salary, that is, reduce it.) It is apparent that the efficiencies depend on how long information would be withheld without insider trading. This includes the question of alternative mechanisms that set incentives to disclose information, the most important legal one being disclosure duties.

15.3 INSIDER TRADING VERSUS DISCLOSURE DUTIES

If disclosure duties were costless and worked perfectly so that all information that should be published was actually published, one could argue that insider trading is superfluous. However, disclosure duties do not work perfectly. In fact, there are inherent problems with disclosure duties and their sanctions.

First, disclosure duties require enforcement actions of some kind. For example, if sanctions include criminal law, then costs involve the work of enforcement agencies that need to find and prosecute violations. If sanctions involve damage payments to plaintiffs under private law, costs include the (private) search for violations, the costs of courts, and so forth. Enforcement actions also include the problem of corruption. Since disclosure duties require enforcement actions by central decision makers, such decision makers may be subject to various forms of "lobbying" which may range from soft intervention to bribery, depending on the country and government. On the contrary, insider trading is market based and thus requires no further enforcement costs. The only cost involved is a potential overinvestment in searching for inside information, as many people have incentives to look for such information.

Second, disclosure duties conventionally include negative incentives for violations, meaning that people acting contrary to the law are threatened with punishment. For example, if they do not disclose a relevant fact under the Securities and Exchange Commission (SEC) rules, they may have to pay a fine. These negative incentives are subject to wealth restraints, which cause a problem of marginal deterrence. If a manager owns $10 million (including future wealth discounted to present values) and knows he or she has to pay $10 million in damages to private plaintiffs, an additional damage payment will not have an incentive effect.

Similar problems arise in the case of imprisonment. Thus, it is not surprising to see managers sometimes take hazardous actions, especially near insolvency when large liability payments and criminal sanctions are pending. Under such circumstances, it makes sense to set positive incentives for disclosing information, that is, to pay insiders by allowing them to trade on the information, since positive rewards are not subject to wealth restraints (for further advantages of rewards versus punishments see Hamdani and Kraakman 2007).

Third, disclosure duties lead to statements that typically express the opinion of only one or very few insiders, whereas insider trading allows the opinions of many to be aggregated. This is an expression of the general idea that a large number of opinions is, on average, more accurate than the opinion of every single participant. The idea was recently recaptured in the form of prediction markets, where people trade on information, like weather forecasts and outcomes of political elections. Prediction markets have also been used for corporate matters, like the expected completion of a production process. In this case, the aggregation of information can function as a disclosure mechanism from employees to managers (Manne 2005). One of the advantages is that traders bear the full gains and losses so they will think about their actions more thoroughly than under any other circumstances. So far, predictions have consistently beaten conventional forecasting methods (Sunstein 2005).

Fourth, and closely connected to the last point, is an argument in favor of disclosure duties. Traditionally, disclosure duties are considered to lead to a "direct" disclosure of information, for example, "the investment has failed," instead of an indirect disclosure by insider trading, for example, a decline in the stock price. This argument seems to run against insider trading as a disclosure mechanism. However, it should be noted that often information is officially disclosed or uncovered as a result of a decline of the stock price.

These arguments do not suggest that insider trading could or should entirely replace disclosure duties. The point is that insider trading can supplement disclosure duties.

15.4 TRADING ON POSITIVE INFORMATION AS MISAPPROPRIATION

One of the main reasons for prohibiting insider trading seems to lie in the intuition that insiders are misappropriating something. The argument seems accurate if positive information is involved. Consider a simple example based on the early case *SEC v. Texas Gulf Sulphur*: A company, engaged in the exploration of natural resources, conducts a geophysical survey and finds copper and other raw materials. Before the news becomes publicly known the members of the exploration group purchase stock and stock options of their own company. After the news is disclosed the stock price increases and the insiders make large profits. What happened in this example? Insiders transferred wealth from the shareholders to their own pockets by (mis)appropriating some of the gains. If this was allowed, shareholders would invest less money than socially optimal in the firm and fewer researches than socially optimal would be conducted. For some exploration projects, only the expected total gains but not the expected gains to shareholders would be larger than the total costs.

Consider a slight alteration of the example: the exploration group simply purchases the land where the company has found raw materials on the basis of inside information. In this case, shareholders would make no gains at all but be left with the search costs. If such (mis)appropriation was possible, they would not invest any money at all in the company and researches would not be conducted. Legal scholarship traditionally captures the second example by the "business opportunity doctrine" but the similarities are evident (Landes 2001). In both cases, shareholders have invested money for the purpose of making a profit. They would not have done so if the profits would go to someone else.

To be sure, profit sharing can be efficient and is typically done through stock option plans. Incentives for managers and other employees to create positive information are set by means of an equity interest. How much incentives need to be set varies from company to company, person to person, and project to project. In any case, rewards through insider trading seem not very well suited as proper incentives. One reason is that it rewards those that first know the information and not the ones that have produced it. Moreover, the rewards are not linked very well to the increase of the firm value because the amount of trading profits depends on a whole different set of factors. Of course, one could argue that stock options and restricted stock are also influenced by other factors (such as overvaluations). However, with insider trading on negative information most of these factors seem to be sorted out.

When negative information is involved, insiders are not misappropriating anything from this point of view. By definition, wealth has already been lost and the corporation is simply overpriced. Of course, the shares could still be sold at an inflated price before negative news is disclosed but this has no incentives for the production of goods (regarding the production of negative information see below). The loss would have to be incurred by another shareholder then.

15.5 PROHIBITING INSIDER TRADING FACILITATES COLLUSIVE AGREEMENTS

The insider trading prohibition causes less information to enter the market. Because insiders are being sanctioned, at least sometimes, they do not trade, or less frequently trade, on that information. One of the important and often overlooked effects of insider trading sanctions is that they facilitate collusive agreements to withhold negative information from the public. Consider a group of managers of a given firm who have recognized that major investments have failed. Under certain given circumstances they will decide not to disclose this fact but to conceal it. They enter into an agreement whereby no one will pass this information on to an outsider and no one will use this information for private trading profits. (As explained above, trading by insiders is likely to cause the information to become public.) If insider trading was allowed, the managers would be more likely to trade on that information, knowing that the other members of the management could deviate from the collusive agreement by trading on and profiting from the information themselves. At times, the collusive agreement will hold even with insider trading, especially when the comparative benefit from concealing the information is larger than the trading profits for every single member of the group. However, at times, some members will be able to make trading profits that outweigh the benefit from concealing the information. Thus, the collusive agreement becomes instable. Like in the prisoner's dilemma, it is likely that the managers trade in order to make at least some profits even though they would be better off concealing the information. Allowing insider trading destabilizes such agreements and causes information to enter the market more quickly.

In turn, if insider trading was prohibited, members who did not abide by the agreement could be sanctioned by the other members who could simply report insider trading activity to the SEC. If those sanctions are high enough, so that the expected loss equals the expected trading benefits, insiders will not deviate from the agreement and information will be concealed. The prohibition allows managers to commit to abiding by the agreement and thus facilitates agreements among insiders to withhold negative information from the public. Clearly, from this point of view, insider trading encourages corporate fraud. For the same reason, a regulation that allowed insider trading on negative information would have to be mandatory if we believe that managers have the power to change the corporate charters or include a clause in the employment contracts.

15.6 INSIDER TRADING VERSUS WHISTLEBLOWING

In the light of the *Dirks* decision and according to efficiency analyses, insiders are considered public agents when they trade on negative information. They perform a task that is otherwise allocated to public authorities (enforcement agencies), the board of directors, especially independent directors (approval or disapproval of certain decisions), auditors (financial statements), private litigants, and so forth. Insiders are public agents because they monitor the firm and uncover negative information by trading on it. They are similar to whistleblowers because they "report" misconduct (Manne 2005; Kobayashi and Ribstein 2006; Macey 2007).

Whistleblowers are conventionally considered to be insiders who report corporate misconduct to the authorities. Legislation in the United States dates back at least to the False Claim Act of 1863 and allowed whistleblowers to bring an action ("qui tam") on behalf of the government. Today, whistleblowers are protected from wrongful dismissal, discriminatory treatment, and they are given monetary incentives to bring an action. Their motives may often be the same as the ones of insiders who trade on negative information, that is, to profit from knowing about misconduct and negative information. To allow and even incentivize whistleblowing while prohibiting insider trading poses somewhat of an inconsistency in the law.

If we wanted to compare whistleblowing with insider trading, the first could be described as the centralized version, where information is collected by a single enforcement agency, the second as the market version, where information exercises influence in the form of dispersed bits. In the case of whistleblowing, whether in the form of a reward for an action on behalf of the government (see False Claim Act, 31 USCS § 3730(d)1: between 15% and 25%) or in the form of monetary rewards for tips to public authorities (SEC "bounties," SEC Act 1934 Sec 21A(e): maximum 10%), there are always circumstances in which both the whistleblower and the wrongdoer are better off "settling" the case, meaning that the wrongdoer bribes the whistleblower into remaining silent.

Consider a simple intuitive example: the whistleblower W knows about managerial fraud by the manager M. The expected damage payments and fines are $d > 0$ [e.g., 100], the expected reward to the whistleblower is rd [e.g., 15], where $0 > r > 1$, that is to say the reward is a fraction of the damage payments or fines. If the whistleblower reports the manager, the two end up with a payoff of rd (W) and $-d$ (M), respectively. The whistleblower receives his or her reward (15) and the manager has to make a payment (–100) to the public authorities. If the manager instead bribed the whistleblower with b [e.g., 50], where $rd < b < d$, the two would end up with the payoffs b (W, 50) and $-b$ (M, –50). Assuming that the whistleblower cannot accept a bribe and then report the manager because the whistleblower would be held liable him- or herself, both would be better off than in the first case. In other words, if the manager paid the whistleblower a bribe larger than the amount that the whistleblower would receive for reporting the manager and smaller than the amount the manager would have to pay in damages and fines, evidently the two will decide for the bribe. The manager prefers –50 over –100 and the whistleblower prefers 50 over 15 (for a similar bribe model, see Grechenig and Sekyra 2006). That is not to say that whistleblower rewards have no effects on the manager's decision whether to conceal information. In fact, the payments the manager may have to make to keep the whistleblower silent may be larger than the losses from disclosing the information lawfully in the first place. Moreover, there may be nonmonetary rewards to the whistleblower, such as the satisfaction of having reported a fraudulent manager. The point is that in most cases there will be a significant bribery problem with whistleblowing as a mechanism to uncover fraud.

With insider trading, the case typically looks different. Consider, again intuitively, there is negative information worth $I > 0$ [e.g., 100], which is known by the manager M and the whistleblower W. Assume that the manger is always first to know the information and thus reaps all the trading profits if the manager decides to trade. (If the manager decides not to trade, the whistleblower can decide to get the trading profits.) M has a larger payoff $P > I$ if M keeps the information secret; otherwise the information will always be disclosed with or without insider trading and whistleblowing. Again, we assume that M and W will abide by their bribery agreement, which is due to the fact that W would be sanctioned for having received a bribe if the information became publicly known. If $P > 2I$ [e.g., $P = 250$], M will bribe the insider with b, where $P - I > b > I$ [e.g., $b = 125$] and both are better off concealing the negative information. On the other hand, if the gains from keeping the information secret are not twice as large as the trading profits, $P < 2I$ [e.g., 150], M will not bribe the whistleblower. M would have to pay at least I (100) to W and end up with $P - I$, which is by definition less than the trading profits I. Thus, M will simply trade on the information in the first place gaining I (100). It is important to note that $P < 2I$ represents the situation the actors *typically* enter first, where an investment has simply failed, and that $P > 2I$ represents the situation where a manager must try to withhold the information by all means because, for example, otherwise the company goes bankrupt and the manager loses his or her job, is subject to lawsuits, and so forth. If information is disclosed in the first place where the manager still has little interest in keeping the information secret, it may be unlikely that we enter into the second set of parameters. Consequently, insider trading does not seem to be subject to the same inherent bribery problem as whistleblowing.

Note that we have assumed that there is only one manager, or a group of fully coordinated managers that act as if they were one person. As suggested above, more managers further destabilize the concealment of information in the case of insider trading.

15.7 PERVERSE INCENTIVES AND "OUTSIDER" TRADING

Maybe, the most serious concern of a regulation that prohibited insider trading on positive information and allowed insider trading on negative information is the incentive effect on the production of information. If insiders were allowed to trade on bad news, they would have incentives to produce this information in the first place (Cox 1986; Fried 1998). Clearly, producing negative information, and then trading on it, is fairly easy. A manager simply has to make a bad investment decision and then sell the stock (or sell it short). Incentives for a deliberate production of negative information result in inefficiencies.

Even though this effect seems to pose a serious problem, at first sight, there appear to be various reasons for managers not to produce negative information deliberately. For example, managers will lose on performance-based compensation, they may be removed in the course of a takeover, they may suffer from a reputational loss which decreases their value on the job market, the company may go bankrupt, especially if product markets are efficient, they may be subject to liability, and so forth (Grechenig 2006). Whether there are incentives to produce bad news deliberately overall is an empirical question. Intuitively, it seems unlikely that the managers will do so; however, it may be possible under certain circumstances.

The problem of perverse incentives could be solved if only certain insiders were allowed to trade, precisely those who have not produced the negative information in the first place. This would simply eliminate the incentives insiders have for the production of negative information. Such an interpretation of the law would be consistent with the *Dirks* decision, where those that profited from the information (Dirks and clients) were different from those who produced the information. It is also consistent with the characterization of insiders as whistleblowers who report fraudulent acts of somebody else (Manne 2005; Kobayashi and Ribstein 2006; Macey 2007). Where insider trading is allowed for whistleblower reasons there seems to be no negative incentive effects on the production of information. Of course, insiders would not have to act for whistleblower reasons, in order to eliminate the negative effects; separating those who trade on the information from those who produce the information would be sufficient.

The idea to remove incentives for the creation of negative information is reflected in Section 16(c) of the Securities and Exchange Act of 1934, though with a slightly different result. Section 16(c) explicitly prohibits short selling by insiders and thus distinguishes between the selling of shares someone owns and the selling of shares someone does not own. Clearly, where someone can sell only shares already owned the person has no incentives to produce negative information. This incentive is given by the possibility of selling shares that the insider can later buy at a lower price. The possibility of simply selling shares will, however, cause less information to enter the market than otherwise. If one believes that there are few incentives for insiders to deliberately produce negative information, then prohibiting short sales must be treated the same as normal sales under these considerations.

Other distortions brought forth against insider trading include incentives to increase the risk in excess of the investors' preferences in order for the managers to increase their profits (Easterbrook 1981) and incentives to make false announcements in order to create trading opportunities. Even though it is not clear whether the level of risk is increased above the optimal level and whether managers can get away with making false announcements, separating those who produce the information (increase the risk, make false announcements) from those who trade on it would eliminate the problem.

Two questions remain. First, do the persons who are allowed to trade always know the information, given that they have not produced it? Likely, they do not always know the information, and thus, less information reaches the market. However, insider trading does not only incentivize traders to distribute information that they have come to know in the course of their regular work. It also incentivizes potential traders to search for this information much like enforcement agencies, newspapers, and others. This involves higher search costs but avoids potential perverse incentives for the production of information. Arguably, employees may find most of the information and then trade on it. In this scenario, employees could be described as special monitors. They seem to be better suited for monitoring the management than an outside enforcement agency that typically has little access to confidential information. Second, if low-tier employees trade, will they cause the price to decline and thereby uncover the negative information? Clearly, this is an empirical question that has not been solved for this specific issue. However, both empirical studies (Meulbroek 1992) as well as anecdotal evidence (Shiller 2005) suggest that investors find out about insider trades.

Similar to the discussion on collusive agreements, a regulation that allowed insider trading on negative information would have to be mandatory if we believed that those who are supposed to be monitored have the power to introduce contractual prohibitions, thereby disposing of the monitoring activity.

15.8 NEGATIVE INFORMATION AND FOREKNOWLEDGE

We have distinguished between negative information and positive information on the basis of an expected delay of disclosure. Because positive information is typically disclosed quickly, trading involves mere foreknowledge. Incentives to profit from foreknowledge lead to a mere redistribution with little or no gains to social welfare (Hirshleifer 1971). We have argued that the disclosure of negative information is sometimes delayed. Whether the positive effects of insider trading on negative information outweigh the costs depends on how far disclosure would be delayed. The comparative social benefit lies in an efficient allocation of the investors' capital between an early disclosure and an otherwise late disclosure, and has to be weighed against the potential losses. Estimating the benefits and losses is a complex empirical question.

The case law reflects this idea in *In re Cady, Roberts and Co.*, 40 SEC 907 (1961), *Diamond v. Oreamuno*, 248 NE 2d 910 (1969), and *United States v. Smith*, 155 F.3d 1051 (1998), where insiders were held to have violated the law for trading on negative information. In *In re Cady, Roberts and Co.*, the stock was sold shortly ahead of the dividend announcement that communicated a sharp reduction in the rate per share. In *Diamond v. Oreamuno*, insiders sold stock in September after having realized at the end of August that the net earnings had sharply declined. The information was finally disclosed in October. In *United States v. Smith*, the insider started selling approximately two months ahead of the disclosure of the relevant information. The court assumed that the information would have become public with or without insider trading. If the courts were right in assuming that insiders added little or nothing to the information flow in such cases, holding insiders liable, even for trading on negative information, seems efficient (Grechenig 2006). However, in many cases it is hard to tell whether information would have become public without insider trading. It could have been that the information was disclosed soon after the trading activity by insiders precisely because of this trading activity. We are again left with an empirical question when asking whether insider trading involved short or long foreknowledge under certain circumstances. It may best be solved on a case by case basis.

15.9 CONCLUSION

Following the contemporary debate, this chapter argues that allowing insider trading on negative information and prohibiting it on positive information may be more efficient than the two extremes discussed in the past. The social benefit lies in the dissemination of information that would have been concealed otherwise. Even insider trading on negative information may at times include short foreknowledge with little gains for the informational efficiency of the market. This idea can be restated as allowing insiders to trade for whistleblowing reasons. Because insiders may be incentivized to produce negative information deliberately, the approach could be refined in order to allow trading by only those insiders who have not produced the information in the first place. This way incentives to produce and incentives to search and distribute the information could be separated. These efficiency arguments are reflected in the current case law.

REFERENCES

Bainbridge, S. 1986. The insider trading prohibition: A legal and economic enigma. *University of Florida Law Review* 38:35–68.

Carlton, D., and D. Fischel. 1983. The regulation of insider trading. *Stanford Law Review* 35:857–95.

Cox, C. C., and K. S. Fogarty. 1988. Bases of insider trading law. *Ohio State Law Journal* 49:353–90.

Cox, J. 1986. Insider trading and contracting: A critical response to the "Chicago school." *Duke Law Journal* 35:628–59.

Dalley, P. J. 1998. From horse trading to insider trading: The historical antecedents of the insider trading debate. *William and Mary Law Review* 39:1289–1354.

Dyer, B. K. 1992. Economic analysis, insider trading, and game markets. *Utah Law Review* 1–66.

Easterbrook, F. H. 1981. Insider trading, secret agents, evidentiary privileges, and the production of information. *Supreme Court Review* 309–65.

Fried, J. 1998. Reducing the profitability of corporate insider trading through pretrading disclosure. *Southern California Law Review* 71:303–92.

Goshen, Z., and G. Parchomovsky. 2001. On insider trading, markets, and "negative" property rights in information. *Virginia Law Review* 87:1229–77.

Grechenig, K. 2006. The marginal incentive of insider trading: An economic reinterpretation of the case law. *University of Memphis Law Review* 37:75–148.

Grechenig, K., and Sekyra,M. 2006. No derivative shareholder suits in Europe—A model of percentage limits, collusion and residual owners. Columbia Law and Economics Working Paper Series 312. http://ssrn.com/abstract=933105.

Hamdani, A., and R. Kraakman. 2007. Rewarding outside directors. *Michigan Law Review* 105:1677–1711.

Hirshleifer, J. 1971. The private and social value of information and the reward of inventive activity. *American Economic Review* 61(4):561–74.

Jalil, J. P. 2003. Proposals for insider trading regulation after the fall of the house of Enron. *Fordham Journal of Corporate and Financial Law* 8:689–716.

Kobayashi, B., and L. Ribstein. 2006. Outsider trading as an incentive device. *UC Davis Law Review* 40:21–84.

Landes, D. R. 2001. Economic efficiency and the corporate opportunity doctrine: In defense of a contextual disclosure rule. *Temple Law Review* 74:837–77.

Leland, H. E. 1992. Insider trading: Should it be prohibited? *Journal of Political Economy* 100(4):859–87.

Macey, J. 2007. Getting the word out about fraud: A theoretical analysis of whistle-blowing and insider trading. *Michigan Law Review* 105:1899–1940.

Manne, H. G. 1966. *Insider trading and the stock market.* New York: Free Press.

Manne, H. G. 2005. Insider trading: Hayek, virtual markets, and the dog that did not bark. *Journal of Corporation Law* 31:167–85.

Meulbroek, L. K. 1992. An empirical analysis of illegal insider trading. *Journal of Finance* 47(5):1661–99.

Pritchard, A. C. 1998. *United States v. O'Hagan*: Agency law and Justice Powell's legacy for the law of insider trading. *Boston University Law Review* 78:13–58.

Shavell, S. 2004. *Foundations of economic analysis of law.* Cambridge, MA: Harvard University Press.

Shiller, R. 2005. *Irrational exuberance.* 2nd ed. Princeton, NJ: Princeton University Press.

Stigler, G. 1970. The optimum enforcement of laws. *Journal of Political Economy* 78(3):526–36.

Sunstein, C. 2005. Group judgments: Statistical means, deliberation, and information markets. *New York University Law Review* 80:962–1049.

Treynor, J. (published under the pseudonym Bagehot, W.) 1971. The only game in town. *Financial Analysts Journal* 27:12–14.

Part 3

Economic Consequences of Insider Trading

CHAPTER 16

The Economic and Financial Features of Insider Trading

François-Éric Racicot and Raymond Théoret

CONTENTS

16.1 INTRODUCTION

It is difficult to study the financial aspects of insider trading because trustworthy statistics are rare on this subject. As we see in this chapter, some authors have studied the returns associated with insider trading but their studies are blurred by specification errors which are not taken into account in their articles. Yet those errors give way to biased estimators of financial performance (Théoret and Racicot 2007).

There are other aspects of insider trading which are more qualitative than quantitative but which are nevertheless important. Indeed, asymmetric information, moral hazard, and agency problems, notions related to the economics of information, must be considered when studying insider trading. These concepts are not new and much has to be done to incorporate them into the theory of insider trading. They have regulatory, institutional, and financial consequences for the analysis of insider trading. This chapter also considers those aspects.

This chapter is organized as follows. Because the economic theory of information is often confusing, we first present the concepts of asymmetric information, moral hazard, and agency costs, which are key notions in this theory. We then show how to integrate these concepts to the theory of insider trading. Thereafter the quantitative aspects of insider trading are studied with the help of a typical paper on the subject. We see that the traditional measures of performance of the insiders are biased and we show how to correct them. Finally, we consider a case of insider trading involving a stock option plan.

16.2 ASYMMETRY OF INFORMATION, MORAL HAZARD, AND AGENCY PROBLEMS: THE CONCEPTS*

In this section, we present the theoretical aspects of three essential notions in corporate finance: asymmetry of information, moral hazard, and agency problems. In the next section, we show how these concepts may help to understand the complexity of insider trading.

* For this section, we borrowed from Coën, Mercier, and Théoret (2004). The article of Arrow (1963) is a classic on the topic of the economics of information. A very well-known reference on this subject is Copeland, Weston, and Shastri (2005).

16.2.1 Asymmetric Information

In a firm, there exists an asymmetry of information because the managers (insiders) are better informed than the shareholders (outsiders) about the evolution of the operating profits of the firm. The managers thus have inside information. Hence they have an incentive to signal to the shareholders the inside information they own in order that the shares be issued at their fair value in such a way that the value of the firm is maximized.

If the firms are not signaling this information, we obtain a pooling equilibrium where the value of their shares will converge to the average firm's one. In other words, a segmented equilibrium will not be obtained, in which the market evaluates firms on the basis of their relative performance. The classical article of Akerlof (1970) is very useful on that matter. Akerlof believes that Gresham's law ends up playing a significant role if there is asymmetric information and no signaling of information is given. The Gresham's law stipulates that bad money drives good money out of circulation. Transposed to finance, this means that nonperforming firms drive performing ones out of business. We thus regress toward a supply of products of very average and even mediocre quality. Akerlof calls such a universe a "lemons one."

The universe of Akerlof, characterized by asymmetric information, gave rise to two new concepts in the economic and financial literature: adverse selection and moral hazard. These concepts are widely used in the insurance field but find also applications in many other research fields. A problem of adverse selection arises when a buyer is unable to distinguish good products from bad ones or when an insurer is unable to discriminate bad risks from good ones. As argued by Akerlof in his example about lemons, asymmetric information gives way to a problem of adverse selection. The poor quality represented by the lemons drives the best cars from the market. Hence a buyer of used cars believing that the chosen car will be a good one is in fact buying a bad one, that is, a lemon. This is an obvious example of adverse selection.

16.2.2 Moral Hazard

Moral hazard is inherent to any insurance program. It comes from the fact that insurers are unable to monitor the behavior of their insured customers. Once insured, the customer has a tendency to take more risk. This is what is called "moral hazard." If people were not insured, they would probably behave more carefully because they know in this case that they could lose all their wealth or belongings if an accident occurs. (A person whose house is insured may be also less vigilant in locking the door than an individual who was not insured.) For instance, this could happen following legal suits. Thus, because of moral hazard, insurance providers may charge an important insurance premium to all their customers. This premium is based on the bad risks. (Thus insurers would not be able to diversify away all unsystematic risk. Therefore they could not fix the premium to its fair actuarial value, that is, its expected discounted value. They must add a premium to it (Arrow 1963). There are other factors that may justify such a premium.) Therefore, low-risk customers, who cannot afford such premiums, will be driven out of the insurance market. This leads to a problem of adverse selection. The insurance company no longer plays its role, which is to diversify risks, and ends up with all the bad risks. It should be understood that at the limit the insurance industry could go bankrupt if the problem of moral hazard becomes too acute. But if the markets are rational, they will implement a mechanism to counter the problems of adverse selection and moral hazard. Therefore, the insurance company will reduce the premium to car drivers who had no accidents, as in the case of large financial institutions in Quebec. To counter adverse selection, the firm will develop trademarks to signal the high quality of its products to the market.

But what are the characteristics of a good system of signaling? To be efficient, a signal must be expensive. If the firm signals forecasts of its profits, for instance, in its periodic financial statements, then this signal is relatively cheap in the sense that it may be imitated by other firms. Yet, a firm can also cheat by sending this kind of signal, which may in turn show up in the price of its stock. But rational shareholders will soon understand that there is something wrong and we will thus return to Akerlof's lemons universe. Hence to be valid, a signal must be expensive and must not be reproduced easily by another firm. How then can a firm signal by modifying its capital structure? Say that a firm increases its leverage. It can therefore signal to the market that its profitability has increased. Indeed, a firm that issues more debt must commit to regular payments, in this case the interest of the debt. Such a signal is expensive because another firm that is not strong enough will not be able to sustain the payment of continuous interest rate. It thus does not have the capacity to imitate a strong firm. A firm that increases its leverage is also exposed to more market discipline. It knows that the interruption of interest payments might be deadly for it or at least that it might lead to an increase of the credit spread on its debt.

In this last case, its credit rating will be downgraded and the firm will be forced to increase the coupon on its debt because of the inflated risk premium. Therefore, by signaling through its leverage, the firm issues a credible signal on its increased performance which will probably have a positive impact on its stock price. However, Miller and Rock (1985) do not consider the leverage increase a favorable signal but a bad one. These authors argue in terms of net financing. A net positive financing, that is, a surplus of dividends over the financing sources, will be considered as good news and will thus have a positive impact on the firm stock price. A net negative financing, say a surplus of debt over dividends, will have a negative impact on the firm value, because it will be considered as an indicator of financial weakness on the side of the firm. To its credit, the theory of Miller and Rock shows that dividend policy cannot be separated from financing policy. A global approach must thus be adopted to analyze firm policies, which considers as endogenous variables the debt, the ownership, the dividends, and other key variables related to the firm value. (This means that we can model the structure of capital by a well-known econometric technique, that is, the simultaneous equation process. On that matter, see Bhagat and Jefferis 2002.)

Another example of an ambiguous signal is the one that is associated to a new stock issue. Indeed, a firm issues new stock when it considers that its stock is overvalued. As the price of its stock will drop anyway, this firm benefits temporarily from the premium incorporated into the price of its stock. On the other hand, it hesitates to issue new stock when it is undervalued, because there is a risk of more dilution. Without the issue of new stock, the stock price should increase sooner or later and the existing shareholders should then benefit from a fair price for the shares they hold. That would not happen if the firm issues new shares when they are undervalued. Myers (1984) explains why a firm will hesitate to issue new shares when they are undervalued. The author assumes that the firm wants to fund a project whose NPV (net present value) is y. To finance this project, the firm must issue stock for an amount of N dollars. But this issue is worth effectively N_1 dollars. The degree of undervaluation or overvaluation of stock is thus given by: $\Delta N = N_1 - N$. If $\Delta N > 0$, the stock is undervalued. To finance its project, assume that the firm has issued stock for an amount of \$10 million and that these shares are effectively worth \$13 million, that is, $\Delta N = N_1 - N = 13 - 10 = \$3\ million$. The degree of undervaluation is thus here equal to \$3 million. Yet, by issuing these shares, the firm has transferred an amount of \$3 million from the old to the new shareholders.

Indeed, it has sold them shares at a price which is too low, thus ipso facto transferring them a capital gain. Sooner or later, the old shareholders would have cashed this capital gain, assuming a market correction for this undervaluation. Following the issue of undervalued stock, there was a dilution of the capital owned by existing shareholders. That is, their net wealth decreased for an amount of $3 million.

16.2.3 Agency Costs

The financial literature provides many definitions for agency costs. A typical agency cost familiar to managers is the perquisites consumed by them, such as plush carpets and company airplanes (On that matter, see Shleifer and Vishny 1997, 742.) Managers who incur such expenses are the typical bad agents who involve themselves in excessive expenses instead of investing the money of the creditors or shareholders in good projects. In the literature, other researchers associate agency costs to the monitoring costs taken for inducing managers to do their work properly. But the best definition of agency costs is focused on the value of the firm. De Matos (2001) defines agency costs as the difference between the value of the firm in a situation of an ideal contract on the one hand and the value which is viable after the negotiation of this contract on the other hand. This definition has the advantage of enlightening one fundamental element of the theory of agency cost, that is, the concept of contract. It has also the advantage of taking as benchmark the optimal value of the firm in the world of perfect markets as defined by Modigliani and Miller (1958). What the creditors and the shareholders really want is for the manager to give them a return that is competitive on their investments. Because of the separation between financing operations and control, this objective is not filled, which gives raise to agency costs.

Let us discuss now some examples contained in the article of Jensen (1986). This author considers the case of a firm having a surplus of free cash flows, which are defined as the funds exceeding the amount required for a project to have a positive NPV. Because of the separation between financing and control, there is no guarantee that the manager will redistribute those cash flows to shareholders. The manager might instead invest them in projects having a negative NPV with the aim of making the firm grow. This is the assumption of overinvestment. The subsequent reduction in the value of the firm is identified as an agency cost. This is one conflict of interest that may appear between shareholders and managers of a firm which owns excess free cash flows and which are not returned to the shareholders. It is well known that dividend policy can decrease the agency costs but we will not dwell on this subject here.

The conflicts of interest that exist between shareholders and creditors create a link between the option theory and the structure of capital. For instance, a project that increases the stock return volatility of a firm leads to an expropriation of the wealth of the creditors, which results in agency costs for them.

Copeland and Weston (1988) present a model of agency costs which aims at evaluating the optimal rate of indebtedness of the firm. We designate agency cost by AC, debt as *B*, and equity by *S*. Then we have:

$$AC = f(B, S)$$

where

$$\frac{\partial AC}{\partial B} > 0; \frac{\partial AC}{\partial S} > 0$$

The agency costs are thus an increasing function of debt and equity. The objective consists in finding the debt ratio of the firm given by $(B/B + S)$ which is associated to the minimum of total agency costs. Figure 16.1 shows how this minimal cost is obtained.

16.3 ASYMMETRY OF INFORMATION, MORAL HAZARD, AGENCY PROBLEMS, AND INSIDE INFORMATION

Obviously, insider trading is a typical case of asymmetric information, moral hazard, and agency problems (Jaffe 1974; Masson and Madhavan 1991; Padilla 2002a, 2002b). As argued by Padilla, there is a "lemons problem" in insider trading. Obviously, the managers of a firm do not know if their agents, who are in charge of trading on behalf of the company, are truthful. Therefore, they cannot discriminate between agents who are involved in insider trading from agents who are not involved in such trading. Yet, as we will see, insider trading can be very detrimental for the value of a firm. Consequently, the managers will decrease the salaries of their agents across the board. Gresham's law will thus be activated: good agents will thus leave this firm in search of better salaries and the bad ones, that is, the lemons, will remain in place. Hence, the reaction of the firm to insider trading has perverse effects: it fosters insider trading instead of reducing it. Yet, there are ways for a company to discourage insider trading

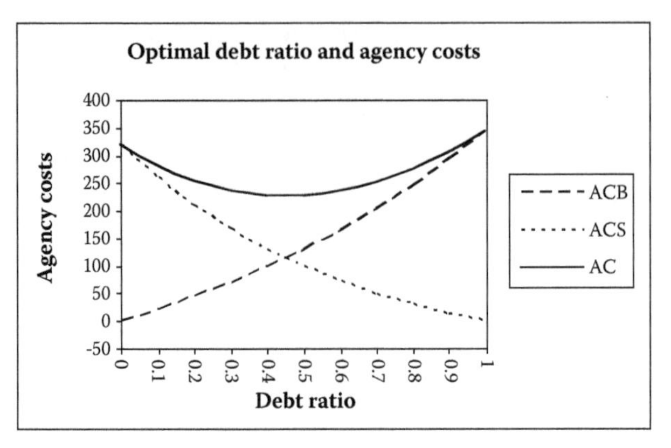

**ACB: agency costs related to debt*
ACS: agency costs related to equity
AC: total agency costs (ACB+ACS)

FIGURE 16.1. The optimal debt ratio with agency costs. ACB, agency costs related to debt; ACS, agency costs related to equity; AC, total agency costs (ACB + ACS).

and encourage its agents to contribute to the maximization of shareholders' wealth. The managers can distribute stock options to their traders. We can argue that in order to see their calls increase in value, the traders would seek to maximize profits instead of devoting themselves to insider trading. But that will not rule out the problem of asymmetric information. Externalities are such that some traders would nevertheless try to cheat because they think that their own transactions have no real impact on the whole transactions of the firm. To discourage insider trading is difficult because of the presence of asymmetric information.

Insider trading also leads to a moral hazard problem (Padilla 2002a, 2002b). We said previously that insurance was the classic example of moral hazard. Insured drivers are more prone to take risk because they will be compensated in case of accidents. The insurance company reacts to this problem by increasing the insurance premium, a procedure that can aggravate the problem instead of solving it. In the context of insider trading, moral hazard is related to a perverse behavior on the side of the traders of a firm which may be very detrimental to its shareholders. Like the insured drivers, the traders do not have to bear the costs of their actions. Actually, their actions may accelerate the decrease of the value of a company.

If they have inside information on an eventual decrease in the value of the firm in which they work, they will short the stock of the firm and that may give way to great instability on the market of the firm shares. There is a counterargument here. The insiders have only anticipated an event which would have happened anyway. They would have thus contributed to the efficiency of the financial markets. But financial markets do not operate smoothly. They can overshoot easily following transactions by the insiders. Perhaps also their operations were only based on rumors without foundations. Anyway, insiders may exacerbate market volatility and that is bad for the markets.

The firms that struggle with insider trading will have to deal with agency problems (Padilla 2002a, 2002b). Actually, shareholders have no enforcement devices to eliminate insider trading. Therefore, agency costs will appear in these companies, which will decrease the value of this firm. As we just said, these costs are related to two other problems: asymmetric information and moral hazard. Agency costs result from problems of governance caused by asymmetric information and moral hazard. These costs are sometimes difficult to compute, like the ones associated to short trading by the insiders. The costs of monitoring that are implemented by a firm for their sake are measurable, and they can decrease other agency costs. Agency costs are communicating vessels and they are thus fussy.

16.4 THE "RETURNS" OF INSIDER TRADING

Several researchers have tried to evaluate the returns of insider trading (Cornell and Sirri 1992; Eckbo and Smith 1998; Jeng et al. 1999). Padilla (2002a) made a survey of these studies from 1976 to 2002. The statistics on insider trading are scarce and they are contaminated by many measurement errors. It is why researchers try to compute the cumulated abnormal return (CAR) on a stock on the days before a signal is issued by a company that has a major impact on its value. As said before, this signal, which is seen as an event, may be an issue of debt, an increase of dividends, or an announcement of a jump in net earnings. It is assumed that insiders were informed about these signals and that they benefited from them. The researchers thus try to evaluate the returns related to this inside information. An indicator of inside transactions is the amount of this return before an "event." If the financial markets are efficient, this return must be low or nonexistent: only on the day of the event would there be a jump in the stock of the firm which has issued the signal because it was unexpected.

Jeng et al. (1999) have studied the returns on insider transactions in the United States from 1975 to 1996. They resort to three methods to compute the abnormal return related to an insider's portfolio. The first method consists in calculating the CAR of this portfolio, a very popular statistic in event studies. The CAR is defined as the insider portfolio return minus the return on the value-weighted market index (the NYSE/AMEX/Nasdaq in this study).

The second method used by Jeng et al. (1999) to compute the abnormal return of an insider's portfolio is based on the alpha computed with two well-known financial models of returns: the capital asset pricing model (CAPM) and the augmented Fama and French model. The unconditional version of the CAPM to compute the abnormal return is:

$$R_{it} - R_{ft} = \alpha_i + \beta_i (R_{mt} - R_{ft}) + \varepsilon_{it}$$

with R_i, the return on the insider portfolio i; R_f, the risk-free rate; α_i, the unconditional alpha of Jensen; β_i, the unconditional beta; R_m, the market return, and ε_i, the residual error. On the other hand, to account for style, they resorted to the augmented Fama and French model, which may be written as follows:

$$R_{it} - R_{ft} = \alpha_i + \beta_{i1}(R_{mt} - R_{ft}) + \beta_{i2} SMB_t + \beta_{i3} HML_t + \beta_{i4} UMD_t + \varepsilon_{it}$$

where SMB is a mimicking portfolio that is long in small cap stocks and short in big cap ones; HML, a mimicking portfolio that is long in stocks having a high ratio of book value to market value and short in stocks having a low ratio of book value to market value; UMD, a mimicking portfolio that is long in stocks having an upward trend and short in stocks having a downward trend.

Finally, they resort to the characteristic selectivity measure (CS) of Daniel et al. (1997) to evaluate the abnormal return of an insider's portfolio. This method matches each insider transaction to a portfolio of similar stocks, and then calculates an excess return relative to this portfolio on each day (Jeng et al., 14). The definition of the CS measure is thus:

$$CS_{it} = R_{it} - Bin_{it}$$

where Bin_{it} is the return of the matching bins of insider's portfolio i in month t. According to Jeng et al. (1999), the CS measure is similar to the

α of Jensen in factor models. The authors divided the insiders' portfolios into purchase portfolios and sale portfolios. They conclude that the purchase portfolio earns abnormal returns of more than fifty basis points per month over the period 1975 to 1996 whereas the sale portfolio did not earn any abnormal return.

The methodology of Jeng et al. (1999) is questionable in some aspects. First, as shown by Théoret and Racicot (2007) and Coën and Racicot (2007), specifications errors which are present in financial models of returns must be removed by resorting to appropriate instruments to arrive at an estimation of alpha corrected for specification errors. Actually, the alpha of Jensen is very sensitive to specification errors contained in a financial model (Coën and Racicot 2007; Théoret and Racicot 2007). Incidentally, Racicot (2003) has shown that we arrive at a new alpha, and therefore at a new version of the CAPM, by correcting this model for its specification errors by using as instruments the moments and co-moments of the explanatory variables of a model. In fact, up to the third order, centralized moments are equal to cumulants. Starting from the fourth order, cumulants become polynomial of centralized moments up to this order. These instruments were originally developed by Dagenais and Dagenais (1997) for cross-section estimation of economic models. Racicot (2003) transposed them to financial time series, especially to models of financial returns like the CAPM.

A second criticism concerns the dynamic aspects of insider trading. Indeed, the strategies followed by insiders are not static, but dynamic. The estimated factor loadings of a financial model of return change day after day during the elapsing time of inside trading. That is especially true for the CAPM beta. When the followed strategies are dynamic, the alpha and the beta become conditional entities. They must be estimated resorting to a "rolling window." Each day, we add to the sample that is used for the estimation process the data of this day and we remove from the data the first day of the previous window. The estimated coefficients thus change day after day. That also reflects the fact that risk changes day after day during the period of a market event following the dissemination of new information.

16.5 A RECENT CASE OF INSIDER TRADING IN THE AERONAUTIC INDUSTRY INVOLVING A STOCK OPTION PROGRAM

Many firms provide stock option plans for their employees. These call options are generally written on the stock issued by these firms. Firms implement those plans to encourage their managers to remain focused on the objective which serves the best the interests of the company shareholders: generally the search for maximum profit. When profits increase, stock price is boosted and the calls which are held by the managers gain in value. Therefore, these stock option plans aim to decrease the agency conflicts between managers and shareholders. But stock option plans may increase agency costs instead of decreasing them because those plans may encourage managers to do insider trading.

Insider trading seems frequent in firms offering stock option plans. Managers have at their disposal inside information which may be used for personal interests instead of collective ones. That gives rise to conflicts between existing shareholders and the managers who benefit from the stock option plan.

Let us suppose that there is a piece of news known only by the managers. The announcement of this information, say, an unexpected decrease of profits, is liable to make the stock price fall. Managers may profit from this inside information by selling their stock before the stock price decreases when the event will occur, which here is the announcement of falling profits. Hence the managers will avoid financial losses associated with the decrease of the price of the call following the event.

Some years ago, there was presumption of inside trading in a large European aeronautic conglomerate. Managers were informed that the board of directors would announce a big increase of the profits of this group. But they were also informed that there would be another announcement, in the days following the first news concerning profits, that there would be delays in the airplane deliveries which would lead to a decrease of the price of the conglomerate shares. Therefore, there were massive sales of options in the days following the profit news and the managers were enriched by their sales. They cashed big profits on their transaction. If they had sold them later, at the occurrence of the second event, that is, the announcement of the delays in deliveries, they would have instead undergone big losses. There was thus here presumption of illegal transactions motivated by inside information.

There is here an obvious conflict of interest between the shareholders who were not informed of the delays and the insiders. The uninformed shareholders underwent big losses. Had they been informed of the delivery delays, they would also have sold their shares to avoid losses. They were misled by the managers of the firm they owned. Managers were bad agents in that case.

There were lawsuits against the managers of this airplane conglomerate. But it is always difficult to prove the existence of insider trading. There are not many cases of suits which give rise to accusations of insider trading, especially in Canada. At the time of writing, this insider trading case is still pending in court.

16.6 CONCLUSION

Much has to be said on insider trading. The regulatory aspects of this phenomenon are particularly controversial. Because of asymmetric information and moral hazard, the efficiency of insider trading regulation is questioned by many authors. It is like legislating on drugs. This legislation gives rise to a black market because the benefits of selling drugs are much higher than the costs. The benefits of insider trading encourage the portfolio managers to cheat because of asymmetric information. Not much can be done to stop that. Even if insider trading leads to big profits, how can these profits be attributed to inside information after the fact? How can it be proved that they were acquired by fraud?

We have also shown in this chapter that a stock option plan designed to decrease agency costs related to conflicts of interests between managers and shareholders may entail perverse effects. Those plans might encourage the managers to involve themselves in insider trading. They can dilute the capital of the existing shareholders. They can also increase the volatility of the stock issued by the firm in which these shareholders invest. Instead of decreasing agency costs, insider trading would thus inflate them.

Because it involves many disciplines, insider trading is therefore a difficult subject to study. Perhaps the unobservable aspects of insider trading might be better dealt with resorting to the econometrics of latent variables. The Kalman filter method would be a way to do so. On this subject, see Racicot and Théoret (2006, Chap. 21).

REFERENCES

Akerlof, G. 1970. The market for lemons: Qualitative uncertainty and the market mechanism. *Quarterly Journal of Economics* 84(3):488–500.

Arrow, K. J. 1963. Uncertainty and the welfare economics of medical care. *American Economic Review* 53(5):941–73.

Bhagat, S., and R. H. Jefferis. 2002. *The econometrics of corporate governance studies*. Cambridge, MA: MIT Press.

Coën, A., and F. E. Racicot. 2007. Capital asset pricing models revisited: Evidence from errors in variables. *Economics Letters* 95(3):443–50.

Coën, A., G. Mercier, and R. Théoret. 2004. *Traité de finance corporative*. Montreal: Presses de l'Université du Québec.

Copeland, T. E., and J. F. Weston. 1988. *Financial theory and corporate policy*. 3rd ed. Reading, MA: Addison Wesley.

Copeland, T. E., J. F. Weston, and K. Shastri. 2005. *Financial theory and corporate policy*. 4th ed. Reading, MA: Addison Wesley and Pearson.

Cornell, B., and E. Sirri. 1992. The reaction of investors and stock prices to insider trading. *Journal of Finance* 47(3):1031–59.

Dagenais, M. G., and D. L. Dagenais. 1997. Higher moment estimators for linear regression models with errors in the variables. *Journal of Econometrics* 76(1):193–221.

Daniel, K., M. Grinblatt, S. Titman, and R. Wermers. 1997. Measuring mutual fund performance with characteristic based benchmarks. *Journal of Finance* 52(3):1035–58.

De Matos, J. A. 2001. *Theoretical foundations of corporate finance*. Princeton, NJ: Princeton University Press.

Eckbo, B. E., and D. C. Smith. 1998. The conditional performance of insiders trades. *Journal of Finance* 53(2):467–98.

Jaffe, J. F. 1974. Special information and insider trading. *Journal of Business* 47(3):410–28.

Jeng, L. A., A. Metrick, and R. Zeckhauser. 1999. Estimating the returns of insider trading. Working paper, The Wharton School.

Jensen, M. C. 1986. Agency costs of free cash flow, corporate finance and takeovers. *American Economic Review* 76(2):323–29.

Masson, R. T., and A. Madhavan. 1991. Insider trading and the value of the firm. *Journal of Industrial Economics* 39(4):333–53.

Miller, M., and K. Rock. 1985. Dividend policy under asymmetric information. *Journal of Finance* 40(4):1031–51.

Modigliani, F., and M. H. Miller 1958. The cost of capital, corporate finance, and the theory of investment. *American Economic Review* 48(3):261–97.

Myers, S. C. 1984. The capital structure puzzle. *Journal of Finance* 39(3):575–92.

Padilla, E. 2002a. The regulation of insider trading as an agency problem. Working paper, University of Law, Economics and Science of Aix-Marseille.

Padilla, E. 2002b. Can agency theory justify the regulation of insider trading? *Quarterly Journal of Austrian Economics* 5(1):3–38.

Racicot, F. E. 2003. On measurement errors in economic and financial variables. In *Three essays on the analysis of economic and financial data*, chap. 3. PhD diss., ESG-UQAM.

Racicot, F. E., and R. Théoret. 2006. *Finance computationnelle et gestion des risques*. Montreal: Presses de l'Université du Québec.

Shleifer, A., and R. W. Vishny. 1997. A survey of corporate governance. *Journal of Finance* 52(2):737–83.

Théoret, R., and F. E. Racicot. 2007. Specification errors in financial models of returns: An application to hedge funds. *Journal of Wealth Management* 10(1):73–86.

CHAPTER 17

Insider Trading, News Releases, and Ownership Concentration*,†

Jana Fidrmuc, Marc Goergen, and Luc Renneboog

CONTENTS

* This chapter was published as Fidrmuc, J., Goergen, M., and L. Renneboog. 2006. Insider trading, news releases and ownership concentration. *Journal of Finance* 61(6):2931–73. Reprinted with permission, Wiley-Blackwell Publishing Ltd (Oxford).

† Acknowledgements: We thank the participants at the 2nd Forum on Corporate Governance organized by Humboldt University and Stanford University in Berlin (2003), the 2003 Workshop on Corporate Governance at the University of Vienna, the 2002 European Finance Association annual meetings in Berlin, the 2003 EEA/ESEM annual meetings in Stockholm, the 2004 IFS Frankfurt Summer School, and the 2004 IPEG Conference on Managerial Remuneration at the University of Manchester for very helpful comments and suggestions. We are also grateful to the participants at seminars at HEC (Paris), CUNEF (Madrid), Tilburg University, and Erasmus University. We are particularly indebted to editor Robert F. Stambaugh, Marco Becht, Arturo Bris, Robert Bushman, Johanna Federmütze, Uli Hege, Rez Kabir, Fred Palomino, Joachim Schwalbach, Henri Servaes, Myron Slovin, Marie Sushka, Greg Trojanowski, Chris Veld, Martin Walker, and an anonymous referee for suggestions that have helped to improve this chapter.

INSIDERS, THAT IS, MANAGERS and members of the board of directors of publicly traded corporations, usually possess more information about their company than do (small) outside shareholders. The main argument in favor of insider trading is that it communicates this superior information to outsiders. For instance, Leland (1992) shows that when insider trading is allowed, share prices incorporate more information, and are higher. However, while an insider purchase conveys positive information about the firm's prospects, it is less clear what information an insider sale conveys. On the one hand, an insider sale may convey unfavorable information about the firm's prospects. On the other hand, an insider sale may be less informative if it is made to meet the liquidity needs of the seller.

Seyhun (1986), Lin and Howe (1990), and Chang and Suk (1998) report positive abnormal returns on insider purchases for the United States. Similarly, several studies, such as Gregory, Matatko, and Tonks (1997), find positive abnormal returns for the United Kingdom over horizons of six to twelve months following directors' purchases.* A more recent U.K. study by Friederich, Gregory, Matatko, and Tonks (2002) on daily share prices corroborates these findings for short-term horizons. In this chapter we analyze the immediate market reaction to directors' transactions (excluding sales after the exercise of options) for companies listed on the London Stock Exchange during the 1990s. Consistent with the findings from

* Other papers that report positive abnormal returns include King and Röell (1988), Pope, Morris, and Peel (1990), and Gregory, Matatko, Tonks, and Purkis (1994).

previous studies, our results suggest that directors' trades convey new information on the firm's prospects.

An interesting aspect of the chapter is that we give a detailed account of both the U.K. and U.S. regulations on insider trading and directors' share dealings. There are marked differences between the two sets of regulations, for example, the regulations with respect to the definition of insiders and (illegal) insider trading, the main aspects of the regulation (e.g., the frequency of information releases and trading bans), the length of the period within which insiders must report their trades, and the level of the enforcement of the regulation. We conclude in our discussion on the regulatory differences between the two countries that directors' trades in the United Kingdom are likely to be more informative and hence trigger larger market reactions.

Our chapter makes two major contributions to the existing literature. First, the chapter is innovative in terms of the event study methodology we use in the context of insider trading. Specifically, we adjust the abnormal returns on insider trades for the release of news during the period preceding the trade and we examine whether the share price reactions to directors' trades remain significant if the trades follow news releases that relate to the firm's prospects, corporate restructuring, changes in capital structure, board restructuring, and other business events. We find that, in general, directors' transactions communicate new information to the market even if they are preceded by news releases. However, the informational content of trades is smaller when news on mergers and acquisitions (and to a lesser extent CEO replacement) precedes the trades. Indeed, in these cases, purchase transactions do not contain new information.

Second, when measuring the market reaction to directors' purchases and sales, we differentiate between the ownership of the directors who trade as well as the ownership held by outsiders. To the best of our knowledge, no other study explores the impact of the presence of different types of blockholders on the announcement effect of directors' transactions. We argue that the market takes into account all available public information—including director and outsider ownership—when reacting to insider transactions. As a result, directors' trades in firms with outside blockholders who monitor the firm may have relatively less informational value than directors' trades in widely held firms that may suffer from higher informational asymmetry. Our analysis therefore provides new evidence on the market's perception of ownership and control.

Our results confirm that the market takes into account the firm's ownership structure when reacting to directors' trades. The market reaction differs significantly depending on the degree of outsider ownership, director ownership, and the type of outsider ownership. In particular, firms controlled by other companies or by individuals or families unrelated to the directors experience significantly lower cumulative abnormal returns (CARs) in absolute value. This suggests that monitoring by these blockholders reduces informational asymmetry and ensures that the management focuses on value maximization, in which case directors' trades convey less information. In contrast, firms whose dominant shareholders are institutional investors have higher CARs on average. This suggests the higher information content of directors' transactions and confirms the findings of Franks, Mayer, and Renneboog (2001) and Faccio and Lasfer (2002), who argue that institutional shareholders in the United Kingdom do not monitor the firms in which they invest and do not mitigate problems of asymmetric information. Further, our evidence is consistent with institutional investors trading on the information signal conveyed by directors' trades. Interviews with fund managers in the City of London confirm that this is indeed the case.

Our results also demonstrate that the market takes into account director ownership when reacting to directors' trades. For firms with little director ownership, the CARs of directors' purchases are strongly positive, which is in line with the precommitment explanation. In contrast, for firms whose directors hold large stakes, the positive news that directors' purchases contain is mitigated by the danger of increased entrenchment. Similarly, the market reacts less negatively when directors with significant stakes sell, as this reduces their dominant position.

For poorly performing firms and those close to financial distress, we find stronger market reactions, with the reaction to directors' purchases (sales) significantly positive (negative) irrespective of the shareholder structure. We fail to find support for the information hierarchy hypothesis (Seyhun 1986). Although CEOs are assumed to have superior knowledge about their company's prospects, the information content of their trades is lower than that of other directors' trades. It is possible that CEOs, who may be subject to greater market scrutiny, trade more cautiously and at less informative moments.

The remainder of the chapter is organized as follows. The next section summarizes the U.K. regulation on directors' dealings and compares it to the U.S. regulation. Section 17.2 develops the hypotheses based on the existing literature. Section 17.3 describes the data and discusses the methodology. Section 17.4 analyses the results and Section 17.5 concludes.

17.1 U.K. AND U.S. REGULATION ON INSIDER TRADING

In the United States, insider trading is regulated by the Securities and Exchange Commission (SEC). The 1934 Securities and Exchange Act and its amendments impose restrictions on insider trading. In the United Kingdom, the 1977 Model Code of the London Stock Exchange (LSE) and the 1985 Companies Act regulate insider trading. There are major differences between the two sets of regulations in terms of (i) the definition of (illegal) insider trading, (ii) the essence of the regulation, (iii) the definition of an insider, (iv) the time within which insiders must report their trades, and (v) the level of the enforcement of the regulation.

The definitions of insider trading and directors' (share) dealings frequently cause confusion. Insider information, according to the U.K. Misuse of Information Act, is information that is "material, current, reliable and not available to the market" and legally qualified as "new and fresh." The Criminal Justice Act makes trading on insider information (information not regularly available and obtained through insiders) a legal offense. This chapter does not deal with illegal insider trading; rather, it focuses on legal trading by directors as defined in the listing rules of the LSE (Source Book August 2002, Chap. 16). Note, however, that while the U.K. code distinguishes between (illegal) insider trading and (legal) directors' dealings, the U.S. regulation does not make such a distinction. Throughout the chapter, we use the term directors' dealings to refer to the U.K. definition of (legal) insider trading or share transactions by directors. We also adopt the U.K. definition of a director, which refers to both nonexecutives and executives. In the United States, in contrast, executives are normally referred to as officers and nonexecutives as directors.

In general, the essence of U.S. rules on insider trading is that insiders must either abstain from trading on undisclosed information or release this information to the public before they trade (Hu and Noe 1997). The U.K. approach is different. In particular, the regulation contained in the 1977 Model Code of the London Stock Exchange, which became effective in April 1979,* and the 1985 Companies Act is stricter than the U.S. regulation (Hillier and Marshall 1998). The directors of companies traded on the LSE cannot trade during the two months preceding a preliminary,

* See Pope, Morris, and Peel (1990, 371).

final, or interim earnings announcement and the one month prior to a quarterly earnings announcement.* Moreover, outside the trading ban periods, directors must obtain clearance to trade from the board's chairman. In general, there are no such restrictions in the U.S. system,† which favors removing possible insider advantages via frequent disclosure rather than banning trades during price-sensitive times.

In the United States, insiders are defined as officers,‡ directors, other key employees, and shareholders holding more than 10 percent of any equity class (Lakonishok and Lee 2001). All these are prohibited from trading on undisclosed "material" information. The U.K. definition of insiders is narrower, including the members of the board of directors (both executives and nonexecutives), but excluding other key employees and large shareholders.

The period within which insider trades must be reported also differs substantially between the United Kingdom and the United States. The U.K. Model Code requires much faster reporting of directors' dealings. The directors must inform their company of the transaction as soon as possible and no later than the fifth business day after a transaction for their own account or on behalf of their spouses and children (Hillier and Marshall 2002). In turn, a company must inform the LSE of the transaction without delay and no later than the end of the business day following receipt of the information.§ Via its Regulatory News Service (RNS), the LSE then disseminates this information immediately to data vendors. The company must also enter this information into its register, which is available for public inspection, within three days of the director's report. In the

---

* In exceptional circumstances in which it is the only reasonable course of action available to a director, the director may be given clearance to sell (but not to purchase) when he or she would otherwise be prohibited from doing so.

† Lustgarten and Mande (1995) show that the volume of U.S. insider trading declines as an earnings announcement approaches but it does not decline to zero. It should be noted, however, that besides the federal regulation, a large fraction of U.S. firms impose additional insider trading restrictions on their directors and officers that in many cases also include trading bans (Bettis, Coles, and Lemon 2000). Further, the Sabanes–Oxley Act of 2002, effective since 2003, imposes insider trading bans during pension fund blackout periods.

‡ The term officer covers the company president, principal financial officer, principal accounting officer, any vice president in charge of any principal business unit, division, or function (such as sales, administration, or finance), and any other person who performs a policy-making function within the company (Bettis, Coles, and Lemon 2000).

§ This implies that information about an insider transaction can reach the market as late as six days after the transaction. However, in practice, this information is disclosed faster: for 85 percent of the directors' dealings in our sample the announcement day coincides with the transaction day or the following day.

United States, insiders only have to report their holdings within the first ten days of the month *following* the month of the trade (Persons 1997). The capital gains U.S. insiders make on short-term swings in prices (formally, within six months) must be repaid to the company. Insider transactions are published in the SEC online *Insider Trading Report.* Chang and Suk (1998) write that trades normally appear in the online report the same day that the SEC is informed. Shortly afterward, the information is published in the *Wall Street Journal* (WSJ) and other publications. Chang and Suk (1998) find that there is a significant share price reaction even after the announcement in the WSJ, suggesting that the SEC online report is read only by a small number of investors whereas the WSJ is read by a larger number of investors.* This implies that not only is the reporting process in the United States slower, but that it also takes time for the information contained in the insider trades to be reflected in the share price.†

The difference in the speed of reporting is also likely to have major implications for the size of the abnormal returns measured around the announcement of insider trading. Given that the period between the trading day and the announcement in the United Kingdom covers up to six days compared to up to forty days in the United States, we expect insider trades in the United Kingdom to be highly informative whereas insider trades in the United States are more likely to be based on stale information.

Although the regulation in the United Kingdom may be stricter than in the United States, what matters is its enforcement. According to Hillier and Marshall (1998), U.K. regulation is well enforced as insider trading is virtually nonexistent during the two-month period prior to the final and interim earnings announcements. Similarly, the regulator, the Financial Services Authority (FSA), argues that past and present regulation has been sufficiently strict and that there have been only a few violations of the trading

* Lakonishok and Lee (2001, 88) report that even after a trade has been reported it may still take several days for outsiders to obtain the information. Further, McConnell, Servaes, and Lins (2005) use a six-day event window for their cumulative abnormal returns because "the information usually does not enter the public domain for several days after it is filed with the SEC."

† Dedman (2004) reports further evidence that in the United States new information is only gradually reflected in stock prices, and in contrast to the United Kingdom, there is leakage of price-sensitive information. Reviewing the existing literature on the United Kingdom and the United States, she finds that, in the United States, stock prices start adjusting five days before a profit warning and that the adjustment takes up to five days after the warning. Conversely, there is no such leakage in the United Kingdom.

bans.* In addition, the Financial Services and Markets Act (FSMA) of 2000 (effective as of December 1, 2001) further refines the definitions of illegal insider trading† and specifies a dual prosecution track that facilitates the procedures to bring insider trading violations to court. Lack of disclosure, violation of trading bans, or misuse of inside information can be prosecuted under the Misuse of Information Act using either a civil law or a criminal law procedure.‡ Given that the new procedures have only recently been introduced and that investigations take time, there has been only one conviction since 2001 (via a civil court procedure), namely, that of Middlesmiss' Company Secretary,§ who traded equity prior to earnings announcements.¶,** In the United States, the Insider Trading and Securities Fraud Enforcement Act (ITSFEA) of 1988 raised the maximum fine for insider trading to $1 million and ten years of imprisonment in response to frequent violations of the existing insider regulation. The Act also placed the liability for illegal insider trading by any of the company's employees with the top management. Garfinkel (1997) documents that the Act has changed the timing of U.S. insiders, trades, with insider trading—especially selling—generally happening after, rather than before, earnings announcements. He also finds that the earnings surprise, that is, the difference between the actual earnings and the median analysts' forecast, has increased since the Act. He states that this "is consistent with less informed

* Based on interviews with several members of the FSA.

† "Any person who does act or engages in any course of conduct which creates a false or misleading impression as to the market in or the price or value of any relevant investments is guilty of an offence if he does so for the purpose of creating that impression and of thereby inducing another person to acquire, dispose of, subscribe for or underwrite those investments or to refrain from doing so or to exercise, or refrain from exercising, any rights conferred by those investments" (FSMA 2000, s. 397).

‡ In 2000, the LSE authority to impose administrative penalties was transferred to the FSA. The LSE passes any information raising the suspicion of insider trading on to the FSA for further investigation.

§ In the United Kingdom, the company secretary is responsible for keeping the minutes at board and general meetings and for generating the various records that must be kept at the registered office. He or she is also responsible for all formal administrative matters.

¶ The conviction occurred in February 2004. With regard to the current state of affairs, the FSA states that "several cases, a mixture of lack of disclosure, violation of trading ban periods and misuse of insider information, are currently being investigated and some of which will be brought to court via the civil or criminal procedure."

** For an alternative view, see Dedman (2004).

trading prior to earnings announcements during the post-Act period and the notion that informed trading encourages price discovery."*

To summarize, there are substantial differences between the U.K. and U.S. regulations on insider trading. The differences pertain to the definition of an insider, the essence of the regulation, the enforcement, and the window within which trades must be reported. We conclude that U.K. insider trades are likely to be more informative on the announcement day than are U.S. trades for the following reasons: (i) A trade must be made public within at most six business days in the United Kingdom, compared to up to forty days in the United States. (ii) Both Lakonishok and Lee (2001) and McConnell, Servaes and Lins (2005) report that the information on insiders' trades enters the public domain in the United States only several days after it is released by the SEC. We show below that no such delay occurs in the United Kingdom. (iii) In the United Kingdom, mandatory reporting by insiders is limited to top management (executive board members) and the nonexecutive directors only. In contrast, U.S. insiders (legally) comprise a much larger group, including large shareholders, (nonexecutive) directors, and managers (officers). The last includes not only the top management with board seats, but also a wider group of managers (e.g., any vice president in charge of any principal business unit, division, or function such as sales, administration, or finance), who may de facto possess less information about their firm's prospects. (iv) The U.K. regulator favors trading bans during price-sensitive periods whereas the U.S. regulator favors more frequent disclosure. All these elements suggest that directors' trades in the United Kingdom are more informative and hence may trigger larger market reactions.

17.2 LITERATURE REVIEW AND HYPOTHESES

The existing empirical literature uses two approaches to measure the effect of insider information on share prices. One strand of the literature argues that the price reaction to insider trading is gradual. This literature measures the price reaction via the CARs earned over the six to twelve months after the transaction. The existence of significant abnormal returns over this period is interpreted as proof of superior insider information (for example, Jaffe 1974; Rozeff and Zaman 1988; Lin and Howe 1990; Gregory, Matatko,

* Although Lakonishok and Lee's (2001) study covers the period before the Act (1975 to 1988) as well as the period after the Act (1989 to 1995), they do not report CARs separately for the two periods.

and Tonks 1997; Lakonishok and Lee 2001). The second strand of the literature assumes that stock markets are (to some degree at least) informationally efficient and that share prices adjust rapidly to insider trades. These studies measure the abnormal return on the date of announcement of the insider trade (Jaffe 1974; Chang and Suk 1998; Friederich et al. 2002). Our research pertains to this second strand of the literature.

We first test the benchmark hypothesis, which conjectures that directors trade on superior information (or at least, that the market believes that the directors trade on superior information). By purchasing shares in their firm, directors communicate a positive signal about the future value of the firm to the market. The signal is costly as the directors put their own wealth at stake and bear the cost of holding less than optimally diversified investment portfolios. Therefore, directors' purchases are credible signals to outsiders. Conversely, directors signal negative news when selling shares. However, the negative signal may be less informative as liquidity needs, rather than changes in expectations about the firm's future cash flows, may force the directors to sell shares (Lakonishok and Lee 2001; Friederich et al. 2002). Thus, given the mixed motivations for sales, we expect that the absolute value of the market reaction to sales is lower than that to purchases.

Hypothesis 1:

(a) The market reaction to the announcement of directors' purchases is positive.
(b) The market reaction to the announcement of directors' sales is negative.
(c) The absolute value of the market reaction to directors' sales is smaller than that to directors' purchases.

Next, we test the information hierarchy hypothesis, which postulates that the information content of the transactions depends on the type of director who trades (Seyhun 1986). According to this hypothesis, directors who are familiar with the day-to-day operations of the company trade on more valuable information. Seyhun (1986) and Lin and Howe (1990) partially confirm this hypothesis using U.S. data.* In particular,

* Seyhun (1986) measures the market reaction to insider trades by the CARs covering the first 50 and 100 days, respectively, following the day of the trade. Lin and Howe (1990) use six- and twelve-month CARs.

Seyhun shows that the cumulative average abnormal returns (CAARs) following the transactions of officers are significantly higher than those of nonexecutive directors, and Lin and Howe (1990) demonstrate that trades of chairmen, directors, officer-directors, and officers contain more information than those of large shareholders. In contrast, Jeng, Metrick, and Zeckhauser (1999) question whether insiders can benefit from their information advantage: "Some insiders are more 'inside' than others. The chief executive, for example, is likely to have better information about the firm's prospects than lesser officers. Since the CEO's trades are likely to be carefully scrutinized, both by shareholders and by regulators, he may be more reluctant to trade on his informational advantage. The net effect of these considerations on the profitability of insider trading is an empirical question." Jeng, Merrick, and Zeckhauser conclude that insiders benefit "handsomely" from their informational advantage, especially from their purchases. However, they do not find any support for the information hierarchy hypothesis, as they report that CEOs realize lower abnormal returns (though not significantly lower) than do other officers and directors.* Their explanation is that CEOs, who are more carefully scrutinized by market participants and regulators, may be more reluctant to trade on an informational advantage. Furthermore, the earlier support of the information hierarchy story that Seyhun (1986) and Lin and Howe (1990) document may be driven by transaction size. In these studies, CEOs' trades are twice as large, on average, as those by other officers or directors, and larger transactions trigger stronger price reactions.

Hypothesis 2:

The abnormal returns associated with purchases and sales depend on the type of director that makes the trade. The positive (negative) abnormal returns following purchases (sales) decrease in absolute value by category of director in the following order: CEO, other executive directors, nonexecutive chairman, and nonexecutive directors.

We proceed by relating informational asymmetries to ownership and control structures. Admati, Pfleiderer, and Zechner (1994) argue that holding a large stake in a firm encourages the owner to monitor the

* The results of Seyhun (1986) and Lin and Howe (1990) are not directly comparable to those of Jeng, Metrick, and Zeckhauser (1999) given the different methodologies they use in calculating the returns.

management. Similarly, Maug (1998) contends that corporations are more closely scrutinized by large shareholders who, given their size, have more incentives and sufficient voting power to intervene. This is what Maug calls the "lock-in effect." Further, Stoughton and Zechner (1998) theorize that large shareholders employ a monitoring technology that increases the expected value of the end-of-period cash flow distribution. Still, the use of this monitoring technology comes at a cost, such that the use of this technology pays off only for sufficiently large shareholders. Here we define monitoring as any activity that creates value that is shared by all shareholders in proportion to their holdings. As monitoring is inherently unobservable and small investors can free-ride on these activities, the incentive for monitoring must be a function of ownership.

Given that monitoring activities are likely to benefit all shareholders, the information asymmetry between management and shareholders is reduced. Hence, directors' dealings are likely to be less important a signal to the market in the presence of a large outsider. This implies that the absolute value of the announcement effect of directors' dealings is likely to be smaller in firms with major outside blockholders.

Holderness and Sheehan (1988) show that the ability and incentives of major shareholders to monitor management depend on their type. Most empirical studies distinguish between three categories, namely, corporations, institutional investors, and individuals or families not related to the management. U.K. institutional investors, such as banks, investment and pension funds, and insurance companies, are not assumed to monitor the companies in which they invest (see, for example, Franks, Mayer, and Renneboog 2001), as they do not usually have sufficient resources to monitor the (many) firms in which they invest. In addition, monitoring would provide them inside information, rendering their investments locked-in (Goergen and Renneboog 2001). Thus, only outsiders such as corporations and individuals or families unrelated to the management are expected to monitor the firms in which they invest.

Hypothesis 3:

(a) The announcement effect of directors' purchases and sales is weakened by the presence of an outside blockholder who monitors the firm (corporations, and individuals or families unrelated to the directors).

(b) The announcement effect of directors' purchases and sales is not influenced by the presence of an institutional blockholder.

Directors not only have direct access to restricted information, but they also have different incentives as compared to outside blockholders (Holderness and Sheehan 1988). For directors, the performance of their shares may be of secondary importance if they derive substantial private benefits of control from their positions in the firm, where these private benefits are not transferable but, rather, are investor specific. For director–owners, such benefits may consist of above-market-rate salaries, perquisites, and prestige or reputation effects (Johnson et al. 2000; Dyck and Zingales 2004; Holmen and Högfeldt 2005). At low levels of control benefits, director ownership may align their incentives with those of the other shareholders (Jensen and Meckling 1976), and increases in ownership may reflect the directors' precommitment to focus on shareholder value creation. Therefore, in widely held firms, the precommitment effect of directors' purchases (sales) may lead to a stronger market reaction. However, at higher levels of director ownership, their purchases may lead to entrenchment such that they may become insulated from disciplinary actions in the case of poor performance (Morck, Shleifer, and Vishny 1988); Franks, Mayer, and Renneboog 2001). Consequently, while the market may react positively to an increase in a director's holding that results in a modest stake, it may respond negatively to a director's purchase if his or her ownership is already substantial. The negative effect of increased entrenchment may even dominate the otherwise positive signal about the firm's prospects.* Similarly, the market may react positively to a director's sale if it views the benefits from reduced managerial discretion as greater than the negative signal.

The entrenchment problem in firms with high director ownership may be less prominent if a large monitoring (outside) shareholder reduces the otherwise high managerial discretion. Conversely, the entrenchment problem may not be reduced by a passive shareholder, that is, an institutional investor.

* The entrenchment effect refers to the fact that directors with substantial voting power may become unaccountable and/or exploit their private benefits at the expense of other shareholders. There is evidence that entrenchment frequently occurs in the United Kingdom. Lai and Sudarsanam (1997), Franks, Mayer, and Renneboog (2001), and Faccio and Lasfer (2000) show that directors with substantial voting power cannot be ousted even in the wake of poor performance.

Hypothesis 4:

(a) In firms with strong director control and without other major shareholders, the positive announcement effect of directors' purchases is weaker when the purchases increase the directors' entrenchment. Likewise, in such firms, the negative announcement effect of directors' sales is weaker when the sales erode the directors' entrenchment.
(b) The market is more concerned about director entrenchment in firms with large share blocks held by both directors and institutional investors, in which case, purchases (sales) trigger a weaker positive (negative) announcement effect. In contrast, the market is less concerned about entrenchment in firms with outside blockholders (families or corporations) that monitor, in which case the announcement effect is stronger.

For poorly performing or financially distressed firms, the probability of insolvency is such that the market awaits new information on the firm's prospects more eagerly. Therefore, we expect a stronger market reaction to directors' transactions. Moreover, given that the costs of an incorrect signal are far more substantial to the directors of such firms, the signal is also more credible to the market. Hence, if directors buy more shares in a poorly performing firm, then the market reaction should be significantly more positive. If directors of poorly performing or financially distressed firms sell shares, this may reflect their loss of confidence in the firm. The CAARs are thus expected to be strongly negative irrespective of the ownership structure.

Hypothesis 5:

For poorly performing or financially distressed companies, directors' purchases and sales trigger stronger announcement reactions.

17.3 DATA SOURCES, DESCRIPTIVE STATISTICS, AND METHODOLOGY

17.3.1 Data Sources

Our data cover directors' dealings, ownership, daily returns, company-specific information such as capital structure changes, number of shares outstanding, industry, accounting data, and news items.

Directors' dealings data cover the period from 1991 to 1998 and come from Hemmington Scott (HS). The original file contains 58,363 entries and includes information on company names, directors' names, directors' shareholdings, directors' positions on the board, transaction and announcement dates, number of shares traded, prices, security types (ninety different types),* and transaction types (twelve different types).† The exclusion of directors' trades in financial firms, duplicate entries, and inaccurate or incomplete transactions records reduces the number of observations by roughly 40 percent.‡ We aggregate multiple purchases (or sales) by the same director on a given day (e.g., we view one sale of 10,000 shares and another of 5,000 shares as one sale of 15,000 shares). Furthermore, when a director purchases and sells shares on the same day, we net the transactions (e.g., we view a purchase of 10,000 shares and a sale of 5,000 shares as a net purchase of 5,000 shares). Following these adjustments, the sample covers 35,439 directors' transactions with respect to 1,498 firms.

The most frequent transactions relate to ordinary shares and the exercise of options, representing 27,416 trades (78 percent of all insider transactions) and 5,885 trades (17 percent), respectively. As the market is likely to ignore very small transactions, we only retain the (net) transactions involving at least 0.1 percent of the shares outstanding. Furthermore, as sales after the exercise of options are likely to be related to the directors' remuneration packages and whether the options are in the money, we expect their information content to be low. Hence, we exclude these sales. These rules eliminate 83 percent of all purchases on ordinary shares (12,019 out of 14,500) and 61 percent of all sales (4,101 out of a total of 6,769 transactions). We analyze the transactions with respect to their

* The ninety security types include ordinary shares, restricted voting shares, options, warrants, and convertibles. The full list of security types is available upon request.

† Transaction types consist of buy, sell, exercise, options granted, post-exercise sale, take up, scrip dividend, inherited, bed & breakfast, gift given, gift recorded, and scrip issue.

‡ The main reason for the reduction in the number of observations is the deletion of duplicate information that results from the fact that directors' transactions are collected from various sources (Regulatory News Service, Reuters, Thomson Financial, and LexisNexis). The number of errors refers to entries for which no code indicates whether the transaction is a sale, post-exercise sale, etc., or to typographical errors in the codes. The number of such omissions is very limited, representing only 253 of 58,363 entries.

relative rather than absolute value because our chapter focuses on relative voting power and changes in control.*

We trace changes in company names using the London Share Price Database (LSPD), which also provides information on the SEDOL number, birth and death dates, and reason for delisting. The number of shares outstanding for each firm-year and the industry code from the LSPD match with the directors' dealings file. We use the number of shares outstanding to calculate the relative size of each transaction.

We obtain ownership data from Worldscope, which records all direct ownership stakes of 5 percent or more of the ordinary shares outstanding. We classify these stakes according to their owner: directors (insiders), corporations, institutional investors, and individuals or families not related to the directors. We use the Stock Exchange Yearbooks to verify whether the individuals reported in the database (around 7,400 persons) are (i) a CEO, (ii) another executive director, (iii) a nonexecutive chairman, (iv) another nonexecutive director, (v) a former director who has recently left the company, or (vi) an individual who is neither a director nor related to a director. The equity stakes held by direct family members (spouses, children, parents) of the directors are added to the ownership stakes of the last category.

We obtain adjusted daily prices, dividends, data on the FTSE All Share Index, market capitalization, after-tax earnings, return on equity, book-to-market, debt-to-equity, and interest coverage from Datastream.

17.3.2 Descriptive Statistics

Table 17.1 reports the summary statistics on our sample trades. Panel A shows the statistics for all the trades (including those resulting from the exercise of options, which we exclude later on). These statistics are directly

* A threshold based on relative size has the disadvantage that the value of the threshold (0.1 percent of market capitalization) varies from company to company. In value terms, our threshold of 0.1 percent amounts to GBP 14,616 (GBP 63,626) for the median (average) purchase transaction, while it amounts to GBP 31,908 (GBP 107,433) for the median (average) sale transaction. Still, a threshold based on absolute transaction value (e.g., GBP 25,000) is also arbitrary. Moreover, the absolute size of the transaction is more likely to be dependent on the director's wealth rather than on company-specific characteristics. The absolute size of the transaction would necessarily have to be standardized by some benchmark of the director's wealth (e.g., by the value of the director's remuneration package). Yet another alternative threshold could be based on the transaction size expressed as a percentage of the director's existing ownership stake. However, the signal emitted by the director's transaction depends on how the relative transaction size relates to the distribution of voting power of the outside blockholders, which would not be captured by such a threshold.

comparable to those of Lakonishok and Lee (2001) for the United States. Panel B shows the statistics on the transaction sizes of net purchases and net sales, respectively, which represent at least 0.1 percent of the market capitalization of a firm. Panel C shows the ownership structure of the firms. According to Panel A, directors of U.K. firms trade less frequently than their U.S. counterparts. Each year, on average, there are only 1.49 (1.09) purchases (sales) per U.K. firm compared to 2.77 (4.74) purchases (sales) per U.S. firm.* We believe that the lower trading activity of U.K. directors compared to U.S. directors is due to the stricter regulation (trading bans) in the United Kingdom. Furthermore, the higher frequency of directors' sales in the United States could be because American directors are awarded more stock options than are their British counterparts (Conyon and Murphy 2000) and sales after the exercise of options are not treated as a separate category in most U.S. studies.

Panel B shows that, on average, directors' purchases are smaller than their sales. The median net purchase is £36,000 compared to £147,000 for the median net sale. The median net purchase (sale) as a proportion of market capitalization amounts to 0.27 percent (0.48 percent). CEOs and chairmen are the most active traders, accounting for 582 and 492 (490 and 350) purchases

---

* Directors' trading activity, measured by the total number of shares traded per firm-year (not shown in the table), increased throughout the beginning of the period, peaked in 1996, and decreased thereafter. During the sample period, U.K. directors sold only two to three times as many shares as they bought compared to seven times for U.S. directors.

TABLE 17.1. Summary statistics for directors' trades, from 1991 to 1998

Panel A: U.K. Sample description (comparable to Lakonishok and Lee, 2001, for directors in United States)

	All	Purchases	Sales	Sales post exercise	Sales and sales post exercise	Exercise
Fraction	0.71	0.51	0.33	0.17	0.50	0.24
No. of trades	4.26	1.49	0.69	0.40	1.09	0.59
% Market capitalization	0.69%	0.24%	0.46%	0.09%	0.48%	0.14%
Number of firms	1,492	1,385	1,119	690	1,203	837

Panel B: Directors' large trades (>0.1% of market capitalization)

	Mean	Median	Minimum	Maximum
Net purchases (2,188 trades)				
Trade value (£'000)	1,076	36	0.019	1,590,000
% Market capitalization	0.96%	0.27%	0.10%	77.45%
% Market capitalization by category of director				
CEO (582 trades)	1.04%	0.31%	0.10%	77.45%
Other top executives (112 trades)	1.29%	0.28%	0.10%	44.29%
Chairman (492 trades)	1.30%	0.36%	0.10%	52.27%
Other incumbent directors (606 trades)	1.34%	0.29%	0.10%	77.45%
Former directors (396 trades)	1.51%	0.31%	0.10%	77.45%
Net sales (2,347 trades)				
Trade value (£'000)	1,305	147	0.032	81,300
% Market capitalization	1.38%	0.48%	0.10%	39.05%
% Market capitalization by category of director				
CEO (490 trades)	1.85%	0.82%	0.10%	18.47%
Other top executives (115 trades)	1.58%	0.54%	0.11%	14.43%
Chairman (350 trades)	2.07%	0.69%	0.10%	39.05%

Other incumbent directors (766 trades)	1.29%	0.46%	0.10%	39.05%
Former directors (626 trades)	1.55%	0.51%	0.10%	23.62%

Panel C: Ownership structure

	% of firms with share stake > 5%	Mean % of ownership	Min. % of ownership	Max. % of ownership	Median % of ownership
Firms with net purchases					
All blockholders with more than 5%	69.1%	41.6%	2.6%	97.0%	40.9%
All outsiders	65.1%	28.2%	2.0%	97.0%	26.0%
Corporations	17.9%	14.5%	1.5%	76.3%	8.7%
Institutional investors	60.0%	22.6%	2.0%	81.8%	19.8%
Individual outsiders	23.8%	9.2%	1.1%	34.2%	7.5%
All directors	42.0%	24.8%	1.0%	77.6%	18.2%
Firms with net sales					
All blockholders with more than 5%	66.1%	34.5%	2.0%	89.0%	32.2%
All outsiders	62.0%	22.0%	1.1%	65.6%	18.6%
Corporations	13.8%	13.6%	0.9%	50.0%	8.5%
Institutional investors	55.6%	18.2%	1.1%	62.3%	16.4%
Individual outsiders	16.2%	10.0%	0.9%	34.7%	7.0%
All directors	39.9%	23.0%	1.1%	88.0%	16.5%

Panel A shows the summary statistics for all directors' trades independent of the trades' size. This panel is comparable to Table 17.1 of Lakonishok and Lee (2001). Fraction is the fraction of firms with at least one trade. No. of trades is the average number of trades per firm per year. % Market capitalization is the average across firms of the number of shares traded by the directors over the number of shares outstanding at the beginning of the year. Panels B and C refer to all purchases and sales of U.K. directors during the 1991 to 1998 period that represent at least 0.1% of a company's market capitalization. Trade value is total number of shares traded by the directors times the share price at the beginning of the year. CEO, Other top executives, Chairman and Other incumbent directors represent the dealings of the CEO/managing director, the deputy CEOs/deputy managing directors/financial directors, the board's chairman, and those of all incumbent directors that are neither executives nor chairmen, respectively. Former directors' dealings are traced up to two months beyond the year in which they left.

(sales), respectively. Former directors who have recently left the company sell more frequently than they purchase (626 versus 396 transactions).*

Directors sell more shares than they purchase in larger firms, more profitable firms, and those with less debt and lower book-to-market ratios (not reported in the table). According to Friederich et al. (2002), directors purchase stock when they believe it is undervalued (as measured by a high book-to-market ratio). Panel C of Table 17.1 reports the ownership structure, measured at the beginning of the transaction year, for firms with net purchases and for those with net sales, respectively: 69 percent of the firms with net purchases and 66.1 percent of the firms with net sales have a blockholder, that is, a shareholder that owns more than 5 percent of the equity. In the firms with blockholders, the outside blockholders jointly hold on average 28.2 percent of the equity in firms whose directors purchase shares, while they control 22.0 percent of the equity in firms whose directors are net sellers. Institutional investors are blockholders in the majority of firms (in 60.0 and 55.6 percent of the firms with net purchases and net sales, respectively), but their blockholdings are more modest as

* Hemmington Scott obtains the list of former directors from the directors' reports in the annual reports. For a firm whose financial year runs from, for example, April to March, the financial-year report for 2001 will list the former directors who were directors for some time during the period of April 2000 to March 2001 but who left at or before the end of March 2001 (for most U.K. firms, the financial year ends in March). The report also shows the former directors' shareholdings and whether there was a change in their equity stake. The trades of these directors are then recorded for the period starting with their departure and ending up to two months after the end of the financial year (in our example, until the end of May 2001) either directly from the directors' report or from the Jordan's database, which is based on information obtained from the company registers. This means that for a director who resigns on March 31, 2001, the trades are recorded for only two months after the director left the firm, whereas for those leaving early in the financial year (e.g., April 2000), the trades may be recorded for more than one year. We estimate for how many months the trades of former directors are traced using the director turnover data of Franks, Mayer, and Renneboog (2001). The vast majority of natural turnover is related to retirement (with a few cases of departure due to illness or death), and most of the retiring directors stay until the end of the financial year. Hence, the average departure date for this type of turnover is the final month of the financial year. This means that, for about 70 percent of the directors who leave, the trades are traced for only two months after their departure. For "conflictual" turnover, the average retirement date lies near the end of month nine. For this category, around one-third (10 percent of the total) leave in the final month of the financial year as this category of turnover also includes the departure of directors reaching the end of their (nonrenewed) contract. Only slightly less than one-fifth of directors leave at various times throughout the year. To conclude, the transactions of 80 percent (88 percent) of all former directors are followed for two (three) months. The transactions of only a small minority of former directors are traced over a longer timeframe. Therefore, it is not implausible that these former directors may be considered "insiders."

they hold 22.6 percent (18.2 percent) on average in firms with net purchases (net sales). Directors are the largest shareholders. Jointly, they own on average around 24.8 percent (23.0 percent) in firms with net directors' purchases (sales). Individuals or families unrelated to the management hold only around 9 percent (10 percent) in firms with net purchases (sales), compared to about 15 percent (14 percent) for corporations.

17.3.3 Methodology

We compute the cumulative (average) abnormal returns (C(A)ARs) by using the market model for a period of forty-one days centered on the announcement day. The market return is proxied by the FTSE All Share Index excluding investment trusts, and the beta is estimated over a period of 200 to 21 days prior to the event day. To verify the robustness of the results, we also calculate market-adjusted returns. Several studies (e.g., Rozeff and Zaman 1988 for the United States; Gregory et al. 1994 for the United Kingdom) highlight the importance of controlling for size when calculating abnormal returns over a long post-event window or when the sample includes a large number of smaller companies.* We use the same size-adjustment method as in Lakonishok, Shleifer, and Vishny (1994), forming ten size portfolios based on market capitalization at the beginning of the calendar year and calculating the equally weighted average return for each portfolio. Each return $R_{i,t}$ is adjusted by return $R_{p(i),t}$ earned on the size portfolio p to which security i belongs.† To test the null hypothesis that the CAARs are equal to zero for a sample of N securities, we use three parametric test statistics: $t_{CAAR,}$ based on Barber and Lyon (1997), and J_1 and $J_{2,}$ both based on Campbell, Lo, and McKinley (1997).‡ We also

* Rozeff and Zaman (1988) argue that abnormal returns are higher for smaller companies. If directors' purchases are concentrated in smaller firms, and if their shares tend to earn positive abnormal returns, then the abnormal returns on directors' trades may be partly attributable to the size effect.

† An alternative method would be the Dimson and Marsh (1986) method that uses betas obtained from size portfolios. However, Gregory, Matatko, and Tonks (1997) report that the difference between the Dimson–Marsh benchmark and the Lakonishok, Shleifer, and Vishny (1994) benchmark is relatively small for U.K. data.

‡ If the true abnormal return is larger for securities with higher variance, then the test statistic should give equal weight to the realized cumulative abnormal returns of each security, which is what J_1 does. If the true abnormal return is constant across securities, then the test statistic should give more weight to the securities with the lower abnormal return variance, which is what J_2 does. In most studies the results are not likely to be sensitive to the choice of J_1 versus J_2 because the variance of the CAR is of a similar magnitude across securities (Campbell, Lo, and MacKinlay 1997, 162).

use Corrado's (1989) nonparametric rank test statistic. We discuss further details on these test statistics as well as information on how we deal with nonsynchronous trading and event clustering in the Appendix.

At a later stage, we also adjust the CARs for the possible release of news prior to a director's transaction.* It is important to make such an adjustment given the nature of the regulation in the United Kingdom. Directors are allowed to trade only after (and not before) the release of corporate information. If a news release precedes a trade, this may influence the market reaction to the trade. Hence, we need to ascertain whether the significant CAR is due to the signal of the director's transaction or, rather, to the release of price-sensitive corporate news. Also, U.K. rules on insider trading proscribe trading prior to earnings announcements. For preliminary, interim, and final earnings announcements, the period during which directors must not trade is as long as two months. In our regression models (see below), we correct for different types of news, for instance, board restructuring, asset restructuring, changes in the capital structure, and earnings announcements. This process allows us to determine whether the market reacts to the news or to the directors' transactions.

We obtain information on news items from two sources. The first source is the RNS of the LSE (information relating to mergers and acquisitions, or M&As, legal disclosure requirements, changes to the board, corporate restructurings, etc.). The second source consists of annual reports, preliminary results, and other corporate announcements as well as the analysis of this information by brokers, journalists, or analysts as covered by Thomson Financial, LexisNexis, Reuters, Bloomberg, and Jordan's Database. After eliminating duplicate news items (e.g., when Reuters disseminates the exact text from the RNS announcements), we categorize these news items into the following classes. The first class of news items relates to changes to the board of directors and/or the audit firm/corporate advisors, namely, (i) a change in the CEO, (ii) the departure/appointment of nonexecutive directors, (iii) the replacement of an executive director (excluding the CEO), and (iv) a change in the firm's advisors such as the auditors, solicitors, registrars, financial advisers, or stockbrokers. The second class covers news relating to corporate and capital restructuring, that is, (i) M&As, (ii) a disposal of a major part of the business, a division, or important assets, (iii) a share repurchase, and (iv) a change in equity capital (including a new stock issue to pay off existing

* We are grateful to the referee for suggesting further analysis along these lines.

debt). A third class covers news on the outlook of the firm, prospects, and other business events, specifically, (i) a forward-looking statement about the company's performance, and (ii) a business event containing any news item that is deemed to be price sensitive but not falling into any of the preceding categories (e.g., a name change, the signing of a new contract, a product launch, a change in accounting policy, a debt rollover, a move to the alternative investment market (AIM), and a change of sector). Given that the archives of the RNS have only been available since 1995, we can only adjust the abnormal returns for news releases as of this year. We collect a total of 15,138 news releases over the four-year period.

Panel A of Table 17.2 indicates that about 27 percent of all news items relate to changes in the board of directors or to the advisors of the firm. Almost 14 percent of the information is related to corporate restructurings such as M&As, the acquisition of a minority stake, the acquisition of a division of another firm, the creation of a joint venture, and so forth. About 4 percent of news items relate to asset disposals and 6 percent to changes in the capital structure. The bulk of the information releases (about 35 percent) relates to information on the firms' prospects.

Panel B of Table 17.2 reports the incidence of trades that are preceded by news items. Of a total of 1,444 purchases, 457 purchases are preceded by news releases during the thirty trading days before the trade, and 251 purchases are preceded by news releases within the week prior to the trade.* Of the 457 purchases preceded by news, 97 are preceded by two news releases whereas 109 are preceded by more than two news releases. All of this suggests that it may be important to correct the CARs for news releases prior to the trade. We make this correction in Section 17.4.5. Panel B also shows whether different types of news items trigger directors' purchases in the periods of two, seven, and thirty days subsequent to the news release, respectively. Overall, we find that there are not many firms with purchases after the release of new information. For instance, in only 2.8 percent of the firms, one or more directors purchase shares subsequent to news on the departure or appointment of a CEO. Even after major asset restructuring (M&A activity, asset disposals), directors' trading remains modest, as in only about 6 percent of the firms these news items trigger a purchase within a week. Purchases are most frequent after news releases that cover the firm's prospects.

* Data on news releases are available only for the 1995 to 1998 period. A table similar to Table 2, relating news items to directors' sales, is available upon request.

TABLE 17.2. Frequency of each type of news announcement

Panel A: Occurrence of news items by type of information content

	1995	1996	1997	1998	Average % for 1995–98
News on changes to the board/advisers					
CEOs	68	80	117	143	2.7%
Executive directors (excluding CEOs)	293	430	641	669	13.4%
Nonexecutive directors	229	327	408	503	9.7%
Advisors, auditors, solicitors	27	38	66	84	1.4%
News on corporate/equity restructuring					
Mergers, acquisitions, joint ventures	228	391	618	870	13.9%
Asset disposals	54	84	227	224	3.9%
Changes in capital structure, seasoned equity offerings, and share buybacks	47	107	243	449	5.6%
News on prospects/other business event					
Prospects	1,123	1,303	1,404	1,461	35.0%
Other business event	127	291	788	976	14.4%
Total	**2,196**	**3,051**	**4,512**	**5,379**	**100%**

Panel B: Incidence of purchases after news releases (1995–1998)

Number of purchases preceded by:		**2 days**	**7 days**	**30 days**
A news release during this period		178	251	457
No news release during this period		1,266	1,193	987
Number of firms with a news item on:		**2 days**	**7 days**	**30 days**
CEO change	215			
Subsample of firms with purchases		6 (2.8%)	6 (2.8%)	11 (5.1%)

Executive director change	464			
Subsample of firms with purchases		17 (3.7%)	30 (6.5%)	68 (14.7%)
Nonexecutive director change	429			
Subsample of firms with purchases		17 (4.0%)	17 (4.0%)	40 (9.3%)
M&A	393			
Subsample of firms with purchases		11 (2.8%)	22 (5.6%)	38 (9.7%)
Asset disposals	268			
Subsample of firms with purchases		2 (0.7%)	2 (0.7%)	6 (2.2%)
Capital structure change	321			
Subsample of firms with purchases		9 (2.8%)	13 (4.1%)	31 (9.7%)
Firm's prospects	513			
Subsample of firms with purchases:		75 (14.6%)	93 (18.1%)	144 (28.1%)
Business event	417			
Subsample of firms with purchases		12 (2.9%)	20 (4.8%)	45 (10.8%)

Panel A shows the incidence of news announcements by category and by year. Prospects is a forward-looking statement on the company's performance. Business event contains news that is deemed to be price sensitive but not included in any of the preceding categories; that is, it relates to a name change, the signing of a new contract, a product launch, a change in accounting policy, a debt rollover, a move to the AIM market, and a change of sector. Panel B shows the incidence of purchases subsequent to news releases.

17.4 RESULTS

We start this section by presenting the CAARs triggered by purchases and sales. As a robustness check we use different measures for the CAARs. We also contrast the market impact of large versus small trades and explain why the CAARs in the United Kingdom are larger than in the United States. We then test the information hierarchy hypothesis in Subsection 17.4.2 and outline the impact of the presence of different types of blockholders in Subsection 17.4.3. In Subsection 17.4.4, we investigate the value of the signal under poor performance and financial distress. We adjust our models in Subsection 17.4.5 for the release of potentially price-sensitive news prior to a trade to determine whether the CARs are caused by the trades or by news releases. Finally, we check our results for the possible effect of thin trading.

17.4.1 The Market Reaction to Directors' Trades

Table 17.3 reports the market reaction to purchases and sales. The table consists of three different panels. Panel A reports the CAARs for large trades, that is, those exceeding 0.1 percent of the firm's market capitalization, Panel B shows the CAARs for all the trades irrespective of their size, and Panel C documents the CAARs for small trades. In the following subsections, we focus on the effect of large trades only. However, we report the figures for all trades and for small trades to allow for a direct comparison of our results with those from U.S. studies.

The results in Table 17.3 strongly support our benchmark hypothesis, Hypothesis 1(a), which states that there is a strong positive market reaction to directors' purchases given their high informational content. For example, Panel A shows that for large trades, the two-day CAAR based on the announcement day and the following day from the market model is 3.1 percent and strongly significantly different from zero regardless of the test statistic used.* Conversely, the CAAR is significantly negative (–1.27 percent) over the 20 days prior to purchases. This suggests that directors are able to time their purchases.

Panel A also shows that the market reacts negatively to the announcements of large sales. The CAR measured over the announcement day and the following day is –0.37 percent and is significantly different from zero. The positive CAAR follows a period of positive abnormal returns of about

* The two-day CAAR from the market-adjusted model is 2.9 percent and that from the size-adjusted model is 2.9 percent. Both are significantly different from zero.

TABLE 17.3. Market reaction to directors' transactions around the announcement day

	CAAR (−20;−1)	CAAR (0;1)	CAAR (0;4)
Panel A: Large trades (>0.1%)			
Large purchases (1,861 trades)			
CAAR	−1.27%	3.12%	4.62%
t_{CAAR}	−2.66	14.84	17.14
J_1	−3.63	28.29	26.46
J_2	−11.81	41.30	39.54
t_{rank}	−2.50	9.17	8.89
Large sales (2,004 trades)			
CAAR	3.07%	−0.37%	−0.53%
t_{CAAR}	8.68	−4.69	−4.51
J_1	14.38	−5.42	−5.01
J_2	22.74	−7.01	−6.16
t_{rank}	7.58	−4.92	−3.95
Panel B: All trades			
	CAAR (−20;−1)	**CAAR (0;1)**	**CAAR (0;4)**
All purchases (10,140 trades)			
CAAR	−2.01%	1.16%	1.65%
t_{CAAR}	−13.38	20.78	21.95
J_1	−18.71	34.15	30.73
J_2	−35.21	42.21	39.15
t_{rank}	−6.73	7.65	7.15
All sales (5,523 trades)			
CAAR	2.29%	−0.26%	−0.49%
t_{CAAR}	13.54	−6.05	−7.96
J_1	20.89	−7.38	−9.01
J_2	29.98	−8.23	−10.94
t_{rank}	8.96	−4.51	−4.98
Panel C: Small trades (<0.1%)			
	CAAR (−20;−1)	**CAAR (0;1)**	**CAAR (0;4)**
Small purchases (8,378 trades)			
CAAR	−2.18%	0.79%	1.07%
t_{CAAR}	−14.30	15.62	15.46
J_1	−20.74	23.82	20.38
J_2	−34.14	28.93	25.93
t_{rank}	−7.56	6.52	5.89
Small sales (3,519 trades)			
CAAR (Market model)	1.84%	−0.25%	−0.55%
t_{CAAR}	10.81	−6.59	−8.59
J_1	15.62	−6.87	−9.30
J_2	20.50	−7.17	−10.47
t_{rank}	8.29	−3.96	−5.14

This table reports the CAARs for directors' purchases and sales for three intervals around the announcement day of the transactions. Panel A covers the trades of at least 0.1% of a company's market capitalization, Panel B covers all the trades irrespective of transaction size, and Panel C reports the abnormal returns of the trades that represent less than 0.1% of market cap. The β_i values are estimated over the (−200;−21)-day window. The test statistics are described in the Appendix and in the methodology section.

3 percent over the twenty days preceding the announcement. As with purchases, directors seem to be able to time their sales very well. We conclude that directors' sales are also information-revealing events, interpreted as negative news. Hence, we fail to reject Hypothesis 1(b). Our results also confirm Hypothesis 1(c), namely, that the absolute market reaction to directors' purchases is larger than that to sales. This is in line with Jeng, Metrick, and Zeckhauser (1999) and Lakonishok and Lee (2001) for U.S. firms, and Friederich et al. (2002) for U.K. firms. For instance, Lakonishok and Lee (2001) report that insider purchases trigger four times larger abnormal returns than do sales. Similarly, for the longer run, Jeng, Metrick, and Zeckhauser (1999) show that purchases yield significantly higher returns than do sales. The reason for this pattern may be that markets attach less informational content to sales because part of the sales may be made due to directors' liquidity needs rather than bad insider news.

The abnormal returns in Panel A of Table 17.3 refer to large transactions only. As we mention above, to facilitate a direct comparison of our results with most U.S. studies, which consider all transactions irrespective of size, Panel B and Panel C present the CAARs for all trades and for the subset of small transactions, respectively. The announcement effect for all purchases is only about one-third of that for large purchases, while the CAAR for all sales is 30 percent smaller than that for large sales. Comparing the U.K. results of Panel B with the U.S. results of Lakonishok and Lee (2001, Table 17.3), the U.K. abnormal returns in absolute terms are three times as high as the U.S. analogs. Over the five-day window the U.K. CAAR for all purchases (sales) is 1.65 percent (–0.49 percent) compared to a U.S. CAAR of only 0.59 percent (0.13 percent).* Finally, in Panel C we observe that the announcement effects of small trades in our sample are much smaller than for large transactions. The abnormal announcement returns triggered by purchases amount to only one-quarter (0.79 percent) of those for large trades and the sales-related returns are one-third smaller (–0.25 percent).

An important question that arises when we compare our results with those from U.S. studies is why the U.K. CAARs are so much higher. Above, we give one explanation, that is, regulation and the speedier reporting of trades in the United Kingdom compared to the United States. In the United

* While Lakonishok and Lee do not report the statistical significance of their findings, they mention in a footnote that "most abnormal returns are significantly different from zero." Still, they suggest that their results are not "economically significant."

Kingdom, directors' transactions are known to the market within six days of the transaction (see Section 17.1). In most cases, the market already knows about a trade within one or two days. We find that for more than 85 percent of our transactions the announcement day coincides or immediately follows the transaction date, and that the information is immediately in the public domain via the RNS, Reuters, and Bloomberg.* In the United States, directors' trades are announced at the earliest ten days and at the latest forty days after the transaction. Furthermore, Lakonishok and Lee (2001) and McConnell, Servaes, and Lins (2005) argue that even after a trade has been reported, it takes several days for outsiders to become aware of it. We calculate the abnormal returns for both the trades whose transaction and announcement dates coincide (36 percent of the purchases and 41 percent of the sales) and the trades whose transaction date precedes the announcement date in order to examine whether reporting speed per se really matters. For the purchases for which the announcement and transaction dates coincide, the CAAR (0;1) is 3.9 percent as compared to only 2.7 percent for the purchases for which the two dates differ. As this difference is statistically significant, it seems that the reporting speed matters for purchases, which suggests that the informational value of insider purchases diminishes over time. Conversely, the difference in the CAAR for sales is not statistically significant.

A second reason why the U.K. CAARs may be higher than those in the United States is that U.K. directors are not allowed to trade over periods that may cover up to six months (prior to earnings announcements; see above), whereas U.S. regulation does not impose such trading bans. As U.K. directors trade less frequently, their transactions may contain more information.

Third, the definition of insiders in the United States is different from that in the United Kingdom. U.S. insiders are officers (comparable to U.K. executives), directors (comparable to U.K. nonexecutives), *as well as* other key employees and large shareholders that own more than 10 percent of the firm's equity.† As some U.S. papers show only aggregate results for

* Throughout the remainder of the paper we focus on returns measured around the announcement date rather than the trading date. We focus on the former, because a director's identity is disclosed only when this information is released via the RNS (and not when the order is placed). The identity of the party trading is not even known by the market maker. This is confirmed by the FSA as well as several investment trust managers.

† Not all large shareholders are considered insiders. Regardless of the size of their holdings, the following shareholders are not viewed as insiders: commercial banks, brokers, insurance companies, investment banks, investment advisers, employee benefit plans, pension funds, and mutual funds.

insiders, it is not surprising that their CAARs are lower than those from U.K. studies because some insiders may have less inside information than officers or directors. Even though some U.S. studies exclude large shareholders, the results still cannot be directly compared to those for the United Kingdom. Indeed, whereas in the United Kingdom insiders are defined as executives and nonexecutives, U.S. insiders comprise (i) officers including the company president, principal financial officer, principal accounting officer, and any vice president in charge of any principal business unit, division, or function (such as sales, administration, or finance), (ii) directors and other persons who perform a policy-making function within the company (Bettis, Coles, and Lemmon 2000), as well as (iii) other key employees. Thus, given the more wide-ranging definition of insiders in the United States, we expect that U.S. insider trades are less informative and hence trigger smaller price reactions.*

Fourth, it is possible that news released prior to the directors' trades contaminates the abnormal returns around the announcement date.† Still, as we point out in the methodology section, the incidence of directors' transactions within a two-day or seven-day period subsequent to the release of news is relatively modest for most types of news (apart from news related to the firm's prospects). In Subsection 17.4.5, a detailed analysis of the impact of news releases on the CARs of directors' trades gives little credence to this fourth explanation, as the contamination effect by the news items is very modest, except for the announcements of M&As and CEO replacements.

In sum, we argue that the main reason "insider" trades trigger higher CAARs in the United Kingdom corresponds to differences in regulation and reporting speed between the United States and United Kingdom.

17.4.2 The Information Hierarchy Hypothesis

Hypothesis 2, the information hierarchy hypothesis, postulates that those directors who are more familiar with the day-to-day operations of the company trade on more valuable information. Our data set distinguishes between five categories of directors: CEOs (including joint CEO-chairmen), other executive directors (the deputy CEO and the financial

* It should also be noted that some U.S. studies include sales after the exercise of options. We exclude such sales from our study as they reveal less information given that the directors may merely sell to release that part of their remuneration (Lustgarten and Mande 1995; Jeng, Metrick, and Zeckhauser 1999; Friederich et al. 2002).

† We are grateful to the referee for suggesting further analysis along these lines.

officer), chairmen (nonexecutives in more than 90 percent of the cases), other incumbent directors (both executive and nonexecutive directors not included in the previous categories), and former directors. Former directors' trades are traced for up to two months after the end of the financial year during which they left the company.* We list the categories in decreasing order with respect to the degree of information superiority they are supposed to possess. As the three most senior executives are already included in the first two categories, that is, "CEOs" and "other top executives" (defined as the deputy CEO and financial director), and on average there are three executives on the board of a U.K. firm, the overwhelming majority of directors in the category of "other incumbent directors" are nonexecutives.†

We test the information hierarchy hypothesis in two ways. First, we compare the average abnormal returns earned after trades by each of the individual categories of directors. Second, we perform a regression analysis with the two-day CAR as the dependent variable and with dummy variables representing the individual categories as explanatory variables. The regressions allow us to control for other factors such as the transaction size, firm size, industry affiliation, simultaneous trading by several directors, and information releases just prior to the transactions.

Panel A of Table 17.4 reports the results of the event studies for purchases made by the different categories of directors. The J-form pattern of the abnormal returns around purchases that obtains for the whole sample also applies to the purchases made by all the individual categories of directors. That is, for all the categories of directors, the CAARs are significantly negative over the twenty days prior to the announcement, but become increasingly positive after the announcement day. In general, the CAARs covering the announcement day and the subsequent day range from 2.4 to 3.8 percent, and are strongly significant. However, there is no support for Hypothesis 2 on the information hierarchy as the differences between the (two-day) CAARs for the different categories of director are not statistically significant (these *t*-statistics are not reported in the table), apart from the differences between the CAARs of CEOs on the one side, and other incumbent (mainly nonexecutive) directors and former

* Former directors are defined in footnote 24.

† The Higgs (2003, 18) report shows that the average board size of all U.K.-listed firms is 6.7 consisting of a chairman (1), the executive directors (3), and the nonexecutive directors (2.7).

TABLE 17.4. Market reaction to directors' purchases according to director categories

Panel A: CAARs by director type

Event window	CAAR (−20;−1)	CAAR (0;1)	CAAR (0;3)	CAAR (0;5)	No. of observations
CEOs	−2.76%	2.38%[a]	3.71%	4.53%	582
t-statistic	−3.76	6.35	8.55	9.37	
All top executive directors	−2.57%	2.71%	4.19%	4.98%	677
t-statistic	−3.87	7.54	9.99	10.81	
Chairmen	−1.40%	3.17%	5.02%	6.26%	493
t-statistic	−1.57	6.98	9.02	9.81	
Other incumbent directors	−2.12%	3.51%[a]	5.17%	5.64%	572
t-statistic	−2.52	7.68	9.53	10.07	
All incumbent directors	−2.40%	2.92%	4.43%	5.14%	1,591
t-statistic	−5.12	11.86	14.81	15.74	
Former directors	−2.50%	3.83%[a]	6.34%	7.21%	396
t-statistic	−2.09	6.47	8.61	8.77	

Panel B: Cross-sectional regression results with CAR(0;1) as the dependent variable

	Model 1		Model 2	
	coef.	*t*-stat.	coef.	*t*-stat.
Constant	0.025	2.76	0.023	2.62
CEOs—multiple purchases	—	—	0.020	1.78
Other top executives	0.024	2.03	0.020	1.57
Other top executives—multiple purchases	—	—	0.034	1.20
Chairman	0.008	1.29	0.010	1.70
Chairman—multiple purchases	—	—	0.003	0.15
Other incumbent directors	0.010	1.66	0.012	2.03

Other current directors—multiple purchases	—	—	0.004	0.13
Former directors	0.020	2.76	0.021	2.75
Former directors—multiple purchases	—	—	0.017	0.81
Multiple purchases	0.015	1.87	—	—
Transaction size	−0.216	−1.40	−0.214	−1.41
Market capitalization	−0.039	−1.21	−0.041	−1.10
Adjusted R^2	1.40%		1.29%	
F	2.59		1.96	

Panel A reports the CAARs of directors' share purchases (of at least 0.1% of the market capitalization) based on the market model. The announcement day is day 0. CEOs and Chairmen stand for the CEOs/managing directors and chairmen of the board, respectively. All top executive directors represents the CEOs, deputy CEOs, and financial directors. Other incumbent directors are all directors not included in the previous categories. All incumbent directors comprise CEOs, top executive directors, chairmen, and other incumbent directors. Former directors refers to former directors whose trades are traced up to two months subsequent to the year in which they left the firm. In Panel B, the dependent variable is CAR(0;1). The number of observations is 1,905. All models include year and industry dummies. All coefficients are adjusted for heteroskedasticity (White procedure). "CEO" equals one if the CEO purchases shares. CEO—multiple purchases is set to one when a CEO and at least one other director, purchase on the same day. Other top executives equals one when a deputy CEO/managing director, or the finance director, purchases while the CEO does not. Chairman equals one if he or she buys while the CEO or other executives do not. Other incumbent directors equals one if directors (excluding the CEO, another executive, or the chairman) buy while no CEO, other executive, or chairman buys. Former directors is set to one if a former director buys while no incumbent director buys. Other top executives (chairmen, other incumbent directors, or former directors)—multiple purchases equals one if at least one director of that category buys while another director also buys. Multiple purchases is set to one if more than one director buys on the same day. Transaction size is the total number of shares bought by directors (over a day) over the total number of shares outstanding at the beginning of the year. Market capitalization is the total number of shares outstanding at the beginning of the year times the share price on the first trading day of that year.

[a] For the (0;1) event window, the difference in CAARs for CEO and former directors is significantly different at the 5% level (t = 2.07), and the differences in CAARs for CEOs and other incumbent directors is significant at the 10% level (t = 1.91).

directors on the other side. Surprisingly, the market reaction is weakest for purchases by CEOs (see below for possible explanations).

Panel B of Table 17.4 summarizes the regression results for directors' purchases. The dependent variable is the CAR covering the announcement day and the subsequent day using the market model as a benchmark. In order to construct mutually exclusive director categories, we use the following algorithm. The "other top executives" dummy is set to one if the deputy CEO or the financial director buys shares, but the CEO does not purchase any shares. The dummies for chairmen, other incumbent directors, and former directors are defined in a similar way. Hence, the constant picks up the effect of the CEO purchasing shares. The coefficients for the other dummy variables then pick up any differential market reaction as compared to the CEO effect. Negative coefficients would indicate that the market reaction to the CEO buying shares is highest, lending support to the information hierarchy hypothesis. In contrast, positive coefficients would indicate that the market reaction to other types of directors buying is higher than that to the CEO buying, challenging this hypothesis. We control for both the (relative) transaction size and firm size (market capitalization at the beginning of the year). We also adjust for the possibility of multiple trades, given that, on some days, more than one director of the same company may buy shares, which would strengthen the signal.* We use two different types of dummies to account for multiple trading. The first is that of multiple purchases, which is set to one if more than one director purchases (with a minimum transaction value of 0.1 percent of the firm's market capitalization), and to zero otherwise. For example, if both the CEO and a former director buy shares on the same day, then the CEO dummy is set to one (as a CEO is higher up the information hierarchy than a former director) and the multiple purchases dummy is set to one. Second, we include interaction dummies in the specification. By using the above example, the CEO dummy is set to one as well as the interactive dummy, CEO—multiple purchases.

Model 1 in Panel B of Table 17.4 shows that the coefficients for all the categories of directors are positive and only one (that for the chairman) is

* For 96 percent of all the days on which directors trade, all transactions are either all purchases or all sales. Hence, for only 4 percent of those days are there simultaneous purchases and sales. In these cases, an event is labeled as a purchase if the size (measured as a proportion of the firm's market capitalization) of the purchase(s) exceeds the size of the sale(s) by at least 0.1 percent. If there is more than one (net) purchase exceeding 0.1 percent of the firm's market capitalization, the multiple purchases dummy is set to one.

not statistically significant at the 10 percent level. The information effect of a CEO purchase (as measured by the constant, which is positive and significantly different from zero) is therefore lowest compared to all the other categories. For example, if a top executive other than the CEO buys shares, the market reaction is 4.9 percent (2.5 + 2.4) compared to only 2.5 percent if the CEO buys. This contradicts Hypothesis 2. Jeng, Metrick, and Zeckhauser (1999) also do not find any support for the information hierarchy view. They explain this as follows: the fact that the market follows CEO transactions more closely may cause CEOs to trade more cautiously and at less informative moments. Another possible explanation is that the positive news associated with purchases of shares is toned down by the negative news that the CEO strengthens his or her control over the firm to a level that causes entrenchment.

The multiple purchases dummy variable in Model 1 picks up the effect of several directors purchasing shares on the same day. The positive and significant coefficient documents that this constitutes a stronger signal for the market. Model 2 in Panel B reports a similar result; namely, multiple purchases make the positive market reaction even stronger. The model includes interaction terms between director-category dummy variables and the dummy for multiple purchases. So, for example, the coefficient on the interaction term CEO—multiple purchases indicates that, when both the CEO and another director purchase shares on the same day, the CAR is on average double (0.23 + 0.20) that when only the CEO buys shares. Note that the coefficients on the other interaction terms are not significantly different from zero (these coefficients refer to cases in which more than two directors of the other categories purchase shares but the CEO does not). Hence, the results suggest that CEO purchases that are accompanied by purchases by other directors have a higher information content than do purchases by the CEO alone.

Table 17.5 gives the market reaction to sales by the different categories of directors. Panel A reports the market reaction measured by the CAARs. The CAARs are negative for all the directors' categories and are significantly different from zero, except for former directors (for the windows of two, four, and six days starting with the announcement day). This suggests that there is no significant market reaction to former directors' sales as their sales are likely to be caused by portfolio diversification motivations and hence are not a signal to the market. Similar to purchases, the market reaction to sales by CEOs tends to be lower than that to sales by

TABLE 17.5. Market reaction to directors' scales according to director categories

	Panel A: CAARs by type of director				
Event window	**CAAR (−20;−1)**	**CAAR (0;1)**	**CAAR (0;3)**	**CAAR (0;5)**	**No. of observations**
CEOs	3.49%	−0.42%	−0.58%	−0.81%	490
t-statistic	5.96	−2.86	−2.66	−2.98	
All top executive directors	3.42%	−0.48%	−0.67%	−0.95%	563
t-statistic	5.88	−3.26	−3.17	−3.60	
Chairmen	3.19%	−0.50%	−0.56%	−0.88%	350
t-statistic	4.72	−3.15	−2.46	−3.17	
Other incumbent directors	3.05%	−0.59%	−0.77%	−1.06%	684
t-statistic	4.97	−4.52	−4.48	−4.97	
All incumbent directors	3.31%	−0.46%	−0.59%	−0.84%	1476
t-statistic	8.76	−5.26	−5.05	−5.73	
Former directors	2.61%	−0.16%[a]	−0.20%	−0.18%	626
t-statistic	3.53	−1.10	−0.98	−0.77	

Panel B: Cross-sectional regression results with CAR(0;1) as the dependent variable

	Model 3	
	coef.	***t*-stat.**
Constant	–0.004	–1.54
Other top executives	–0.004	–0.74
Chairman	–0.001	–0.43
Other incumbent directors	–0.001	–0.36
Former directors	0.002	0.83
Multiple sales	–0.005	–2.12
Transaction size	0.068	1.05
Market capitalization	–0.700	–1.50
Adjusted R^2	0.52%	
F	1.20	

Panel A reports the market model-based CAARs of directors' share sales (of at least 0.1% of the market capitalization). The announcement day is day 0. CEOs and Chairmen stand for the CEOs/managing directors and chairmen of the board, respectively. All top executive directors stands for CEOs, deputy CEOs, and financial directors. Other incumbent directors are all directors not included in the previous categories. All incumbent directors comprise CEOs, top executive directors, chairmen, and other incumbent directors. Former directors refers to former directors whose trades are traced up to two months subsequent to the year in which they left the firm. In Panel B, the dependent variable is CAR(0;1). Model 3 includes time and industry dummies. It has 1,993 observations. Multiple purchases is set to one if more than one director buys on the same day. Transaction size is the total number of shares sold by directors (over a day) over the total number of shares outstanding at the beginning of the year. Market capitalization is the total number of shares outstanding at the beginning of the year times the share price on the first trading day of that year.

[a] For the (0;1) event window, the difference in CAARs for current and former directors is significantly different at the 10% significance level. All other pair-wise tests on differences of CAARs (0;1) are not statistically significant.

other directors. Still, the differences are not statistically significant, which implies that our results do not support Hypothesis 2.

Panel B of Table 17.5 shows the regression results for sales. Note that Model 3 is similar to Model 1 for purchases in Panel B of Table 17.4.* The regression has very low explanatory power and none of the coefficients on the types of directors is significantly different from zero. The only coefficient that is significantly different from zero (although small in economic terms) is that on multiple sales. This suggests that the market interprets directors' sales as negative news if several directors sell simultaneously. Conversely, if only one director sells, the market seems to treat this as a sale due to liquidity needs rather than bad news. In line with the regressions for purchases, the regression for sales does not support Hypothesis 2 on information hierarchy.

17.4.3 The Effect of Outside Ownership

In what follows we test the impact of ownership concentration on the information content of directors' trades (Hypotheses 3 and 4). The two-day CARs are regressed on a set of ownership variables that measure the possible information content of directors' transactions in firms with different categories of blockholders, namely, corporations, individuals or families unrelated to the directors, institutional investors, and directors. A specific ownership concentration dummy is set to one if a shareholder of that category owns at least 5 percent of the equity (the definition of blockholder that we use).† We also control for other determinants that may influence the information content of directors' transactions, that is, simultaneous trading by several directors, transaction value, firm size, book-to-market, profitability, and leverage.

Table 17.6 contains the regression results for directors' purchases, whereas Table 17.7 gives the results for sales.‡ The results from Model 4 in Panel A of Table 17.6 for purchases provide strong support for Hypothesis 3(a). The coefficients measuring the information effect of blockholders who are likely to monitor the management, again, corporations and individuals or families, are both negative. However, only the coefficient on corporations

* We do not report the equivalent of Model 2 for sales as the model is not significant.

† Dispersed ownership is the base case.

‡ There are five very large purchases involving more than 30 percent of the equity. As such large acquisitions trigger a mandatory tender offer for all shares outstanding, we also run the regressions without these trades. However, this does not change any of the results in Models 4 through 8.

TABLE 17.6. Market reaction to directors' purchases and control structure

Panel A: Regressions with dominant blockholders				
	Model 4		Model 5	
	coef.	***t*-stat.**	**coef.**	***t*-stat.**
Constant	0.050	2.16	0.044	1.95
Other top executives	0.016	1.51	0.015	1.35
Chairmen	0.002	0.36	0.003	0.52
Other incumbent directors	0.009	1.16	0.009	1.18
Former directors	0.015	2.00	0.016	2.15
Concentrated blockholder dummies				
Corporations	-0.021	-2.84	—	—
Institutional investors	0.013	2.29	—	—
Individuals/families	-0.010	-1.58	—	—
Directors	-0.014	-2.59	—	—
Dominant blockholder dummies				
Dominant corporations	—	—	0.007	0.28
with institutional investors present	—	—	-0.016	-0.69
with individuals/families present	—	—	0.021	1.04
with directors present	—	—	-0.027	-1.21
Dominant institutional investors	—	—	0.027	3.08
with corporation present	—	—	-0.029	-2.90
with individuals/families present	—	—	-0.013	-1.15
with directors present	—	—	-0.026	-3.10
Dominant individuals/families	—	—	-0.021	-2.28
with institutional investors present	—	—	0.019	0.94
Dominant directors	—	—	0.011	1.26
with corporation present	—	—	-0.058	-2.20
with institutional investors present	—	—	-0.006	-0.67
with individuals/families present	—	—	-0.017	-1.75
Other variables				
Multiple purchases	0.014	1.56	0.014	1.62
Transaction value	-0.001	-0.67	-0.002	-0.70
Size	-0.001	-0.20	0.000	-0.16
B/*M* ratio	-1.609	-0.86	-2.289	-1.22
ROE	1.687	2.41	1.644	2.29
Leverage	0.002	0.94	0.003	0.98
Adjusted R^2	3.35%		4.57%	
F	2.15		2.06	

(continued)

TABLE 17.6. Market reaction to directors' purchases and control structure (*continued*)

Panel B: Regressions with concentrated blockholders and loss dummies						
	Model 6		Model 7		Model 8	
	Negative earnings		Low interest coverage		Dividend decrease	
	coef.	***t*-stat.**	**coef.**	***t*-stat.**	**coef.**	***t*-stat.**
Constant	0.045	1.95	0.044	1.87	0.048	2.04
Other top executives	0.012	1.07	0.015	1.17	0.009	0.89
Chairman	−0.004	−0.58	−0.006	−0.89	0.002	0.34
Other incumbent directors	0.007	0.94	0.010	1.28	0.007	0.85
Former directors	0.024	2.66	0.023	2.47	0.017	2.02
Concentrated blockholder dummies						
Corporations	−0.020	−3.08	−0.020	−2.96	−0.021	−2.38
Institutional investors	0.012	1.96	0.014	2.21	0.012	1.97
Individuals/families	−0.010	−1.38	−0.018	−2.61	−0.011	−1.52
Directors	−0.011	−1.72	−0.010	−1.61	−0.010	−1.70
Interaction term: director category × loss dummy						
CEO	0.052	2.11	0.038	1.79	0.001	0.05
Other top executives	0.063	1.97	0.041	1.50	0.027	0.70
Chairman	0.071	2.60	0.056	2.45	−0.003	−0.13
Other incumbent directors	0.062	2.24	0.035	1.41	0.010	0.44
Former directors	0.023	0.87	0.016	0.68	−0.013	−0.55
Interaction term: blockholder × loss dummy						
Corporations	−0.011	−0.66	−0.011	−0.73	0.002	0.14
Institutional investors	−0.031	−1.53	−0.022	−1.30	0.006	0.40
Individuals/families	−0.008	−0.48	0.018	1.18	0.008	0.48
Directors	−0.028	−1.72	−0.017	−1.25	−0.014	−0.94
Dispersed	−0.056	−2.22	−0.032	−1.46	0.016	0.63
Other variables						
Multiple purchases	0.013	1.48	0.013	1.44	0.013	1.41
Transaction value	−0.001	−0.53	−0.001	−0.49	−0.001	−0.58

Size	−0.001	−0.21	0.000	−0.19	−0.001	−0.24
B/*M* ratio	−2.456	−1.18	−2.594	−1.24	−1.602	−0.82
ROE	1.454	2.05	1.717	2.36	1.518	2.29
Leverage	0.002	0.99	0.002	0.83	0.002	0.79
Adjusted R^2	4.62%		4.44%		3.74%	
F	2.13		2.11		1.69	

The dependent variable is CAR(0;1). Other top executives equals one when a deputy CEO/managing director or the finance director purchases while the CEO does not. Chairmen stands for chairmen of the board, respectively. Other incumbent directors are all directors not included in the previous categories. Former directors refers to former directors whose trades are traced up to two months subsequent to the year in which they left the firm. Concentrated blockholder dummies—corporations, institutional investors, individuals/families, and directors are dummy variables. All these dummy variables equal one if a blockholder of the corresponding type holds a stake of at least 5% of the equity, and is zero otherwise. Dominant blockholder dummies—corporation, institutional investor, individual, and insider are dummy variables set to one if the sum of all the blocks of that type of blockholder is the largest compared to the combined stakes of other blockholder types. With corporation, institutional investor, individual, or directors present is an interaction term between the "dominant" blockholder dummy and a "concentrated' blockholder dummy of another type. Multiple purchases is set to one if more than one director buys on the same day. Transaction value is the value of the share block purchased by the director and it is defined as the natural log of the total number of shares transacted by a director times the price per share at the beginning of the calendar year. Size is the natural log of the total number of employees at the beginning of the year. B/M ratio is the book value divided by the market value, both measured at the beginning of the year. ROE and Leverage are the return on equity and the debt-to-equity ratio at the beginning of the year, respectively. All models include year and industry dummies. The number of observations in Panel A is 1,428 and in Panel B is 1,481.

TABLE 17.7. Market reaction to insider sales and control structure

Panel A: Regressions with dominant blockholders	Model 9		Model 10	
	coef.	***t*-stat.**	**coef.**	***t*-stat.**
Constant	0.007	1.27	0.005	0.99
Other top executives	0.001	0.15	–0.001	–0.05
Chairmen	0.001	0.43	0.001	0.49
Other incumbent directors	0.001	0.32	0.001	0.32
Former directors	0.004	1.48	0.004	1.59
Concentrated blockholder dummies				
Corporations	–0.001	–0.32	—	—
Institutional investors	–0.002	–0.99	—	—
Individuals/families	–0.004	–1.63	—	—
Directors	0.004	2.01	—	—
Dominant blockholder group dummies				
Dominant corporations	—	—	–0.003	–0.57
with institutional investors present	—	—	0.007	0.96
with individuals/families present	—	—	0.020	1.36
with directors present	—	—	–0.009	–1.03
Dominant institutional investors	—	—	–0.004	–1.83
with corporation present	—	—	–0.009	–1.82
with individuals/families present	—	—	–0.005	–1.21
with directors present	—	—	0.005	1.97
Dominant individuals/families	—	—	–0.008	–1.24
with institutional investors present	—	—	0.016	1.65
with directors present	—	—	–0.009	–0.85
Dominant directors	—	—	0.002	0.61

with corporation present	—	—	0.010	1.15
with institutional investors present	—	—	0.002	0.42
with individuals/families present	—	—	−0.007	−1.69
Other variables				
Multiple sales	−0.004	−1.61	−0.004	−1.64
Size	−0.002	−2.36	−0.001	−2.01
B/*M* ratio	−0.714	−0.80	−0.880	−0.95
ROE	3.410	1.09	3.710	1.15
Leverage	0.348	0.79	0.505	1.07
Adjusted R^2	2.02%		3.30%	
F	1.55		1.57	

Panel B: Regressions with Concentrated Blockholders and Loss Dummies

	Model 11		Model 12		Model 13	
	Negative earnings		Low interest coverage		Dividend decrease	
	coef.	***t*-stat.**	**coef.**	***t*-stat.**	**coef.**	***t*-stat.**
Constant	0.008	1.54	0.008	1.55	0.007	1.46
Former directors	0.002	1.00	0.002	0.96	0.004	1.72
Concentrated blockholder dummies						
Corporations	0.000	0.09	-0.001	-0.31	0.000	-0.01
Institutional investors	-0.001	-0.70	-0.002	-1.03	0.000	-0.11
Individuals/families	-0.002	-0.79	-0.003	-1.14	-0.005	-1.70
Directors	0.004	1.96	0.004	1.97	0.003	1.54
Interaction term: director category x loss dummy						
Incumbent directors	-0.038	-3.10	-0.031	-2.71	-0.015	-0.96
Former directors	-0.031	-2.41	-0.023	-1.98	-0.018	-1.27
Interaction term: blockholder x loss dummy						
Corporations	0.005	0.61	0.009	1.15	-0.010	-0.59
Institutional investors	0.020	1.88	0.019	1.88	-0.006	-0.49
Individuals/families	-0.010	-1.18	-0.006	-0.69	0.010	0.98
Directors	0.015	1.93	0.009	1.18	0.021	1.84
Dispersed	0.041	3.16	0.029	2.41	0.025	1.53
Other variables						
Multiple sales	-0.004	-1.87	-0.004	-1.91	-0.004	-1.79
Size	-0.002	-2.34	-0.002	-2.31	-0.002	-2.36
B/M ratio	-0.663	-0.73	-0.394	-0.44	-0.642	-0.69
ROE	2.666	0.99	2.546	0.91	3.418	1.12
Leverage	0.527	1.19	0.448	1.02	0.315	0.70
Adjusted R^2	3.32%		2.79%		3.03%	
F	1.94		1.61		1.62	

Panel B: Regressions with Concentrated Blockholders and Loss Dummies

The dependent variable is CAR(0;1). Other top executives equals one when a deputy CEO/managing director or the finance director purchases while the CEO does not. Chairmen stands for the chairmen of the board, respectively. Other incumbent directors are all directors not included in the previous categories. Former directors refers to former directors whose trades are traced up to two months subsequent to the year in which they left the firm. Concentrated blockholder dummies—corporations, institutional investors, individuals/families, and directors are dummy variables. All these dummy variables equal one if a blockholder of the corresponding type holds a stake of at least 5% of the equity, and is zero otherwise. Dominant blockholder dummies—corporation, institutional investor, individual, and insider are dummy variables set to one if the sum of all the blocks of that type of blockholder is the largest compared to the combined stakes of other blockholder types. With corporation, institutional investors, individuals/families, or directors present is an interaction term between the "dominant" blockholder dummy and a "concentrated" blockholder dummy of another type. Multiple purchases is set to one if more than one director buys on the same day. Size is the natural log of the total number of employees at the beginning of the year. B/M ratio is the book value divided by the market value, both measured at the beginning of the year. ROE and Leverage are the return on equity and the debt to equity ratio at the beginning of the year, respectively. All models include year and industry dummies. The number of observations is 1,681.

is significantly different from zero (at the 1 percent level of significance). Our results confirm that directors' purchases convey less new information when other corporations own a considerable stake in the firm.

Hypothesis 3(b) postulates that the presence of institutional blockholders has no effect on the signal of directors' transactions. Our findings do not support the hypothesis but support the notion that institutional blockholders do not act as monitors. The coefficient on the institutional investor dummy is positive and highly significant (at the 1 percent level). This implies that the market reaction is higher for firms with institutional ownership. Thus, institutional owners do not act as monitors and hence do not lower the informational asymmetry. Moreover, the fact that institutional investors do not monitor gives them the opportunity to trade on publicly available signals; institutional investors seem to follow directors' purchases in order to rebalance their portfolios as their trades strengthen the positive (negative) signal of directors' purchases (sales).*

Hypothesis 4(a) postulates that the positive informational effect of directors' purchases is weakened by the danger of (more) entrenchment. Panel A of Table 17.6 supports this hypothesis. The coefficient on directors' block ownership is negative and statistically significant. In the presence of substantial director ownership, directors' purchases convey two important counteracting signals: (i) the positive news about the firm's prospects, and (ii) the negative news associated with increased entrenchment. Our results suggest that the latter effect is quite strong (within the 1 percent level of significance). The adjusted R^2 for Model 4 is more than double that for Models 1 and 2 without the control dummies. However, Models 4 and 5 do not reject the existence of a precommitment effect: in widely held firms (the base case), directors' purchases trigger strongly significant

* We interview six fund managers based in the City of London (from Schroder Investment Management, Morgan Stanley, Credit Suisse, Knox d'Arcy, and Deutsche Bank London). Five of these managers indicate that some of the funds they manage consider the quality of management, changes in the ownership by directors, the reasons directors leave, and the change in the equity position of directors who leave when making their investment decisions. Some funds (e.g., Credit Suisse Insider Strategy and funds managed by Knox d'Arcy) use directors' transactions (including those made by directors leaving the firm) to create trading rules. The purchases by former directors are taken to be a signal of their confidence in the remaining incumbent management of the firm. The fund managers confirm that the fact that a director who leaves the firm or who has recently left liquidates his or her equity stake does not constitute a signal to the market. However, if such a director increases his or her share stakes, this sends important information to the market.

abnormal returns. That is, in cases in which director ownership is low, the danger of entrenchment is low.

Model 5 in Panel A tests for the impact of the relative power of the different categories of blockholders on the CARs. Here we focus on the effect of the *dominant* blockholder type as opposed to the effect of the presence of a blockholder type regardless of the relative size of its holding. We regard a particular type of blockholder as dominant if the sum of the shareholdings of this category is larger than that of any other category.* Since this set of dummy variables is mutually exclusive, only one dummy variable is equal to one at a time, and the dummy variables for all the other categories are equal to zero. Once we determine which specific category of shareholder dominates a firm, we also use interaction terms that indicate whether the other categories of owners are among the firm's blockholders.†

Model 5 fails to support Hypothesis 3(b) as we find that the presence of dominant institutional investors strengthens the positive market reaction to purchases. Moreover, the interaction term of dominant institutional investors with corporate blockholders shows that the above effect is largely neutralized if corporations are present as blockholders. This provides further evidence in support of Hypothesis 3(a): monitoring by blockholders reduces the information value of directors' purchases. Moreover, Model 5 confirms the findings of Model 4 in support of Hypothesis 3(a): the information gap is reduced as the positive market reaction to directors' purchases is less strong when individuals or families are the dominant blockholders.‡

Again, Model 5 also presents a test of Hypothesis 4(b), which states that there is a danger of potential entrenchment by the directors in the presence of passive outside blockholders, that is, institutional investors. The coefficient on the dummy of a dominant institutional investor is positive and highly significant, providing incremental support to what we obtain from Model 4. More importantly, the interaction term between the dominant institutional investor dummy and the dummy that equals one if the directors are blockholders is negative and highly significant. This implies

* When we consider the largest blockholder by category of owner rather than the sum of the category's shareholdings, the results remain largely similar. This makes sense as in most companies there is at most only one large blockholder within each specific category.

† We multiply the dominant blockholder dummy by the dummies for individual blockholder categories.

‡ The coefficient on dominant directors in Model 5 of Table 17.6 is not statistically significant, which is caused by the facts that there are few companies with dominant directors and the negative effect of directors' blockholdings.

that when directors are already large shareholders in the presence of a dominant institutional investor, the market no longer perceives their purchases as a signal of good news.

Panel A of Table 17.7 confirms that the information content of sales is much lower than that of purchases. As we stated in Hypothesis 1(c), directors' sales are less informative as some of the sales may be due to liquidity needs. Further, the negative signal of directors' sales is much stronger in smaller firms. This may be due to the higher uncertainty and the lower availability of information about smaller firms as, for instance, they are followed by fewer analysts. The improved liquidity of the stock as a result of the sale may also cause a stronger market reaction. Lakonishok and Lee (2001) observe the same pattern for the United States. Panel A of Table 17.7 also shows that with the exception of directors, the presence of specific categories of blockholders has little impact on the CARs (Model 9). When directors are blockholders, a reduction in their control (and hence a reduced potential for the private benefits of control) is positively received by the market and reduces the negative signal of directors' sales. This finding supports Hypothesis 4(a).* Model 10 is similar to Model 5 in Table 17.6 and is also based on relative ownership. We find that in the presence of dominant institutional investors, the market reaction to directors' sales is significantly negative. However, when the dominance of institutional investors is accompanied by strong director ownership, the market reaction to sales is neutral. This provides further support for Hypothesis 4(b) on entrenchment.

17.4.4 The Effect of Bad Performance and Financial Distress

We expect that poor financial performance and near insolvency also influence the information content of directors' trades. For the case of purchases, Models 6 through 8 in Panel B of Table 17.6 are similar to those in Panel A, but include additional regressors that consist of interaction terms between director categories and blockholder types on the one side, and poor performance and/or financial distress on the other side. We measure poor performance and financial distress by dummy variables that are set to one if there are earnings losses (Model 6), low interest coverage (Model 7),† and decreased or omitted dividends (Model 8), respectively. Poor

* Model 9 does not include the sales transaction value as it is highly correlated with firm size and the book-to-market ratio.

† Interest coverage becomes dangerously low when it falls below two. At this stage, a firm's bonds typically lose their investment grade rating (Copeland, Koller, and Murrin 1995).

performance and near-insolvency are expected to trigger more intensive shareholder and/or creditor monitoring. We find that purchases generate positive CARs, which are substantially higher when the company incurs losses or is financially distressed (see the interaction terms between the directors' types and losses/interest coverage in Models 6 and 7). Thus, in situations of poor performance and near insolvency, the market interprets purchases as strong positive signals. This supports Hypothesis 5.

The signs and significance levels for the coefficients on the blockholder dummies in Models 6 through 8 of Panel B are similar to those in Panel A. However, the interaction terms between ownership concentration and poor performance (measured by earnings losses and dividend reductions) or between ownership and near insolvency (low interest coverage) are not significant. In poorly performing companies with strong outsiders and with directors who can facilitate corporate recovery the directors' trading signal is not stronger; this suggests that the market does not expect the blockholders to turn around the firm. This result is not at all surprising as poor performance may be the consequence not only of poor management but also of poor past blockholder monitoring. To conclude, given poor performance, the signal of directors' purchasing shares is important irrespective of the shareholder structure.

For directors' sales (Panel B of Table 17.7), we also use a set of interaction terms between director categories and blockholder types on the one side, and losses (Model 11), low interest coverage (Model 12), and dividend decreases/omissions (Model 13), respectively, on the other side. The results of Models 11 and 12 again strongly support Hypothesis 5 as the interaction term between the incumbent directors' dummy* and poor performance is highly significant. This suggests that for poorly performing or financially distressed companies, directors' sales trigger more negative CARs. Table 17.8 provides an overview of the hypotheses and summarizes the results.

17.4.5 The Impact of News Releases prior to Directors' Transactions

In the methodology section, we mention that it may be important to account for news releases that precede trades as they may be one of the reasons the CARs we find are larger than those documented by studies

* Using interaction terms based on the individual directors' categories as we use in Table 17.6 yields less significant results as the number of observations for individual directors' categories is small.

TABLE 17.8. Summary of findings

Directors' share dealings	Directors' share purchases		Directors' share sales	
Hypothesis	Expected announcement effect	Hypothesis confirmed (Yes/No)	Expected announcement effect	Hypothesis confirmed (Yes/No)
H1(a)/(b) Announcement effect	Positive	Yes	Negative	Yes
H1(c) The absolute value of the market reaction to purchases is higher than that to sales	Higher absolute reaction to purchases	Yes	Lower absolute reaction to sales	Yes
H2 Information hierarchy	More strongly positive for executives	No	More strongly negative for executives	No
H3(a) Monitoring outsider blockholders reduces informational asymmetry	Less positive	Yes	Less negative	No
H3(b) Institutional investors do not have any impact on the market reaction	No impact	No, institutions follow directors	No impact	No, institutions follow directors
H4(a) Director entrenchment reduces the market reaction	Less positive in firms with strong directors	Yes	Less negative	Yes
H4(b) Market is more concerned about director entrenchment in the presence of passive shareholders and is less concerned in the presence of monitoring shareholders	Weaker positive effect in the presence of institutional investors	Yes	Weaker negative effect in the presence of institutional investors	Yes
H(5) Poor performance/financial distress triggers stronger market reactions	Strongly positive	Yes	Strongly negative	Yes
Other findings:				

With multiple transactions	More strongly positive	Yes	More strongly negative	Yes
With larger transaction size	More strongly positive	No	More strongly negative	No
With smaller corporate size	More strongly positive	No	More strongly negative	Yes
Immediately after news releases	no impact	No; only reduced market reaction for news on M&A and CEO changes	No impact	No; only reduced negative reaction to exec. director change
In firms with thin trading	Less strongly positive	Yes	Less strongly negative	Yes

TABLE 17.9. The impact of news releases prior to directors' purchases

Timing of news	Up to 2 days prior to purchase		Up to 7 days prior to purchase		Up to 30 days prior to purchase	
	coef.	***t*-stat.**	**coef.**	***t*-stat.**	**coef.**	***t*-stat.**
Constant	0.069	2.36	0.068	2.30	0.071	2.41
Other top executives	0.015	1.20	0.015	1.20	0.013	1.01
Chairmen	0.002	0.27	0.003	0.42	0.003	0.41
Other incumbent directors	-0.008	-0.93	-0.008	-0.98	-0.008	-0.99
Former directors	0.015	1.24	0.016	1.32	0.018	1.49
Concentrated blockholder						
Corporations	-0.028	-2.58	-0.028	-2.57	-0.028	-2.57
Institutional investors	0.006	0.99	0.006	1.03	0.007	1.17
Individuals/families	-0.016	-1.77	-0.016	-1.69	-0.017	-1.89
Directors	-0.011	-1.85	-0.011	-1.87	-0.011	-1.80
News items on:						
CEOs	-0.102	-1.33	-0.114	-1.41	-0.058	-1.98
Executive directors	0.002	0.05	0.001	0.03	0.001	0.08
Nonexecutive directors	0.028	0.74	0.013	0.60	0.012	0.87
Mergers and acquisitions	-0.060	-1.48	-0.041	-2.26	-0.035	-3.04
Asset disposals	0.007	0.15	-0.016	-0.42	-0.024	-1.05
Capital structure changes	0.004	0.11	-0.006	-0.32	-0.012	-0.91
Equity buybacks	-0.015	-1.32	-0.015	-1.28	-0.018	-1.79
Firm's prospects	0.008	0.53	0.008	0.79	0.008	1.14
Business events	-0.072	-1.64	-0.034	-1.35	-0.016	-0.95
Other variables						
Multiple purchases	0.005	0.46	0.005	0.44	0.007	0.57
Transaction value	-0.004	-1.19	-0.004	-1.12	-0.004	-1.23
Size	-0.001	-0.17	-0.001	-0.22	-0.001	-0.27

B/M ratio	−2.506	−1.03	−2.574	−1.06	−2.649	−1.12
ROE	1.733	2.32	1.780	2.36	1.882	2.47
Leverage	0.003	0.72	0.004	0.80	0.004	0.88
Adjusted R^2	5.81%		5.19%		5.31%	
p-value of *F*-test	0.00		0.00		0.00	

This table shows Model 4, which now also includes the news releases. The dependent variable is CAR(0;1) around the announcement date. The number of observations is 873. Other top executive equals one when a deputy CEO/managing director or the finance director purchases while the CEO does not. Chairmen stands for chairmen of the board. Other incumbent directors are all directors not included in the previous categories. Former directors refers to former directors whose trades are traced up to two months subsequent to the year in which they left the firm. Concentrated blockholder dummies—corporations, institutional investors, individuals/families, and directors are dummy variables. All these dummy variables equal one if a blockholder of the corresponding type holds a stake of at least 5% of the equity, and is zero otherwise. The variables on the news items are dummies that equal one for CEOs, Executive directors, and Nonexecutive directors if there is news on changes to each of these types of board positions, respectively. A second set of dummy variables equals one in the case of news relating to corporate and capital restructuring, that is, mergers and acquisitions, asset disposals, capital structure changes (including a new stock issue to pay off existing debt), and equity buybacks, respectively. A third set of dummies equals one when there is news on the firm's prospects and on other business events (e.g., a name change, the signing of a new contract, a product launch, a change in accounting policy, a debt rollover, a move to the AIM market, and a change of sector), respectively. Multiple purchases is set to one if more than one director buys on the same day. Transaction value is the value of the share block purchased by the director and it is defined as the natural log of the total number of shares transacted by a director times the price per share at the beginning of the calendar year. Size is the natural log of the total number of employees at the beginning of the year. B/M ratio is the book value divided by the market value, both measured at the beginning of the year. ROE and Leverage are the return on equity and the debt-to-equity ratio at the beginning of the year, respectively. All models include year and industry dummies.

based on U.S. data. Indeed, the announcement effect may not be due to the directors' transactions but rather to the release of news. Thus, while Panel B of Table 17.2 shows that a relatively small percentage of news items are followed by directors' transactions (apart from the announcements relating to the firm's prospects), the CARs may still be significantly influenced by specific types of price-sensitive information.

We rerun Models 1 to 13 and include dummy variables that capture the release of news two, seven, and thirty days, respectively, prior to the directors' transactions. Table 17.9 reports the results for purchases. On the whole, our previous findings hold. We also show that most news releases prior to directors' transactions do not have any impact on the value of the signal, not even frequent announcements about a firm's prospects.

There is one type of news release that does have a significant impact on the CARs. Table 17.9 shows that if news regarding a merger or acquisition is released within the seven or thirty days prior to a purchase, the market reaction is close to zero. This suggests that directors' purchases do not contain much additional information after an M&A announcement.* We also find (weaker) evidence that the information value of directors' trades is reduced when the trade occurs within a month following news concerning the replacement of the CEO. These two types of news reduce or even cancel out the otherwise positive market reaction to purchases.

Betzer and Theissen (2004) investigate the market reaction to executives' and nonexecutives' trades in German firms prior to news releases on the firm's prospects. They conclude that "their results also provide a rationale for the U.K. type of regulation that prevents insiders from trading prior to earnings announcements. Trades that occur during the blackout period do have a larger price impact. This is consistent with informational asymmetries between corporate insiders and the capital market being larger prior to earnings announcements." The authors argue in favor of trading bans because insiders trading on inside information in Germany seem to benefit from their informational advantage. However, this violates the principle of equal treatment of shareholders.

As Subsection 17.4.1 shows, directors' sales are less informative than are their purchases. There is little impact of news releases on the market effect of sales. Only when directors sell equity immediately after the replacement of executive directors is the negative sales signal strengthened.

* We do not report the estimation results of Models 5 through 8 with the news dummies in a table. However, all the tables are available upon request from the authors.

17.4.6 Thin Trading

Although the abnormal returns are corrected for nonsynchronous trading (Dimson and Marsh 1986), our results may still be biased because of a correlation between the CARs and thin trading.* No or limited trading over specific periods may prevent the information conveyed by directors' transactions from being incorporated in the share price. We therefore set up a simple test. Specifically, we record the number of nontrading days for each firm and classify firms into two categories, "firms with thin trading" (the number of nontrading days is above the median) and "firms without thin trading" (the number of nontrading days is below the median). We find that the announcement effect of directors' transactions is negatively related to thin trading. The purchase announcement effect (CAAR(0;1)) amounts to 3.5 percent for firms without thin trading whereas it is only 2.7 percent for firms with thin trading. For sales, the announcement effect for firms with more thin trading is stronger (–0.6 percent) than that for firms with less thin trading (–0.2 percent). When we include a thin trading dummy (which equals one if the number of nontrading days is above the median) in Models 1 through 13, we find that the market is more receptive to signals conveyed by directors' trades in firms that suffer less from thin trading. However, even when correcting for possible thin trading, all the results from Section 17.4 hold.†

17.5 CONCLUSIONS

This study provides a major contribution to the literature on the information content of executive and nonexecutive directors' trading by analyzing the impact of ownership and control. To avoid a contamination of the signal conveyed by the directors' transactions, we adjust the market reaction for recent releases of corporate news related to board and asset restructuring (such as M&A activity and asset disposals), to the firm's prospects, and to other important business events. Several important conclusions emerge. First, consistent with most existing U.K. and U.S. studies, directors' purchases and sales trigger significant immediate market reactions of 3.12 and –0.37 percent, respectively, measured over the two-day

* We are grateful to the referee for bringing this to our attention.

† All tables are available upon request. The correlation coefficients between share illiquidity and control concentration by category of shareholder are positive but below 0.1. As expected, the correlation between illiquidity and the free float is negative (but only 0.08). The inclusion of a Herfindahl index, which captures the distribution of the ownership concentration, in all our models does not change the results and its coefficient is not statistically significant.

window starting with the announcement day. The lower market reaction to sales suggests that the market associates a lower informational content with sales, as sales may be motivated in part by liquidity needs. Given that directors are banned from trading prior to earnings announcements, it is likely that they trade on additional information relative to that contained in the earnings announcement. Alternatively, earnings announcements do not convey all available information on the company. The existence of trading bans does not appear to curtail the value of the signal.

Second, when several directors trade on the same day, the announcement reaction is stronger. Clearly, multiple trades give more credibility to the signal conveyed to the market.

Third, we do not find support for the information hierarchy hypothesis. Although CEOs are assumed to have the best knowledge about their company's prospects, the information content of their trades is lower than that of other directors' trades. The most plausible explanation for this result is that the FSA and the market may follow CEO transactions more closely, which causes CEOs to trade more cautiously and at less informative moments.

Fourth, there is a strong relation between the presence of specific categories of blockholders and the price reaction to directors' transactions. It is important to distinguish between blockholdings held by directors and different types of outsiders. Additionally, it is important to distinguish between blockholders who are likely to monitor the management (i.e., corporations, and individuals or families unrelated to the directors) and those who are not (i.e., institutional investors). We find that, if corporations or individuals/families are blockholders, then the price reaction to directors' purchases is reduced. The presence of institutional investors generates the opposite effect. The evidence is consistent with institutional investors trading on the directors' trade signals. Thus, while the presence of institutional investors strengthens negative sales signals, the result is less strong than for purchases.

Fifth, the market reacts to changes in director entrenchment. In general, the positive impact of directors' purchases is reduced when the directors already own substantial stakes. Similarly, the market reaction resulting from directors' sales is less negative.

Sixth, the share price reactions to directors' transactions as well as the above effects caused by the firm's control structure remain valid when the transactions are preceded by news on board changes, corporate restructuring, changes in the capital structure, and the firm's prospects. However, it is crucial to adjust for news regarding mergers and acquisitions (and to a lesser extent CEO replacements) as these news items mitigate and even cancel the significant share price reactions to directors' purchases.

Finally, although in general the ownership and control structure has a strong impact, it does not matter for poorly performing or financially distressed firms, as in these firms, directors' trades always convey stronger signals about prospects (perhaps even about the likelihood of survival) irrespective of any potential monitoring or entrenchment effects.

All in all, this chapter provides strong evidence that the market takes into account the firm's control structure, the level of director entrenchment, and whether several directors trade when it reacts to directors' trades.

17.6 APPENDIX: TEST STATISTICS

To test the null hypothesis that the CAARs are equal to zero for a sample of N securities, we use three parametric test statistics:

$$t_{CAAR} = \frac{\frac{1}{N}\sum_{i=1}^{N} CAR_i}{s(CAR)/\sqrt{N}}$$

$$J_1 = \frac{CAAR}{s(CAAR)}$$

$$J_2 = \sqrt{\frac{N(L_i - 4)}{L_i - 2}}\frac{1}{N}\sum_{i=1}^{N}\frac{CAR_i}{s(CAR_i)}$$

where CAR_i is the cumulative abnormal return for security i, $CAAR$ is the cumulative average abnormal return, $s(CAR_i)$ is the sample standard deviation of the individual cumulative abnormal returns and equals

$$s(CAR_i) = \sqrt{\sum_{t=t_1}^{t_2} s_i^2},$$

$s(CAAR)$ is the standard deviation of the cumulative average abnormal returns and equals

$$s(CAAR) = \sqrt{\frac{1}{N^2}\sum_{i=1}^{N}\sum_{t=t_1}^{t_2} s_i^2},$$

and

$$s_i = \sqrt{\frac{1}{L_i - 2} \sum_{t=T_{0i}}^{T_{1i}} \left(R_{i,t} - \hat{\alpha}_i - \hat{\beta}_i R_{m,t}\right)^2},$$

which is the usual sample standard error from the market model regression over the estimation window.

The t_{CAAR} is the test statistic as in Barber and Lyon (1997). It is Student-t distributed with $N - 1$ degrees of freedom and approaches the normal distribution as N increases. The variables J_1 and J_2 are based on Campbell, Lo, and McKinley (1997). The choice between these two statistics depends on the hypotheses regarding the variance of the abnormal returns. If the abnormal return is larger for securities with higher variance, J_1 is preferable as it gives equal weight to the realized cumulative abnormal return of each security. If the true abnormal return is constant across securities, J_2 is preferable as it gives more weight to the securities with the lower abnormal return variance (Campbell, Lo, and MacKinlay 1997). For most studies, Campbell, Lo, and MacKinlay argue, the results are expected not to be sensitive to the choice of the above test statistics because the variance of the CAR is usually of a similar magnitude across securities.

The above test statistics are based on the assumption that returns are jointly normally, independently, and identically distributed. Below, we discuss the following robustness checks: (i) nonnormality of abnormal returns, (ii) nonsynchronous trading, and (iii) event clustering. To check the robustness of our results with respect to nonnormality, we use the nonparametric t_{rank} test, that is, Corrado's (1989) nonparametric rank statistic. This nonparametric rank statistic does not require that abnormal returns be normally distributed. Moreover, Campbell and Wasley (1993) document that, compared to both the (parametric) standardized test statistic and the (parametric) portfolio test statistic, this rank statistic is consistently the best specified and most powerful test statistic across numerous event conditions. It is robust to multiday event periods, clustered event dates, and increases in variance on the event day.

The nontrading (or nonsynchronous trading) effect arises when prices are assumed to be recorded at time intervals of one length when in fact they are recorded at time intervals of other, possibly irregular lengths (MacKinlay 1997). This can lead to biased betas in the market model. Scholes and Williams (1977) and Dimson (1979) present a consistent estimator of beta in the presence of nontrading that adjusts the beta estimates upwards. This results in smaller abnormal returns for thinly traded securities. However, Jain (1986) shows that, in general, the adjustment for thin trading is not substantial. Campbell and Wasley (1993) also conclude that adjustment according to Scholes and Williams (1977) does not improve the Type-I error or the power of parametric test statistics. Furthermore, they show that the rank statistic using the abnormal returns obtained from the market model performs best. Therefore, we also rely on the rank test for the robustness checks of the test statistics of firms that suffer from thin trading.

The above expressions for the standard deviation of the CARs assume that the event windows of individual securities do not overlap. This assumption of an absence of clustering allows us to calculate the variance of the sample's cumulative abnormal returns without concern about the covariances across securities as they are zero (MacKinlay 1997). If this assumption is incorrect, then the parametric tests may be biased. Still, Brown and Warner (1985) conclude that, in general, the use of daily or weekly data makes clustering of events on a single day much less severe than the use of monthly data. Also, diversification across industries further mitigates the problem (Bernard 1987). The rank statistic takes care of the event clustering problem as it takes cross-sectional dependence into account via the aggregation of the abnormal returns on an individual security into a time series of portfolio mean ranks. Campbell and Wasley (1993) show that the rank test is again well specified, and in particular for multiday event periods. Therefore, the rank test is a good robustness check in the case of event clustering. It should also be noted that event clustering is not a serious problem in this study as the average number of insider transactions per firm over the eight-year period of 1991 to 1998 is 2.86 purchases and 2.77 sales with medians of 2 for both.

Furthermore, for hypothesis tests over intervals of more than one day, the autocorrelation of the abnormal returns should be taken into consideration. Failure to do so may result in misspecification of the estimated variance of the CAARs. However, Brown and Warner (1985) show that, even though autocorrelation is present, the benefits from autocorrelation adjustments appear to be limited. Campbell and Wasley (1993) draw a similar conclusion: they show that test statistic specifications are not significantly affected by serial dependence. A shift in the variance and the mean of the returns on the event day resulting from the release of new information may cause another type of misspecification, namely, event-induced variance. Still, Campbell and Wasley (1993) show that the rank test is not liable to such misspecification.

REFERENCES

Admati, A., P. Pfleiderer, and J. Zechner. 1994. Large shareholder activism, risk sharing, and financial market equilibrium. *Journal of Political Economy* 102(6):1097–1130.

Barber, B., and J. Lyon. 1997. Detecting long-run abnormal stock returns: The empirical power and specification of test statistics. *Journal of Financial Economics* 43(3):341–72.

Bernard, V. 1987. Cross-sectional dependence and problems in inference in market-based accounting research. *Journal of Accounting Research* 25(1):1–48.

Bettis C., J. Coles, and M. Lemmon. 2000. Corporate policies restricting trading by insiders. *Journal of Financial Economics* 57(2):191–220.

Betzer, A., and E. Theissen. 2004. Insider trading and corporate governance—The case of Germany. Working paper, University of Bonn.

Brown, S., and J. Warner. 1985. Using daily stock returns: The case of event studies. *Journal of Financial Economics* 14(1):3–31.

Campbell, C., and C. Wasley. 1993. Measuring security price performance using daily NASDAQ returns. *Journal of Financial Economics* 33(1):73–92.

Campbell, J., A. Lo, and A. MacKinlay. 1997. *The econometrics of financial markets*. Princeton, NJ: Princeton University Press.

Chang, S., and D. Suk. 1998. Stock prices and secondary dissemination of information: The Wall Street Journal's "Insider Spotlight" column. *Financial Review* 33(3):115–28.

Conyon, M., and K. Murphy. 2000. The price and the pauper? CEO pay in the U.S. and U.K. *Economic Journal* 110(467):640–71.

Copeland, T., T. Koller, and J. Murrin. 1995. *Valuation: Measuring and managing the value of companies*. New York: John Wiley & Sons.

Corrado, C. 1989. A nonparametric test for abnormal security-price performance in event studies. *Journal of Financial Economics* 23(2):385–95.

Dedman, E. 2004. Discussion of reactions of the London Stock Exchange to company trading statement. *Journal of Business Finance and Accounting* 31(1–2):37–47.

Dimson, E. 1979. Risk measurement when shares are subject to infrequent trading. *Journal of Financial Economics* 7(2):197–226.

Dimson, E., and P. Marsh. 1986. Event study methodologies and the size effect. *Journal of Financial Economics* 17(1):113–42.

Dyck, A., and L. Zingales. 2004. Private benefits of control: An international comparison. *Journal of Finance* 59(2):537–600.

Faccio, M., and M. Lasfer. 2000. Do occupational pension funds monitor companies in which they hold large stakes? *Journal of Corporate Finance* 6(1):71–110.

Faccio, M., and M. Lasfer. 2002. Institutional shareholders and corporate governance: The case of U.K. pension funds. In *Corporate governance regimes: Convergence and diversity,* ed. J. McCahery, P. Moerland, T. Raaijmakers, and L. Renneboog. Oxford: Oxford University Press.

Franks, J., C. Mayer, and L. Renneboog. 2001. Who disciplines management in poorly performing companies? *Journal of Financial Intermediation* 10(3–4):209–48.

Friederich, S., A. Gregory, J. Matatko, and I. Tonks. 2002. Short-run returns around the trades of corporate insiders on the London Stock Exchange. *European Financial Management* 8(1):7–30.

Garfinkel, J. 1997. New evidence on the effects of federal regulation on insider trading: The Insider Trading and Securities Fraud Enforcement Act (ITSFEA). *Journal of Corporate Finance* 3(2):89–111.

Goergen, M., and L. Renneboog. 2001. Strong managers and passive institutional investors in the U.K. In *The control of corporate Europe,* ed. F. Barca and M. Becht. Oxford: Oxford University Press.

Gregory, A., J. Matatko, and I. Tonks. 1997. Detecting information from directors' trades: Signal definition and variable size effects. *Journal of Business Finance and Accounting* 24(3):309–42.

Gregory, A., J. Matatko, I. Tonks, and R. Purkis. 1994. U.K. directors' trading: The impact of dealings in smaller firms. *Economic Journal* 104(422):37–53.

Higgs, D. 2003. The review of the role and effectiveness of the non-executive director. U.K. Department of Trade & Industry, London. http://www.ecgi.org/codes/code.php?code_id=124.

Hillier, D., and A. Marshall. 1998. The timing of directors' trades. *Journal of Business Law* 454–67.

Hillier, D., and A. Marshall. 2002. Are trading bans effective? Exchange regulation and corporate insider transactions around earnings announcements. *Journal of Corporate Finance* 8(4):393–410.

Holderness, C., and D. Sheehan. 1988. The role of majority shareholders in publicly held corporations. *Journal of Financial Economics* 20:317–46.

Holmen, M., and P. Högfeldt. 2005. Pyramidal discounts: Tunneling or agency costs? Working paper, European Corporate Governance Institute.

Hu, J., and T. Noe. 1997. The insider trading debate. *Federal Reserve Bank of Atlanta Economic Review* Fourth Quarter:34–45.

Jaffe, J. 1974. Special information and insider trading. *Journal of Business* 47(3):410–28.

Jain, P. 1986. Analyses of the distribution of security market model prediction errors for daily returns data. *Journal of Accounting Research* 24(1):76–96.

Jeng, L., A. Metrick, and R. Zeckhauser. 1999. The profits to insider trading: A performance-evaluation perspective. Working paper 6913, National Bureau of Economic Research.

Jensen, M. C., and W. Meckling. 1976. Theory of the firm: Managerial behavior, agency costs, and ownership structure. *Journal of Financial Economics* 3(4):305–60.

Johnson, S., R. La Porta, F. Lopes-de-Silanes, and A. Shleifer. 2000. The near crash of 1998—Tunneling. *American Economic Review* 90(2):22–27.
King, M., and A. Röell. 1988. Insider trading. *Economic Policy* 6(3):163–93.
Lai, J., and S. Sudarsanam. 1997. Corporate restructuring in response to performance decline: Impact of ownership, governance and lenders. *European Finance Review* 1(2):197–233.
Lakonishok, J., and I. Lee. 2001. Are insiders' trades informative? *Review of Financial Studies* 14(1):79–112.
Lakonishok, J., A. Shleifer, and R. Vishny. 1994. Contrarian investment, extrapolation, and risk. *Journal of Finance* 49(5):1541–1678.
Leland, H. E. 1992. Insider trading: Should it be prohibited? *Journal of Political Economy* 100(4):859–87.
Lin, J. C., and J. Howe. 1990. Insider trading in the OTC market. *Journal of Finance* 45(4):1273–84.
Lustgarten, S., and V. Mande. 1995. Financial analysts' earnings forecasts and insider trading. *Journal of Accounting and Public Policy* 14(3):233–61.
MacKinlay, C. 1997. Event studies in economics and finance. *Journal of Economic Literature* 35(1):13–39.
Maug, E. 1998. Large shareholders as monitors: Is there a trade-off between liquidity and control? *Journal of Finance* 53(1):65–98.
McConnell, J., H. Servaes, and K. Lins. 2005. Changes in equity ownership and changes in the market value of the firm. Working paper, London Business School.
Morck, R., A. Shleifer, and R. Vishny. 1988. Management ownership and market valuation: An empirical analysis. *Journal of Financial Economics* 20:293–315.
Persons, O. 1997. SEC's insider trading enforcements and target firms' stock values. *Journal of Business Research* 39(3):187–94.
Pope, P., R. Morris, and D. Peel. 1990. Insider trading: Some evidence on market efficiency and directors' share dealings in Great Britain. *Journal of Business Finance and Accounting* 17(3):359–80.
Rozeff, M., and M. Zaman. 1988. Market efficiency and insider trading: New evidence. *Journal of Business* 61(1):25–44.
Scholes, M., and J. Williams. 1977. Estimating betas from nonsynchronous data. *Journal of Financial Economics* 5(3):309–27.
Seyhun, N. H. 1986. Insiders' profits, costs of trading and market efficiency. *Journal of Financial Economics* 16(2):189–212.
Stoughton, N., and J. Zechner. 1998. IPO-mechanisms, monitoring and ownership structure. *Journal of Financial Economics* 49(1):45–77.

CHAPTER 18

Incentives to Acquire Information

Philippe Grégoire

CONTENTS

18.1 INTRODUCTION

A trader has just heard a rumor about a stock. She can either trade on this noisy information or she can dig further to unveil the exact story behind the rumor. She is also in a position where she can communicate information to the market (as an analyst, a broker, or a journalist, for example). She then has two decisions to make, the first being how well informed she wants to be and the second being how much of her information to disclose to the market. Sending information to the public attracts liquidity trades, which in turn helps camouflage the informed trader's order to the market maker. If liquidity traders are risk averse, then their reaction to disclosure depends on the precision of the informed trader's information. If, for instance, a perfectly informed trader completely discloses her information, then her asset is riskless, which is not the case with an imperfectly informed trader. Hence the information acquisition decision influences the disclosure decision.

Rumors and speculation drive stock markets. Countless times we have seen, for instance, companies jump due to takeover or merger rumors. As nowadays information flows easily and rapidly, it is important to develop theoretical models that help understand the information acquisition process in an environment where disclosure plays a central role. In the present chapter, we investigate the incentives to acquire information by traders who are expected to make public appearances and reveal what they know to the public.

To do this, we develop a static trading model as in Kyle (1985), with many assets and one informed trader per asset. Each informed trader is endowed with noisy information at the start of the game and has the option to remove the noise from this information at no cost. These traders have also the opportunity to communicate information to the market through a public signal that consists of a noisy transformation of their private information. When information is released, the other market participants know whether the trader is perfectly or imperfectly informed. There is one liquidity trader with negative exponential utility over wealth who can freely allocate his trades across the different assets. The more information released by an informed trader about a stock, the smaller the expected loss and the risk from trading that stock and thus the greater the fraction of liquidity trades allocated to the stock. Orders are simultaneously sent to competitive market makers who cannot differentiate informed trades from liquidity ones and thus the latter serve as camouflage for the former. Market makers set prices such that the market-clearing trades they provide yield them a zero expected payoff.

We find that if two or more informed traders decide to become perfectly informed, then at least two of them completely disclose their information, all the liquidity trades are allocated to the assets with full disclosure, and insiders do not make any profit. If, on the other hand, at most one informed trader decides to become fully informed, then all insiders retain some information, there are liquidity trades allocated to all stocks, and all insiders make positive profits. Any equilibrium is then such that at most one trader decides to become perfectly informed. This result suggests that traders expected to communicate information to the public are better off limiting their knowledge to hazy information than doing thorough investigations to clarify as much as possible the information they have.

As has been demonstrated by Admati and Pfleiderer (1988), Foster and Viswanathan (1990), Bhushan (1991), Chowdhry and Nanda (1991), and Huddart, Hughes, and Brunnermeier (1999), risk-neutral liquidity traders are attracted to stocks with smaller information asymmetry as the latter have prices that are less sensitive to market orders and this reduces the liquidity trader's expected loss. When liquidity traders are risk averse, then they worry not only about their expected loss but also about the risk remaining in trading an asset given the public information about it. As we show in the present chapter, the risk remaining after disclosure is related to the precision of the information held by the trader privy to the stock and this in turn affects the information gathering decision of that trader. This is our main departure from Huddart et al. (1999), which is the paper we mainly build on. In Huddart et al., insiders who can choose between different exchanges to list their stock select the one with the most demanding disclosure requirements as this is where liquidity flows. We modify Huddart et al.'s framework by giving noisy information to the informed trader instead of the exact value of the firm, we give informed traders the opportunity to acquire more information, and we leave the disclosure decision to the discretion of the informed traders instead of modeling it as an exchange requirement.

Grossman and Stiglitz (1980) show that a rational expectations trading model may not have an equilibrium when all traders can become informed of the asset value due to the revelation of that information through the asset price. Verrechia (1982) shows that an equilibrium always exists in a rational expectations model when information can be gradually acquired. Goenka (2003) shows that traders may not be willing to acquire costly information if some of it leaks to the market before trade begins, and Morrison and Vulkan (2005) show that not knowing the exact number of informed traders in a market may limit the entry of informed traders and this creates a rent to traders who acquire costly information. Regarding the link between information acquisition and disclosure, Verrechia (1990) shows that the greater the precision of a manager's information, the more likely he is to disclose it, as withholding it raises more suspicion than with less precise information. Our model provides a similar result as the presence of many perfectly informed traders leads to full disclosure but the drivers are different. That is, we obtain such a result through a competition for liquidity between different assets rather than in a single asset model wherein the manager's compensation is linked to the asset price.

The chapter is structured as follows: the next section describes the model, Section 18.3 characterizes the liquidity allocation, Section 18.4 derives the disclosure decisions, Section 18.55 analyzes the information acquisition process, and Section 18.6 concludes.

18.2 THE MODEL

Consider a one-period trading model with M firms, denoted $m = 1, 2, \ldots, M$. The liquidation value of firm m's stock is given by $v_m \sim N\left(0, \sigma_v^2\right)$ and firm values are not correlated. There is one informed trader per asset, one discretionary liquidity trader, and one market maker per asset. Trading is as in Kyle (1985), where traders simultaneously submit market orders to market makers who set prices such that their expected payoff is zero. There are no restrictions on short sales.

At the start of the game, informed traders are endowed with private information represented by

$$i_m = v_m + \delta_m,$$

where $\delta_m \sim N\left(0, \sigma_{\delta_m}^2\right)$. The δ_m s are independently distributed and the variance $\sigma_{\delta_m}^2$ can take on two values, namely, $\sigma_{\delta_m}^2 = 0$ or $\sigma_{\delta_m}^2 = \sigma_\delta^2 > 0$. The choice of the variance of δ_m is the information acquisition decision of the trader. We will refer to traders with $\sigma_{\delta_m}^2 = 0$ as perfectly informed and to traders with $\sigma_{\delta_m}^2 = \sigma_\delta^2 > 0$ as imperfectly informed. Let $\Sigma_\delta = \left(\sigma_{\delta_1}, \sigma_{\delta_2}, \ldots, \sigma_{\delta_M}\right)$. The vector Σ_δ is common knowledge; all market participants know whether a trader is partially or perfectly informed.

Before trade begins, a signal

$$s_m = i_m + \varepsilon_m$$

is sent for all m, where $\varepsilon_m \sim N\left(0, \sigma^2_{\varepsilon_m}\right)$, all ε_m terms being independently distributed. An informed trader's disclosure decision consists of choosing the standard deviation σ_{ε_m}, where $\sigma_{\varepsilon_m} = 0$ corresponds to full disclosure, $\sigma_{\varepsilon_m} = \infty$ corresponds to no disclosure at all, and any value in between corresponds to partial disclosure. This way of modeling disclosure is similar to Admati and Pfleiderer (1986), Shin and Singh (1999), and Huddart et al. (1999). Note the implicit assumption that informed traders tell the truth on average; i.e., they never intentionally mislead the market. Let $S = \left(s_1, s_2, \ldots, s_M\right)$ and let $\Sigma_\varepsilon = \left(\sigma_{\varepsilon_1}, \sigma_{\varepsilon_2}, \ldots, \sigma_{\varepsilon_M}\right)$, both being common knowledge.

For the model to be solvable, the events related to information happen in the following order for each informed trader:

The informed trader chooses between $\sigma_{\delta_m} = 0$ and $\sigma_{\delta_m} = \sigma_\delta > 0$.

She chooses a disclosure policy σ_{ε_m}.

She receives her private signal i_m and the public signal s_m is sent.

Informed traders trade only in the shares of the company they have information on and their objective is to maximize their trading profits. The market order of informed trader m is denoted x_m and her ex-post payoff is given by $\pi_m = \left(v_m - p_m\right) x_m$, with p_m the price set by market maker m once orders are submitted.

There is one discretionary liquidity trader whose overall market order is exogenously given by $u \sim N\left(0, \sigma^2_u\right)$. The reasons behind this liquidity shock are not modeled here. The liquidity trader can allocate his trades across the different assets in order to maximize his utility over the monetary payoff of his transactions. The fraction of u allocated to asset m is denoted g_m, where $g_m \in \left[0,1\right]$ and $\sum_m g_m = 1$. The liquidity trader's utility function is given by $-e^{-rw}$, where r is the risk aversion coefficient and $w = \sum_m g_m u\left(v_m - p_m\right)$. Let $g = \left(g_1, g_2, \ldots, g_M\right)$.

Market maker m determines the price of asset m after seeing the aggregate order flow $y_m = x_m + g_m u$. The asset price is given by $p_m = E\left[v_m \mid y_{m,} s_m\right]$, which assumes that the market maker provides zero-expected-profit market-clearing trades.

The timing of events is as follows:

Each informed trader decides how precise her private information will be.

Each informed trader decides how precise her public signal will be.

Private signals are received and public signals are sent.

Market orders are submitted to market makers.

Market makers set asset prices and payoffs are realized.

We solve the game backwards, commencing with the trading outcome given information acquisition and disclosure policies. We restrict our attention to equilibria where the asset prices are linear functions of their respective order flow. With all variables normally distributed, the projection theorem gives us (details in Appendix)

$$p_m = E\left[v_m \mid s_m\right] + \lambda_m y_m,$$

where

$$\lambda_m = \frac{\text{var}\left(v_m \mid s_m\right) - \text{var}\left(v_m \mid i_m\right)}{2\bar{g}_m u},$$

with $\bar{g}_m$ being the correct anticipation of g_m.

To simplify the notation, $\text{var}\left(v_m \mid s_m\right)$ and $\text{var}\left(v_m \mid i_m\right)$ will be represented by $\sigma^2_{m|s}$ and $\sigma^2_{m|i}$, respectively. We will also denote $f_m = \sqrt{\sigma^2_{m|s} - \sigma^2_{m|i}}$, which represents the information advantage of trader m, or the information asymmetry in asset m. Hence an informed trader who completely discloses her private information has $f_m = 0$ and an informed trader who does not disclose any information at all has $f_m = \sqrt{\sigma^2_v - \sigma^2_{m|i}}$. As f_m is continuous and strictly increasing in σ_{ε_m}, we will use f_m as the control variable for disclosure as this simplifies the notation. Let $f = \left(f_1, f_2, \ldots, f_M\right)$.

As in Kyle (1985), traders take λ_m as given but consider the impact of their trades on the stock price. Given the inverse supply curve $p_m = E\left[v_m \mid s_m\right] + \lambda_m y_m$, informed trader m's trading strategy x_m is linear in her own private information. The trading equilibrium found is unique when we restrict ourselves to linear strategies. Kyle (1985) shows the existence and uniqueness of the informed trader's trading strategy and the market maker's price schedule in the class of linear functions. This need not be the case when nonlinear pricing is allowed. Rochet and Vila (1994) show uniqueness of equilibrium in the version of Kyle's (1985) one-shot market-order game when the insider observes liquidity trading, assuming the liquidity shock and the asset value to have compact supports.

The parameter λ_m captures relevant information in y_m that cannot be inferred from s_m. When an informed trader completely reveals her private information, the aggregate order flow y_m is uninformative and thus $\lambda_m = 0$. In absence of liquidity trading in asset m, market maker m must set $\lambda_m = \infty$ in order to prevent losses from his market-clearing trades.

Given the precision of her information and her disclosure policy, the expected payoff to an informed trader before she receives her private signal is given by

$$E\left[\pi_m\right] = \frac{\sigma_u f_m \bar{g}_m}{2}$$

where, as before, $\bar{g}_m$ is the correct anticipation of g_m.

18.3 LIQUIDITY ALLOCATION

The negative exponential utility function of the liquidity trader and the normality assumption on the distribution of the random variables gives us the following problem for the liquidity trader (details in Appendix):

$$\min_{\{g_1, g_2, \ldots, g_M\}} \sum_{m=1}^{M} g_m^2 \left(\lambda_m + \frac{r}{8}\left(f_m^2 + 4\sigma_{m|i}^2\right)\right)$$

$$\text{s.t. } g_m \geq 0 \text{ for all } m, \text{ and } \sum\nolimits_m g_m = 1.$$

The liquidity trader's objective function has two main components: the first is the expected loss due to the presence of informed traders and the second component arises from the trader's risk aversion. This second element is what eventually drives our results as it induces the liquidity trader to diversify across the different assets when informed traders have imperfect information.

If there are some perfectly informed traders ($\sigma_{m|i}^2 = 0$) who completely disclose their private information to the market ($f_m = 0$), then the liquidity trader only trades in such assets; that is, all assets with either $f_m > 0$ or $\sigma^2_{m|i} > 0$ do not attract any liquidity trades.

If, on the other hand, $f_m^2 + 4\sigma^2_{m|i} > 0$ for all m, then the liquidity trader allocates a positive fraction of his trades to all assets with sufficiently small information asymmetry. Since corner solutions will not arise in equilibrium, we only mention interior liquidity allocations (i.e., such that $g_m > 0$ for all m). The details of the following lemma are in the Appendix.

Lemma 1

Take Σ_δ as given and let $g_m(f)$ denote the liquidity trades allocated to asset m given $f = \left(f_1, f_2, \ldots, f_M\right)$. If $f_m^2 + 4\sigma^2_{m|i} > 0$ for all m, then an interior solution is such that

$$g_m(f) = \frac{1 + \dfrac{4}{r\sigma_u}\displaystyle\sum_{k=1}^{M}\left(f_k - f_m\right)\left(f_k^2 + 4\sigma^2_{k|i}\right)^{-1}}{\left(f_m^2 + 4\sigma^2_{m|i}\right)\displaystyle\sum_{k=1}^{M}\left(f_k^2 + 4\sigma^2_{k|i}\right)^{-1}} \quad \text{for all } m.$$

What stands out from Lemma 1 is that if two informed traders have the same advantage ($f_m = f_k$ for two assets m, k) and one of them is perfectly informed ($\sigma_{m|i} = 0$) while the other is imperfectly informed ($\sigma_{k|i} > 0$), then the asset with a perfectly informed trader receives a greater share of the liquidity trades. If two informed traders have the same level of information precision, then the asset with the greatest disclosure receives a greater share of the liquidity trades.

18.4 DISCLOSURE DECISIONS

In this section and throughout the rest of the chapter, we alternatively use f_m and $\sigma_{m|s}$ in reference to insider m's disclosure policy. When making her disclosure decision, an insider does not know the liquidation value of her asset and thus her decision is based on the expected payoff

$$E\left[\pi_m\right] = \frac{\sigma_u f_m \bar{g}_m}{2}$$

given the vector of information precision Σ_δ. As shown in the Appendix, the payoff function of insider m is concave in f_m (strictly concave as long as $f_m^2 + 4\sigma_{m|i}^2 > 0$ for all m) and thus an equilibrium in pure strategies always exists. We only consider pure strategy equilibria in this chapter. The following lemma is proved in the Appendix.

Lemma 2

Given Σ_δ, disclosure decisions are as follows:

If there are at least two perfectly informed traders, that is, if $\sigma_{m|i} = 0$ for at least two assets, then at least two of the perfectly informed traders completely disclose their information (i.e., they choose $f_m = 0$). There can be many solutions in this case if there are more than two perfectly informed traders.

If there is at most one perfectly informed trader, then all traders retain some information, that is, $f_m > 0$ for all m. The solution is unique in this case and is such that all imperfectly informed traders choose the same disclosure policy.

Hence unless two or more traders are perfectly informed, there is only partial disclosure of information by all informed traders. Huddart et al. (1999), who have only perfectly informed traders, find a "race toward full disclosure," which induces insiders to list their shares in the exchange with the most stringent disclosure requirements. Even though the framework is different, our result is similar to Verrechia (1990) in that the greater the precision of the traders' information, the higher the level of disclosure.

As partial information leads to less than perfect disclosure which in turn entails greater profit, a trader's incentives to seek information are weak. The following result is proved in the Appendix and will be useful when analyzing information acquisition decisions.

Corollary 1

When there is exactly one perfectly informed trader, disclosure policies are such that more than one half of the liquidity trades are allocated to the perfectly informed trader's asset. That is,

$$g_m > \frac{1}{2}$$

for the asset such that $\sigma_{m|i} = 0$ when $\sigma_{k|i} = 0$ for all $k \neq m$.

18.5 INFORMATION ACQUISITION

Since the presence of two perfectly informed traders leads to full disclosure and eliminates informed trading profits, a trader has no incentive to become perfectly informed when another trader is perfectly informed. As such, there will always be at most one perfectly informed trader in equilibrium. The main result of this chapter is stated in the next proposition.

Proposition 1

With respect to information acquisition, the model described above suggests the following:

In all cases, at most one trader chooses to become perfectly informed.

When the number of assets is sufficiently large, there is always one trader who becomes perfectly informed, and this regardless of the parameter values r, σ_u, σ_δ, and σ_v. This is always the case when $M \geq 6$.

When $M < 6$, whether or not one trader becomes perfectly informed depends on the parameter values.

Figure 18.1 to Figure 18.6 show the disclosure policies and expected payoffs that prevail for different values of

$$\sigma_\Delta = \frac{\sigma_v \sigma_\delta}{\sqrt{\sigma_v^2 + \sigma_\delta^2}},$$

which represents the variance of the asset conditional on the information of an imperfectly informed trader, and for different values of the risk aversion coefficient r. Note that $\sigma_\Delta \to 0$ as $\sigma_\delta \to 0$ and $\sigma_\Delta \to 1$ as $\sigma_\delta \to \infty$. The reason disclosure policies and expected payoffs fall as σ_Δ becomes large is the boundary condition

$$f_m \leq \sqrt{\sigma_v^2 - \sigma_{m|i}^2} \quad \text{for all } m.$$

As shown in Figure 18.1, there are cases when $M = 2$ where all traders remain imperfectly informed. This is true as long as the disclosure policy prevailing in the all-imperfectly informed case is not limited by the boundary condition. Once this point is reached, which is around $\sigma_\Delta = .625$,

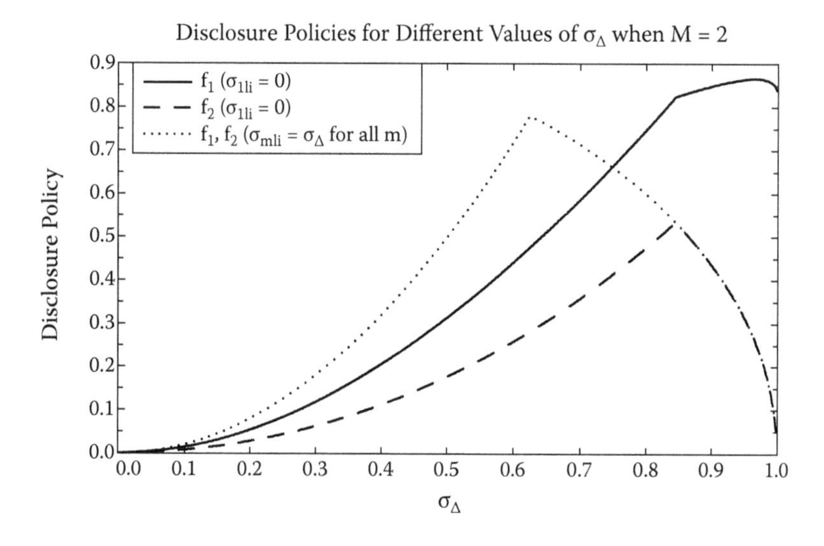

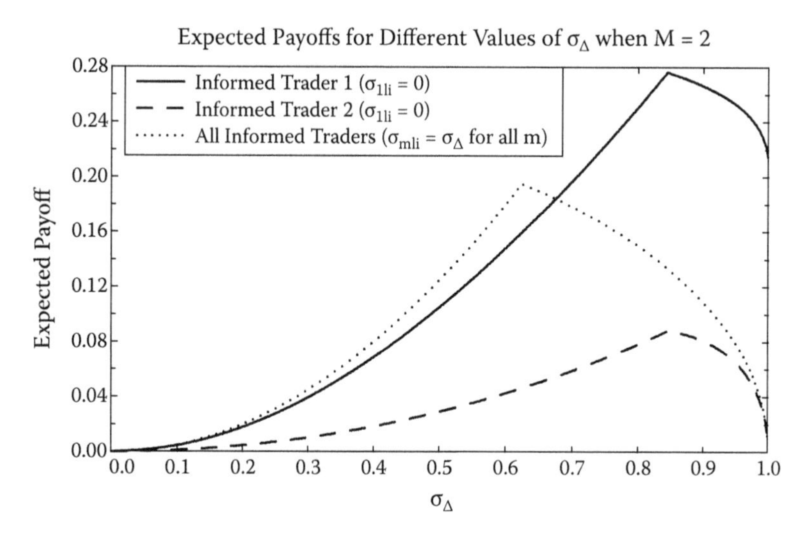

FIGURE 18.1. *Top,* disclosure policies for different values of σ_Δ when $M = 2$. *Bottom,* expected payoffs for different values of σ_Δ when $M = 2$.

expected payoffs in the all-imperfectly informed case start falling and are eventually surpassed by the expected payoff of a trader who chooses to become perfectly informed. What we can also see on this figure is that the greater the variability of the unknown component, that is, the greater σ_Δ, the greater the amount of information retained by informed traders and the greater their expected profits. In Figure 18.2, we can see that the greater the

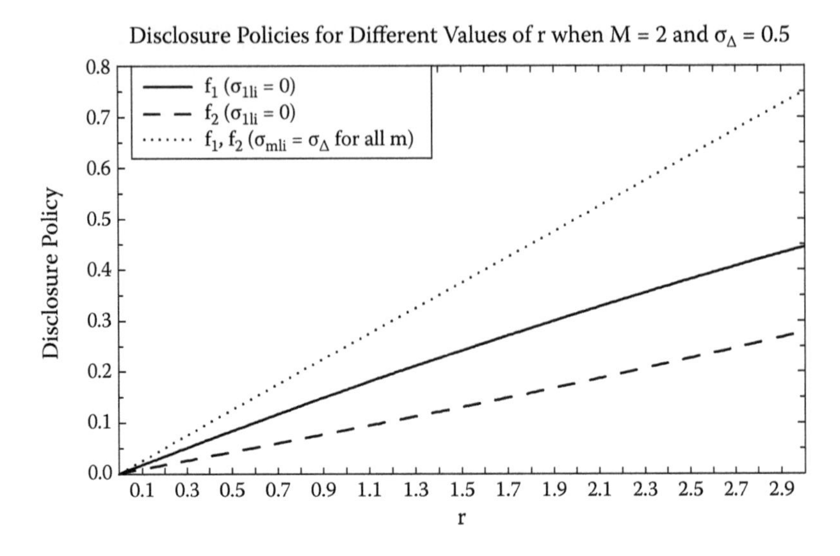

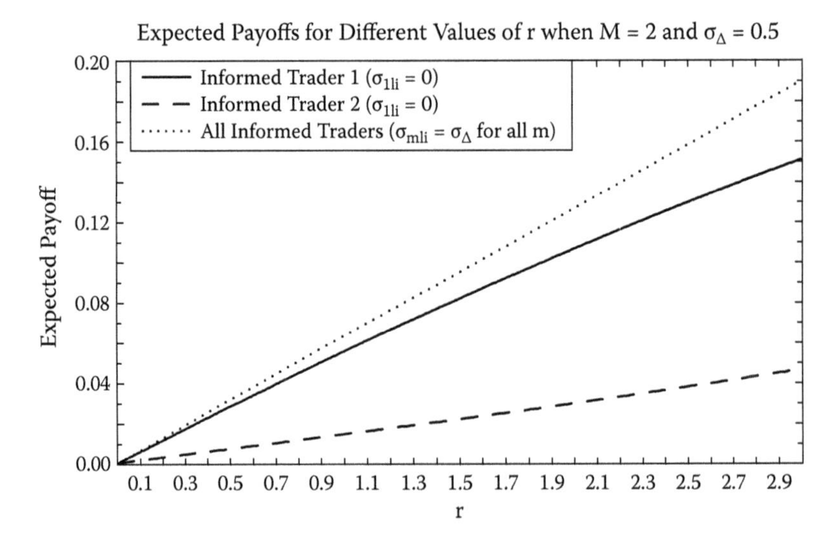

FIGURE 18.2. *Top,* disclosure policies for different values of r when $M = 2$ and $\sigma_\Delta = 0.5$. *Bottom,* expected payoffs for different values of r when $M = 2$ and $\sigma_\Delta = 0.5$.

risk aversion of the liquidity trader, the smaller the amount of information disclosed by informed traders and the greater the latter's expected profits. This result obtains as a more risk averse liquidity trader has a greater need to diversify and is thus less responsive to disclosure. Note that there is generally more information disclosed when there is one perfectly informed trader as the competition for liquidity is fiercer in this case.

Figure 18.3 and Figure 18.4 show the disclosure policies and expected payoffs when $M = 4$. In this case, given the parameter values we have chosen, there is always one trader who chooses to become perfectly informed as her share of liquidity trades then increases from 0.25 to more than 0.5. Note that increasing the number of assets intensifies the competition for liquidity and this leads to more disclosure by informed traders and thus lower expected payoffs.

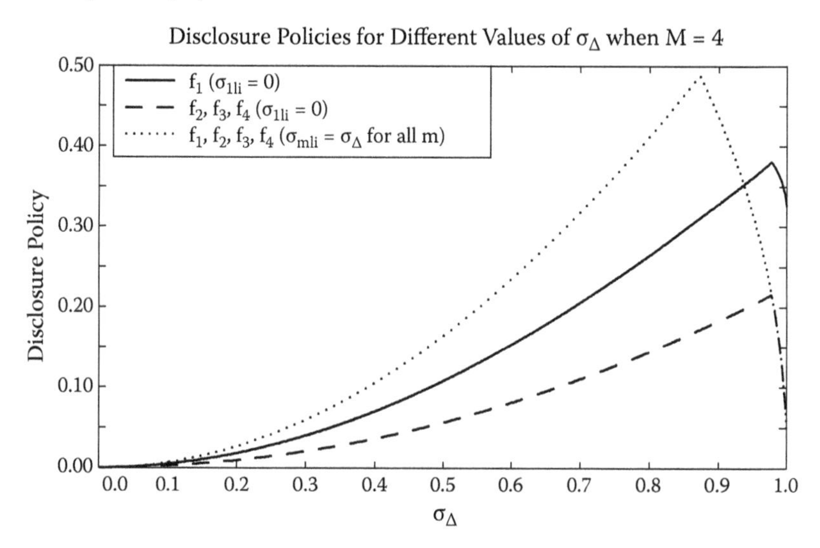

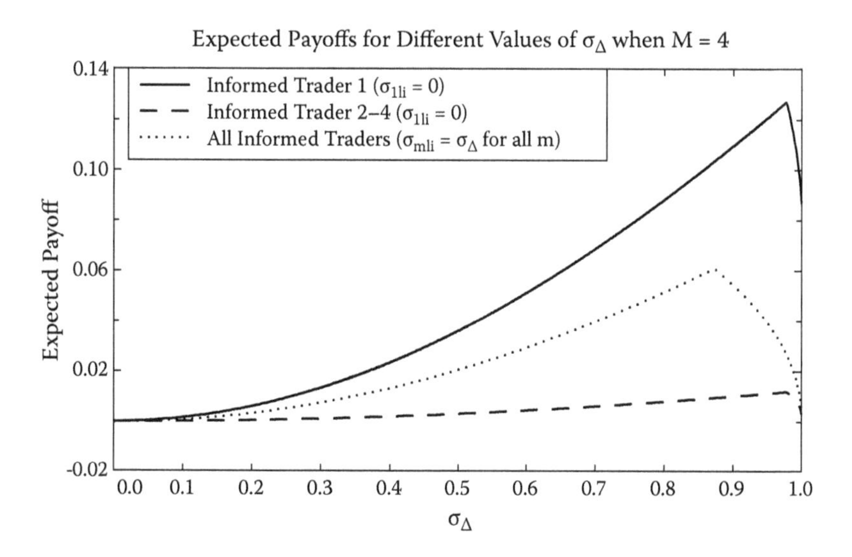

FIGURE 18.3. *Top*, disclosure policies for different values of σ_Δ when $M = 4$. *Bottom*, expected payoffs for different values of σ_Δ when $M = 4$.

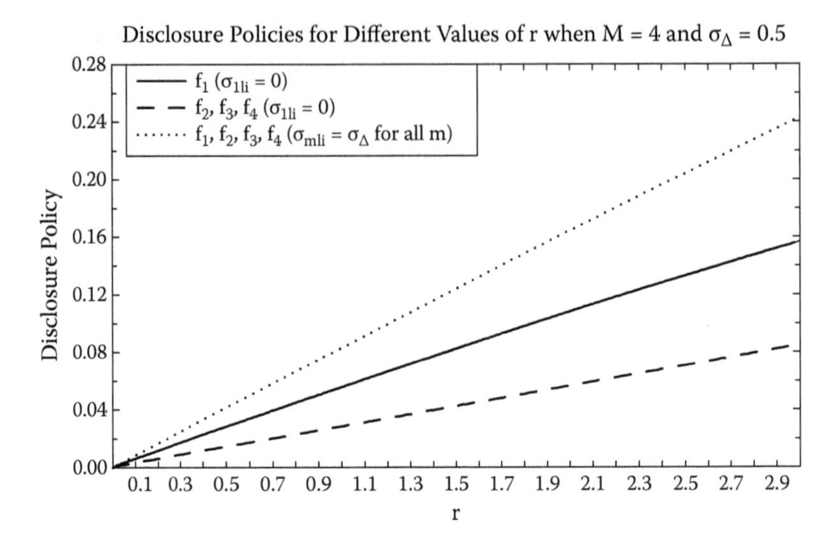

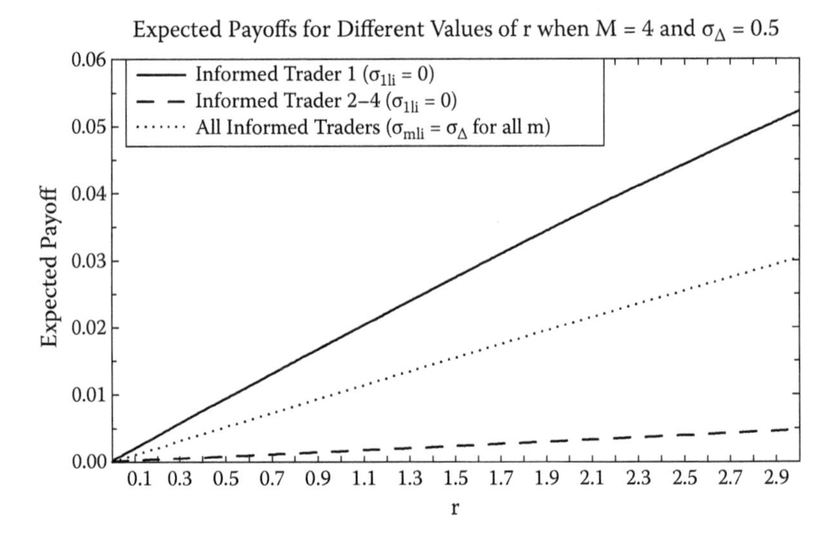

FIGURE 18.4. *Top*, disclosure policies for different values of r when $M = 4$ and $\sigma_\Delta = 0.5$. *Bottom*, expected payoffs for different values of r when $M = 4$ and $\sigma_\Delta = 0.5$.

Figure 18.5 and Figure 18.6 show the disclosure policies and expected payoffs when $M = 6$. We can see on these figures that the improvement in expected payoff to a trader becoming informed is amplified compared to the case $M = 4$. Note also that expected payoffs are lower compared to the previous cases.

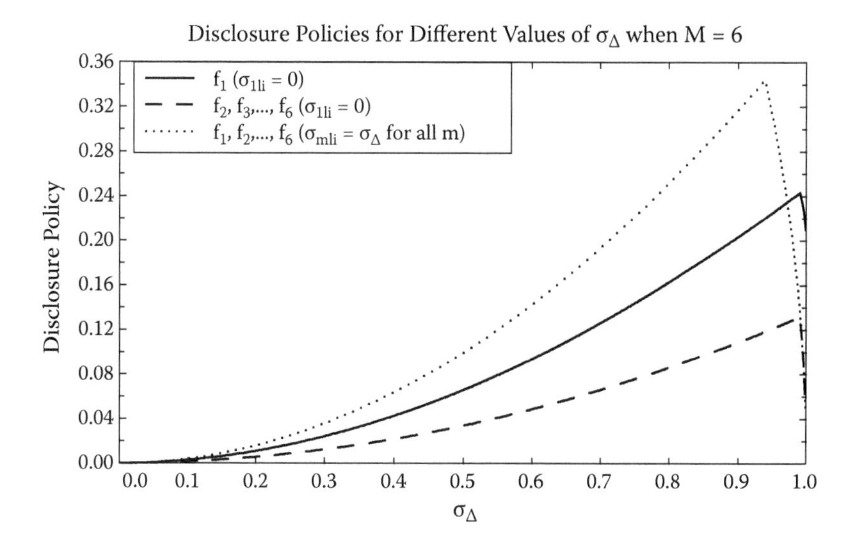

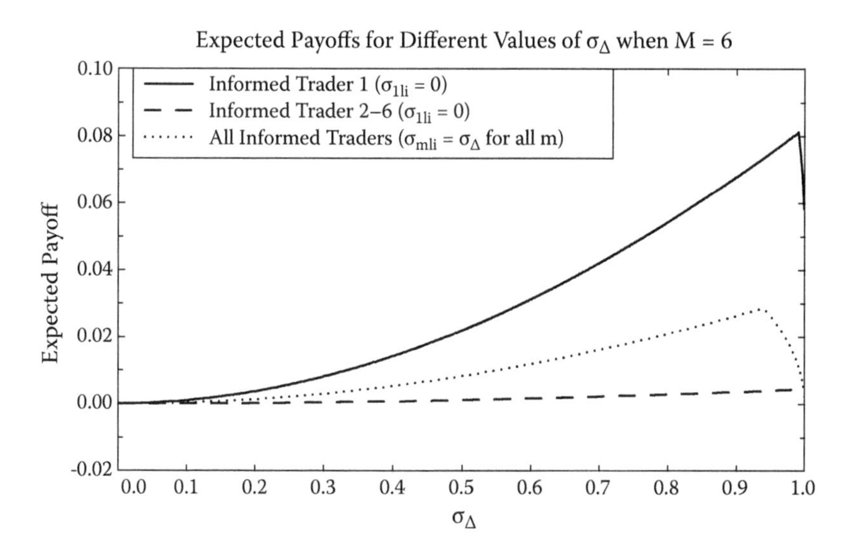

FIGURE 18.5. *Top,* disclosure policies for different values of σ_Δ when $M = 6$. *Bottom,* expected payoffs for different values of σ_Δ when $M = 6$.

18.6 CONCLUSION

A model has been developed where informed traders compete to attract the trades of a discretionary liquidity trader. These trades are attracted by disclosing private information, and this is done through a public signal which is a noisy transformation of the informed trader's private sig-

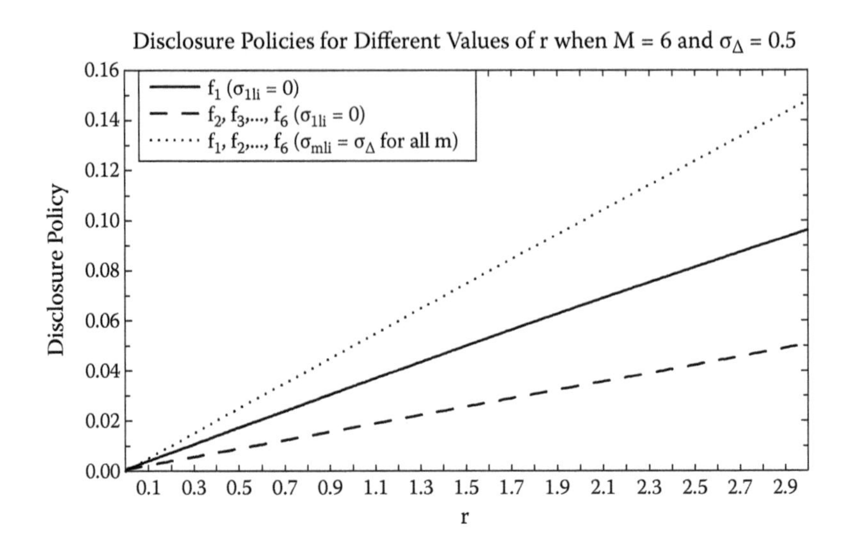

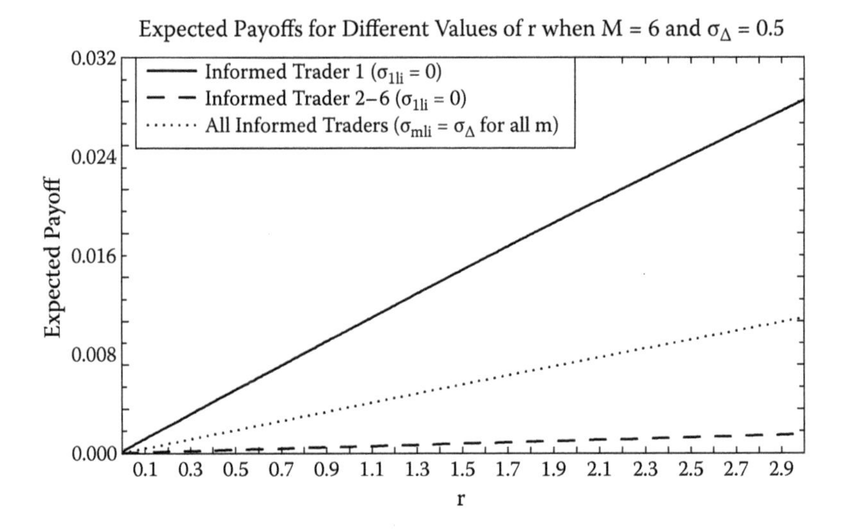

FIGURE 18.6. *Top*, disclosure policies for different values of r when $M = 6$ and $\sigma_\Delta = 0.5$. *Bottom*, expected payoffs for different values of r when $M = 6$ and $\sigma_\Delta = 0.5$.

nal. It has been shown that the precision of a trader's information affects the efficacy of disclosure in attracting liquidity trades. The main result is that at most one trader ever chooses to become perfectly informed as the presence of two or more perfectly informed traders generates a race toward full disclosure and then all traders make a zero expected profit. When the number of assets is small, it is possible that all traders remain partially informed, but there is always one trader who becomes perfectly informed when the number of assets is sufficiently large. The presence of a perfectly informed trader increases the level of disclosure of all traders and this effect is amplified as the number of assets increase.

These results shed some light on the incentives to do thorough research when a trader is expected to make public announcements divulging some of their findings. These traders can be analysts, mutual funds, brokerage firms, rating agencies, and so forth. Whether these individuals are provided with the proper incentives to gather and disclose information is an important question. What is shown in this chapter is that these types of traders have incentives to remain partially informed even when information can be acquired at no cost.

A next step to this chapter would be to compare how different trading mechanisms can affect traders' incentives to gather information. Markets would clearly benefit from giving all analysts the incentives to use all the resources available to them to collect the information needed for their research.

18.7 APPENDIX

18.7.1 Pricing Functions and Insiders' Expected Payoff

Suppose Insider m's trading strategy is linear in her private signal, that is,

$$x_m = \alpha_m + \beta_m v_m,$$

where α_m and β_m are constant. Since v_m and s_m are multivariate normal, the variable x_m given s_m is also normally distributed. Let then $E[x_m \mid s_m] = u_{x|s}$ and a variance $\mathrm{var}(x_m \mid s_m) = \sigma^2_{x|s}$. The aggregate order flow to asset m, $y_m = x_m + g_m u$, is therefore normally distributed with $E[y_m \mid s_m] = E[x_m + g_m u \mid s_m] = E[x_m \mid s_m] = \mu_{x|s}$, and thus

$$E[v_m \mid s_m, y_m] = E[v_m \mid s_m] + \lambda_m (y_m - u_{x|s}),$$

where

$$\lambda_m = \frac{\mathrm{cov}(v_m, y_m \mid s_m)}{\mathrm{var}(y_m \mid s_m)}.$$

Since $p_m = E[v_m | s_m, y_m]$ and $E[g_m u | i_m, s_m] = 0$, Insider *m*'s order flow is obtained from

$$\max_x E\left[(v_m - (E[v_m | s_m] + \lambda_m (x - u_{x|s})))x | i_m, s_m\right],$$

which gives

$$x_m = \frac{1}{2\lambda_m}\left(E[v_m | i_m] - E[v_m | s_m] + \lambda_m u_{x|s}\right),$$

where $E[v_m | i_m] = i_m$ and $E[v_m | s_m] = \frac{\sigma_v^2 s_m}{\sigma_v^2 + \sigma_{\varepsilon_m}^2}$. Since

$$E[x_m | s_m] = u_{x|s} = E\left[\frac{1}{2\lambda_m}\left(E[v_m | i_m] - E[v_m | s_m] + \lambda_m u_{x|s}\right)\right] = \frac{1}{2} u_{x|s},$$

we must have $u_{x|s} = 0$ and thus

$$x_m = \frac{1}{2\lambda_m}\left(i_m - E[v_m | s_m]\right).$$

x_m is therefore effectively a linear function of i_m, with

$$\beta_m = \frac{1}{2\lambda_m} \text{ and } \alpha_m = -\frac{1}{2\lambda_m} E[V_m | s_m],$$

and thus our initial conjecture was correct.

It is now possible to find the distribution of x_m given s_m, which is

$$x_m | s_m \sim N\left(0, \frac{\sigma_{m|s}^2 - \sigma_{m|i}^2}{4\lambda_m^2}\right),$$

where

$$\sigma_{m|s}^2 = \text{var}(v_m \mid s_m) = \frac{\sigma_v^2\left(\sigma_{\delta_m}^2 + \sigma_{\varepsilon_m}^2\right)}{\sigma_v^2 + \sigma_{\delta_m}^2 + \sigma_{\varepsilon_m}^2}$$

and

$$\sigma_{m|i}^2 = \text{var}(v_m \mid i_m) = \frac{\sigma_v^2 \sigma_{\delta_m}^2}{\sigma_v^2 + \sigma_{\delta_m}^2}.$$

Note that $\sigma_{\delta_m}^2 = 0$ for a perfectly informed trader. Since $\text{cov}\left(v_m, i_m \mid s_m\right) = \sigma_{m|s}^2 - \sigma_{m|i}^2$ and since $\text{cov}\left(v_m, u \mid s_m\right) = 0$, we have

$$\lambda_m = \frac{\text{cov}(v_m, y_m \mid s_m)}{\text{var}(y_m \mid s_m)} = \frac{\text{cov}(v_m, x_m \mid s_m)}{\text{var}(y_m \mid s_m)} = \frac{\frac{1}{2\lambda_m}\left(\sigma_{m|s}^2 - \sigma_{m|i}^2\right)}{\frac{\sigma_{m|s}^2 - \sigma_{m|i}^2}{4\lambda_m^2} + g_m^2 \sigma_u^2},$$

which gives

$$\lambda_m = \frac{\sqrt{\sigma_{m|s}^2 - \sigma_{m|i}^2}}{2 g_m \sigma_u}.$$

Letting $f_m^2 = \sigma_{m|s}^2 - \sigma_{m|i}^2$, Insider m's expected payoff before knowing $E\left[\pi_m\right] = \frac{f_m^2}{4\lambda_m} = \frac{1}{2} f_m g_m \sigma_u$ is given by

$$E\left[\pi_m\right] = \frac{f_m^2}{4\lambda_m} = \frac{1}{2} f_m g_m \sigma_u.$$

18.7.2 Derivation of the Liquidity Trader's Problem

The liquidity trader's problem is

$$\max_{\{g_m\}_{m=1}^M} E\left[-\exp\left(-r \sum_{m=1}^{M} g_m u\left(v_m - p_m\right)\right) \middle| s, u\right].$$

For all m, we have

$$v_m - p_m = v_m - E\left[v_m \mid s_m\right] - \frac{1}{2}\left(E\left[v_m \mid i_m\right] - E\left[v_m \mid s_m\right]\right) - \lambda_m g_m u,$$

and thus

$$\left(v_m - p_m\right) \mid s_m, u \sim N\left(-\lambda_m g_m u, \frac{1}{4}\left(\sigma^2_{m|s} + 3\sigma^2_{m|i}\right)\right).$$

With asset values uncorrelated, the discretionary liquidity trader's portfolio, given u and s, has an expected return equal to

$$-\sum_m \lambda_m g_m^2 u^2$$

and a variance

$$\frac{1}{4}\sum_m g_m^2 u^2 \left(\sigma^2_{m|s} + 3\sigma^2_{m|i}\right).$$

As the liquidity trader's problem represents the moment generating function of a normal variable with a coefficient r, the liquidity trader's problem can be rewritten as

$$\max_{\{g_m\}_{m=1}^{M}} \left\{-\sum_{m=1}^{M} \lambda_m g_m^2 u^2 - \frac{r}{8}\sum_{m=1}^{M} g_m^2 u^2 \left(\sigma^2_{m|s} + 3\sigma^2_{m|i}\right)\right\}.$$

18.7.3 Proof of Lemma 1

Using μ as the Lagrange multiplier in the liquidity trader's problem, we obtain the first-order conditions

$$-2g_m\left(\lambda_m + \frac{r}{8}\left(\sigma^2_{m|s} + 3\sigma^2_{m|i}\right)\right) + \mu = 0 \text{ for all } m$$

and $\sum_m g_m = 1$, which gives

$$g_m = \frac{4}{r\left(\sigma^2_{m|s} + 3\sigma^2_{m|i}\right)}\left(\mu - \frac{\sqrt{\sigma^2_{m|s} - \sigma^2_{m|i}}}{\sigma_u}\right) + \mu$$

Using $\sum_m g_m = 1$, we solve for μ and find

$$g_m = \frac{1 + \frac{4}{r\sigma_u}\sum_{k=1}^{M}\frac{\sqrt{\sigma^2_{k|s} - \sigma^2_{k|i}} - \sqrt{\sigma^2_{m|s} - \sigma^2_{m|i}}}{\sigma^2_{k|s} + 3\sigma^2_{k|i}}}{\left(\sigma^2_{m|s} + 3\sigma^2_{m|i}\right)\sum_{k=1}^{M}\frac{1}{\sigma^2_{k|s} + 3\sigma^2_{k|i}}}.$$

Replace $< \sqrt{\sigma^2_{m|s} - \sigma^2_{m|i}}$ by f_m; this gives

$$g_m = \frac{1 + \frac{4}{r\sigma_u}\sum_{k=1}^{M}\frac{f_k - f_m}{f_k^2 + 4\sigma^2_{k|i}}}{\left(f_m^2 + 4\sigma^2_{m|i}\right)\sum_{k=1}^{M}\frac{1}{f_k^2 + 4\sigma^2_{k|i}}}.$$

18.7.4 Proof of Lemma 2

Let

$$\Psi_m(f) = \frac{\partial E\left[\pi_m\right]}{\partial f_m}$$

for all m. If $f_k^2 + 4\sigma^2_{k|i} > 0$ for all $k \neq m$, then

$$\Psi_m(f) = g_m(f)\left(1 - \sum_{k \neq m}\frac{f_m^2 - 4\sigma^2_{m|i}}{f_k^2 + 4\sigma^2_{k|i}}\right) - \frac{4}{r\sigma_u}\sum_{k \neq m}\frac{f_m}{f_k^2 + 4\sigma^2_{k|i}}.$$

Note that $\Psi_m(f)$ is strictly decreasing in f_m and thus each insider's payoff function is strictly concave in her own strategy when $f_k^2 + 4\sigma^2_{k|i} > 0$ for all

$k \neq m$, which implies that an equilibrium in pure strategies always exists in this case.

18.7.5 There Are at Least Two Perfectly Informed Traders

Suppose $\sigma_{m'|i} = \sigma_{m''|i} = 0$ for two assets m', m'', suppose that $f_k^2 + 4\sigma_{k|i}^2 > 0$ for all $k \neq m'$, m'', and suppose that $f_{m'}, f_{m''} > 0$. Then, for asset m', we have

$$\Psi_{m'}(f) = g_{m'}(f)\left(1 - \frac{f_{m'}^2}{f_{m''}^2} - \sum_{k \neq m', m''} \frac{f_{m'}^2}{f_k^2 + 4\sigma_{k|i}^2}\right) - \frac{4}{r\sigma_u}\sum_{k \neq m'} \frac{f_{m'}}{f_k^2 + 4\sigma_{k|i}^2}.$$

To have $\Psi_{m'}(f) = 0$, we need $f_{m'} < f_{m''}$, which then implies that $\Psi_{m''}(f) < 0$, and thus this cannot be an equilibrium. The only possible equilibrium disclosure policies in this case are such that $f_{m'} = f_{m''} = 0$.

18.7.6 There Is Exactly One Perfectly Informed Trader

Suppose there is exactly one perfectly informed trader, trader 1, say. That is, suppose that $\sigma_{1|i} = 0$ and $\sigma_{m|i} > 0$ for all $m \neq 1$. Then

$$\Psi_1(f) = g_1(f)\left(1 - \sum_{k \neq 1} \frac{f_1^2}{f_k^2 + 4\sigma_{k|i}^2}\right) - \frac{4}{r\sigma_u}\sum_{k \neq 1} \frac{f_1}{f_k^2 + 4\sigma_{k|i}^2}.$$

If $f_1 = 0$, then $g_1 = 1$ and $\Psi_{m''}(f) < 0$, inciting informed trader 1 to choose a higher f_1. Hence in this case informed trader 1 always chooses an $f_1 > 0$. For the remaining traders, we have

$$f_m = 0 \Rightarrow \Psi_m(f) = g_m(f)\left(1 + \sum_{k \neq m} \frac{4\sigma_{m|i}^2}{f_k^2 + 4\sigma_{k|i}^2}\right) > 0$$

since $g_m > 0$ when $f_1 > 0$ and $f_m = 0$. An imperfectly informed trader will always choose $f_m > 0$ in this case. As is shown below, all imperfectly informed traders choose the same disclosure policy in this case. Moreover, since

$$\frac{\partial \Psi_m(f)}{\partial f_k} > 0.$$

for all k, m, $k \neq m$, the solution is unique.

Hence all insiders retain some information when at most one of them is perfectly informed and there is a unique vector of equilibrium disclosure policies.

18.7.7 All Traders Are Imperfectly Informed

Suppose that $\sigma_{\delta_m} = \sigma_\delta > 0$ for all m. Let in this case $\sigma_{m|i} = \sigma_\Delta$ for all m. If we single out two assets m', m'', we have

$$\Psi_{m'}(f) = g_{m'}(f)\left(1 + \frac{4\sigma_\Delta^2 - f_{m'}^2}{f_{m''}^2 + 4\sigma_\Delta^2} - \sum_{k \neq m', m''} \frac{f_{m'}^2 - 4\sigma_\Delta^2}{f_k^2 + 4\sigma_\Delta^2}\right) - \frac{4}{r\sigma_u} \sum_{k \neq m'} \frac{f_{m'}}{f_k^2 + 4\sigma_\Delta^2}.$$

To have both $\Psi_{m'}(f) = 0$ and $\Psi_{m''}(f) = 0$, we must have $f_{m'} = f_{m''}$. Suppose, for instance, that $f_{m'} > f_{m''}$. Then

$$g_{m'}(f) < g_{m'}(f),$$

$$\frac{4\sigma_\Delta^2 - f_{m'}^2}{f_{m''}^2 + 4\sigma_\Delta^2} < \frac{4\sigma_\Delta^2 - f_{m''}^2}{f_{m'}^2 + 4\sigma_\Delta^2},$$

$$\sum_{k \neq m', m''} \frac{f_{m'}^2 - 4\sigma_\Delta^2}{f_k^2 + 4\sigma_\Delta^2} > \sum_{k \neq m', m''} \frac{f_{m''}^2 - 4\sigma_\Delta^2}{f_k^2 + 4\sigma_\Delta^2},$$

and

$$\sum_{k \neq m'} \frac{f_{m'}}{f_k^2 + 4\sigma_\Delta^2} > \sum_{k \neq m'} \frac{f_{m''}}{f_k^2 + 4\sigma_\Delta^2},$$

which implies that if $\Psi_{m'}(f) = 0$, then $\Psi_{m''}(f) = 0$, and thus this cannot be possible in equilibrium. Hence we must have $f_{m'} = f_{m''}$ for all m', m'', when $\sigma_{m|i} = \sigma_\Delta$ for all m. Note that even when there is one perfectly informed trader, imperfectly informed traders choose the same disclosure policy.

If all the informed traders choose the same disclosure policy $\bar{f}$, say, then $g_m = 1/M$ for all m and the first-order conditions give us

$$\Psi_m(f) = \frac{1}{M}\left(1 - \sum_{k\neq m} \frac{\bar{f} - 4\sigma_\Delta^2}{\bar{f} + 4\sigma_\Delta^2}\right) - \frac{4}{r\sigma_u} \sum_{k\neq m'} \frac{\bar{f}}{\bar{f} + 4\sigma_\Delta^2} = 0$$

for all m. Solving for $\bar{f}$, we obtain

$$\bar{f} = \begin{cases} r\sigma_u \sigma_\Delta^2 & \text{if } M = 2, \\ \sqrt{\left(\frac{2M(M-1)}{r\sigma_u(M-2)}\right)^2 + \frac{4M\sigma_\Delta^2}{M-2}} - \frac{2M(M-1)}{r\sigma_u(M-2)} & \text{if } M > 2. \end{cases}$$

Since $\bar{f}$ is bounded above by $\sqrt{\sigma_v^2 - \sigma_\Delta^2}$, the disclosure policies will be

$$f_m = f^* = \min\left\{\bar{f}, \sqrt{\sigma_v^2 - \sigma_\Delta^2}\right\}$$

for all m. It is clear here that this equilibrium is unique.

18.7.8 Proof of Corollary 1

Suppose informed trader 1 is perfectly informed and the remaining traders are partially informed. In an interior solution, we have, for all m,

$$\Psi_m(f) = \left(\frac{1 + \frac{4}{r\sigma_u} \sum_{k\neq m}^{M} \frac{f_k - f_m}{f_k^2 + 4\sigma_{k|i}^2}}{1 + \left(f_m^2 + 4\sigma_{m|i}^2\right) \sum_{k\neq m}^{M} \frac{1}{f_k^2 + 4\sigma_{k|i}^2}}\right)\left(1 - \sum_{k\neq m} \frac{f_m^2 - 4\sigma_{m|i}^2}{f_k^2 + 4\sigma_{k|i}^2}\right)$$

$$- \frac{4}{r\sigma_u} \sum_{k\neq m} \frac{f_m}{f_k^2 + 4\sigma_{k|i}^2} = 0,$$

which can be rearranged as

$$f_m^2 + \frac{8\left(1+4\sigma_{m|i}^2 B_m\right)}{r\sigma_u + 4A_m} f_m - \frac{1+4\sigma_{m|i}^2 B_m}{B_m} = 0$$

where

$$A_m = \sum_{k \neq m} \frac{f_k}{f_k^2 + 4\sigma_{k|i}^2} \text{ and } B_m = \sum_{k \neq m} \frac{1}{f_k^2 + 4\sigma_{k|i}^2}.$$

We have shown earlier that all imperfectly informed traders choose the same disclosure policy, $\bar{f}$, say. Letting $\sigma_{m|i} = \sigma_\Delta$ for all $m \geq 2$, we have

$$B_1 = \frac{M-1}{\hat{f}^2 + 4\sigma_\Delta^2}$$

and thus

$$f_1^2 = \frac{\hat{f}^2 + 4\sigma_\Delta^2}{M-1} - \frac{8f_1}{r\sigma_u + 4A_1}.$$

Writing the first-order condition for trader 1 as

$$g_1(f)\left(1 - \frac{(M-1)f_1^2}{\hat{f}^2 + 4\sigma_\Delta^2}\right) - \frac{4(M-1)f_1}{r\sigma_u\left(\hat{f}^2 + 4\sigma_\Delta^2\right)} = 0,$$

we obtain

$$g_1(f) = \frac{4(M-1)f_1}{r\sigma_u\left(\hat{f}^2 + 4\sigma_\Delta^2 - (M-1)f_1^2\right)}$$

$$= \frac{4(M-1)f_1}{r\sigma_u\left(\hat{f}^2 + 4\sigma_\Delta^2 - (M-1)\left(\dfrac{\hat{f}^2 + 4\sigma_\Delta^2}{M-1} - \dfrac{8f_1}{r\sigma_u + 4A_1}\right)\right)}$$

$$= \frac{1}{2} + \frac{2A_1}{r\sigma_u}.$$

Since

$$\frac{2A_1}{r\sigma_u} > 0,$$

this means that $g_1(f) > \frac{1}{2}$.

18.7.9 Proof of Proposition 1

Suppose informed trader 1 is perfectly informed, and all other informed traders choose a disclosure policy $\hat{f}$ and $\sigma_{m|i} = \sigma_\Delta$ for all $m \geq 2$. Using

$$f_1^2 = \frac{\hat{f}^2 + 4\sigma_\Delta^2}{M-1} - \frac{8f_1}{r\sigma_u + 4A_1}$$

and

$$g_1(f) = \frac{1}{2} + \frac{2A_1}{r\sigma_u},$$

we can write

$$g_1(f)f_1 = \left(\frac{1}{2}+\frac{2A_1}{r\sigma_u}\right)\left(\sqrt{\left(\frac{4}{r\sigma_u+4A_1}\right)^2+\frac{\hat{f}^2+4\sigma_\Delta^2}{M-1}}-\frac{4}{r\sigma_u+4A_1}\right)$$

$$= \sqrt{\left(\frac{2}{r\sigma_u}\right)^2+\left(\frac{1}{2}+\frac{2A_1}{r\sigma_u}\right)^2\frac{\hat{f}^2+4\sigma_\Delta^2}{M-1}}-\frac{2}{r\sigma_u}.$$

Note that

$$\frac{2}{r\sigma_u}<\frac{2(M-1)}{r\sigma_u(M-2)}$$

for all M and

$$\left(\frac{1}{2}+\frac{2A_1}{r\sigma_u}\right)^2\frac{\hat{f}^2+4\sigma_\Delta^2}{M-1}>\frac{4\sigma_\Delta^2}{4(M-1)}>\frac{4\sigma_\Delta^2}{M(M-2)}$$

whenever $M \geq 6$. Therefore,

$$g_1(f)f_1 = \sqrt{\left(\frac{2(M-1)}{r\sigma_u(M-2)}\right)^2+\frac{4\sigma_\Delta^2}{M(M-2)}}-\frac{2(M-1)}{r\sigma_u(M-2)}=\frac{1}{M}\hat{f}$$

whenever $M \geq 6$, and this for any value of r, σ_u, and σ_Δ. That is, there is always one informed trader who chooses to become perfectly informed when M is sufficiently large.

REFERENCES

Admati, A. R., and P. Pfleiderer. 1988. A theory of intraday patterns: Volume and price variability. *Review of Financial Studies* 1(1):3–40.

Bhushan, R. 1991. Trading costs, liquidity, and asset holdings. *Review of Financial Studies* 4(2):343–60.

Chowdhry, B., and V. Nanda. 1991. Multimarket trading and market liquidity. *Review of Financial Studies* 4(3):483–511.

Foster, D. F., and S. Viswanathan. 1990. A theory of the interday variations in volume, variance, and trading costs in securities markets. *Review of Financial Studies* 3(4):593–624.

Goenka, A. 2003. Informed trading and the leakage of information. *Economic Theory* 109(2):360–77.

Grossman, S., and J. E. Stiglitz. 1980. On the impossibility of informationally efficient markets. *American Economic Review* 70(3):393–408.

Huddart, S., J. S. Hughes, and M. Brunnermeier. 1999. Disclosure requirement and stock exchange listing choice in an international context. *Journal of Accounting and Economics* 26(1):237–69.

Kyle, A. S. 1985. Continuous auctions and insider trading. *Econometrica* 53(6):1315–35.

Morrison, A. D., and N. Vulkan. 2005. Making money out of publicly available information. *Economic Letters* 89(1):31–38.

Rochet, J. C., and J. L. Vila. 1994. Insider trading without normality. *Review of Economic Studies* 61(1):131–52.

Shin, J., and R. Singh. 1999. Corporate disclosure: Strategic donation of information. Working paper, Carlson School of Management, University of Minnesota.

Verrechia, R. E. 1982. Information acquisition in a noisy rational expectations economy. *Econometrica* 50(6):1415–30.

Verrechia, R. E. 1990. Information quality and discretionary disclosure. *Journal of Accounting and Economics* 12(4):365–80.

CHAPTER 19

Insider Trading in Emerging Stock Markets

The Case of Brazil

Otavio Ribeiro de Medeiros

CONTENTS

19.1 INTRODUCTION

In the past twenty years or so, the development of the Brazilian stock market has raised concerns among investors and investment professionals, as well as among academic and legal professionals and the media about the use of insider information in Brazil. In recent years, a number of insider trading cases have been documented, such as the *Ambev Brewery* case, investigated by CVM (Comissão de Valores Mobiliários, the Brazilian version of the SEC, Securities and Exchange Commission in the United States), regarding trades executed by major shareholders and executive directors before Ambev merged with Belgian brewery Interbrew in 2004 (Proença 2005). Another recent example is CVM Administrative Sanctioning Lawsuit no. 18/01, which investigated the use of inside information related to the relevant event of an equity issue by Copel Parana Energy disclosed on July 25, 2001. The CVM considered the possibility of applying the same trading conditions to owners of this company's ordinary shares as the State of Paraná (one of the states of the Brazilian federation, which comprises 26 states and the federal district) applied to Bndespar. The disclosure of the relevant fact caused the price of the company's ordinary shares to appreciate by about 14 percent. This process led to the imposition of sentences on investment fund managers who traded Copel's stocks using insider information, based on CVM instruction no. 31/84 (Rochman and Eid 2006).

More recently, the Brazilian Federal Justice blocked a suspicious deal of BRL 4 million involving the sale of Ipiranga Group shares before the announcement of the acquisition of the group by Petrobras, Braskem, and Ultra Group. The injunction order, solicited by CVM and by the Federal Public Ministry, was issued on March 21, 2007 and it impeded two investors—a foreign fund and a physical person—from receiving the money obtained from the sale of the shares. However, the Justice's decision does not hinder the sale of the company. The press reported that two days prior to the announcement of the sale of Ipiranga, the price of the company's stocks traded in Bovespa appreciated by 33 percent. For this reason, immediately after the disclosure of the sale of Ipiranga, CVM stated that it suspected insider trading and would launch an investigation. The federal judge, Mauro da Costa Braga, of the 1st Federal Court of Justice of Rio de Janeiro, said he understood that the trade affected market credibility. CVM asked for clarification concerning the strong appreciation of the stock price. Ipiranga replied that assessors of the controlling shareholders had held confidential negotiations with assessors of potential buyers. For the first time, a legal act blocked a stock market trade based on suspicious insider information. If insider trading is confirmed during the investigation, CVM can impose a fine and demand that the investors involved be banned from the financial market for up to twenty years (Revista Consultor Jurídico, March 22, 2007). By the time this chapter was finished, no decision had yet been taken by CVM or by the Federal Justice.

We support the idea that insider trading is a form of corruption. As such, the higher the corruption level in a country, the more intense insider trading practices in that country are likely to be. In 2006, in a global ranking of perceived corruption, Transparency International, ranked Brazil 70th among 160 countries. The first positions are occupied by less corrupt countries, and the last positions by the most corrupt ones. This means Brazil is in an intermediate zone, along with Saudi Arabia, China, Egypt, Ghana, India, Mexico, Peru, and Senegal. In Brazil, there is a widespread feeling among the public that the authorities do not fight insider trading efficiently enough and that the market tolerates it.

Some authors argue that corruption is inversely related to education. For example, Mauro (1995) points out that unstable and corrupt governments tend to invest less in public education, which corroborates the idea advanced by Shleifer and Vishny (1993) that investment in education would provide fewer opportunities for corruption. Pereira (2004) suggests that the struggle against corruption requires the control of public administrators, which can be exerted through voting, controlling congress, and bureaucratic procedures. He adds that the set of actions necessary to fight corruption involves educating the citizens, besides ensuring transparency in the administration of public goods. On the other hand, the poor educational performance of Brazilians is well documented (Birdsall and Sabot 1996). De Barros, Henriques, and Mendonça (2002) assert that "the comparison of the Brazilian reality with the international experience confirms the weak performance of our educational system in the last decades. Brazil presents a delay, in terms of education, of about one decade in relation to a typical country with a pattern of similar development to ours." Several authors have pointed out systematic underinvestment in education in Brazil (De Barros and Mendonça 1997), while others have maintained that investments in education in Brazil are not low. The problem is that they are badly spent (Schwartzman 2004).

De Araujo (2005) suggests that effective combating of corruption must destroy the rent-seeking microeconomic rationale that uses institutional weaknesses to expropriate wealth. He also affirms that in the long term, investment in public education, especially at a fundamental level, is one of the key factors in the struggle against corruption. On the one hand, such investments are less susceptible to corruption; on the other hand, they assure that civil society will have its human capital sufficiently developed to best monitor public institutions, as long as they are transparent. The author concludes by saying that transparency and education are instrumental in ensuring the strengthening of the institutions and reducing corruption.

19.2 CORRUPTION AND INSIDER TRADING

Insider trading is often regarded as a form of corruption. Although Shleifer and Vishny (1993) define government corruption as the sale by government officials of government property for personal gain, Braguinsky (1996) points out that "corruption is commonly defined as misappropriation of government property or revenues made possible through government regulation." However, he also argues that "insider trading is a more common sort of corruption that does not involve government property, and that both corruption and insider trading are a consequence of asymmetry of information. Valuable information exists, but not everyone has access to it" (Braguinsky 1996).

Du and Wei (2004) consider the extent of corruption in a country's judicial system. The authors maintain that, "on an ex-ante basis, it is plausible to expect that legal corruption and insider trading are positively correlated: if the judges can be influenced by bribery, then it is highly probable that the laws regarding the prohibition of insider trading are not vigorously or fairly enforced. Furthermore, they find that legal corruption is positively and significantly associated with insider trading: countries with a higher degree of legal corruption are also likely to have more prevalent insider trading" Du and Wei (2004, 11).

We share the idea, advanced by Braguinsky (1996) and tested empirically by Du and Wei (2004), that insider trading is akin to corruption. Power and González (2003) carried out a cross-country empirical study aimed at determining economic and cultural factors that influence corruption. They noticed that in the 1990s, theorists of social capital were attempting to set up associations between political culture and the behavior of governmental institutions. Thus, they argue that if such relationship exists, then it should be possible to detect the cultural factors that are linked to corruption. Power and González (2003) test this hypothesis across a sample of countries. For this, they developed multivariate models designed to capture the impact of cultural factors—such as religious tradition, interpersonal confidence, and law abidance—on levels of corruption in several countries. They used a reputation index taken as a proxy for corruption developed by Transparency International (a nongovernmental organization devoted to fighting corruption worldwide) as a dependent variable. The paper shows that the level of corruption in a particular country is essentially an attribute of the type of political regime and the level of economic development in that country. However, certain cultural characteristics provide explanatory power to these models, thus contextualizing macroeconomic and macropolitical interpretations of corruption. In this paper, Power and González (2003) show that Brazil presents a high level of perceived corruption both when evaluated by Transparency International and when measured by an index of the people's perception of corruption. Additionally, they find that Brazilians share a low level of interpersonal confidence.

As Filgueiras (2006) suggested, there is a culturalist vision, adopted by anthropologists, according to whom corruption arises from an extension of the public to the private sphere as a result of cultural patterns that approximate the individual to the person; that is, the public sphere is permeated by personal relationships that lead to ad hoc authorities, which is directly proportional to the set of personal relationships the individual has in society. Hence, according to Bezerra (1995), corruption derives from existing personal relationships of the members of the state bureaucracy, implying illicit gains using public resources. Therefore, personal relationships in Brazil are established and socially institutionalized day-to-day practices, which are then not questioned or fought. Besides, contravention networks are related to personal networks, such as kinship or friendship (Bezerra 1995).

19.3 BRAZILIAN INSIDER TRADING LAWS AND REGULATIONS

The disclosure of relevant facts, loyalty, and the duty to provide information to prevent the use of privileged information were introduced in the Brazilian market by Laws 6404 and 6385 of 1976. With the new Anonymous Societies Law, Law 10303 of 2001, the use of privileged information began to be considered a crime liable to punishment:

> Article 27-D: To utilize relevant information not yet disclosed to the market, of which one has knowledge and about which one should keep confidential, capable of providing to oneself or to a third party undue advantage by means of trading with securities, in one's own name or in the name of others:
>
> Penalty—confinement, from 1 (one) to 5 (five) years, and a fine of up to 3 (three) times the amount of the illicit advantage obtained in consequence of the crime.

Concerns with trading transparency and equity of rights led CVM to issue Instruction 358 in 2002, which updated Instruction 31 of 1984, enforcing the disclosure of trades carried out by those related to the firm's administration who possess or might have access to privileged information, such as described in Article 11:

> Article 11: The directors, the members of the board, of the fiscal board and of any bodies with technical or consulting functions, created by statute, are obliged to communicate to the CVM, to the company, and, if it is the case, to the stock exchange and to the entity responsible for the over the counter market where the securities issued by the company are listed for trading, the quantity, the characteristics and the form of acquisition of the securities issued by the company and by its controlled or controlling companies that are public listed companies or that are referred to them, of which they are owners, as well as changes in their positions.
>
> Paragraph 1: The communication must contain at least the following information:
>
> I—name and qualification of the informer, indicating his registration number in the National Register of Legal Persons or the National Register of Physical Persons;
>
> II—quantity, by type and class, in the case of stocks, and additional characteristics in the case of other securities, besides the identification of the issuing company; and
>
> III—type, price, and date of transactions.
>
> 2nd Paragraph: The directors, members of the board, members of the fiscal board and members of any bodies with technical or consulting function, created by statute must provide the communication referred to in the above caption immediately after the investiture in the position or by the time of presentation of the documentation for the registration of the company as publicly listed and within the maximum period of 10 (ten) days after the end of the month when the change of positions owned by them has occurred, indicating the balance of the position in the period.

> 3rd Paragraph: The natural persons mentioned in this article will also indicate the securities under the property of husband/wife neither of whom are legally separated, of a companion, of any dependent included in his annual income tax statement, and of companies directly or indirectly controlled by them.

It should be noticed that according to the third paragraph of Article 11 of CVM Instruction 358/02, family members of the administrators and board members also become compelled to disclose trades carried out with securities, thus contributing to the improvement of the company's transparency and its corporate governance. CVM has demonstrated to the market its strong position against insider trading that brings losses to the other market participants, as one can perceive from the statement pronounced by CVM CEO, Mr. Marcelo Fernandez Trindade, during the judgment of the Sanctioning Administrative Process 18/01:

> Insider trading, the act of trading with privileged information, not available to the regular agents, is among the gravest infractions in the capital market, exactly because it undermines the market in its most important fundamental aspect, which is the trust on the agents and on the information available. Therefore, as a matter of proportionality, and being very grave conduct, it is correct to apply serious punishment, such as the maximum pecuniary fines and the penalty of suspension of the authorization for exerting the role of portfolio manager, proposed by the vote of the Reporting Director and, for that reason, supported by the remaining members of the Board.

19.4 EVIDENCE OF INSIDER TRADING IN BRAZIL

There are several empirical studies on insider trading practices in Brazil. In one of these studies, Leal and Amaral (2000) test the existence of insider trading in Brazil using the event study methodology (for a detailed description of event study methodology, see MacKinlay 1997). They argue that when striving to attract new investors, firms intending to issue stocks become more aggressive in the secondary market and in the media by attempting to improve the quality of the information on the firm and its perspectives. Therefore, they argue that we should expect positive abnormal returns on stocks in a period prior to the announcement of an equity issue. They studied twelve stocks from January 1981 to December 1985, and they found significant evidence of insider trading.

Da Costa (2002) argues that CVM tries to justify the regulation of the capital market based on ad hoc criteria of justice and equity, thus following the model proposed by IOSCO (International Organization of Securities Commissions). However, in his view, the perspective of property rights is also rooted in the logic of the regulatory apparatus that the commission exerts for the enforcement of the regulation, since the concerns of CVM with the system of incentives to the undue appropriation of firms' corporate property are widely known. Da Costa (2002) analyzes the model of regulation of insider trading in the Brazilian stock market, by evaluating the impact of Law 7913 of 1989 on the stock returns of trades registered in CVM in the period from 1989 to 1991. The study, which was carried out before the CVM issued its instruction 358/02, concluded that corporate investors did not obtain statistically significant excess returns for having a monopoly of privileged information. The analysis of the legal system focuses on CVM Instructions No. 8 of October 1979, No. 31 of February 1984, and No. 202 of December 1993, as well as on Articles 155 to 159 of Law No. 6404 of December 1976 and Law No. 7913 of December 1989. The last was assumed as the regulatory mark for the empirical assessment of buying and selling trades of stocks with possible inside information. The econometric model Da Costa used to assess the effectiveness of insider trading regulation in the Brazilian stock market is the Black, Jensen, and Scholes (1972) version of the so-called market model. The model attempts to explain whether the implementation of the law was capable of eliminating or reducing returns from insider trading operations during the period after its implementation. Due to insufficient data available, the analyzed period included the years 1989, 1990, and 1991 only. Based on the results obtained in this study, Da Costa (2002) affirms that, during the sample period, it is possible to observe a clear trend of falling differential returns obtained with buying or selling insider trading operations in the Brazilian stock market. He argues that even during 1989 this trend was already noticeable, although differential returns were statistically more significant. With the sanctioning of Law 7913 in December 1989, the incentive for the undue appropriation of the company's property became even lower, provoking a drop in differential returns. The author adds that, although the law has not completely eliminated these returns, since the existence of differential returns obtained from the undue use of insider trading during the sample period is considerable, one could say that the law was efficient in achieving its purposes, as the differential returns became statistically less significant after its implementation. Of course, Da Costa's (2002) study presents several problems, such as a small sample size and a reduced period of analysis. Nevertheless, it seems quite possible that Law 7913 of December 1989 has, indeed, been effective in curbing differential returns of insider traders. This possibility does not overcome the idea that insider-trading activities are still practiced on a significant scale in Brazil, despite the laws and CVM efforts. Several other empirical studies performed in the country have shown evidence of insider trading activities and profiting in the Brazilian stock market.

In another empirical study, De Medeiros and Matsumoto (2006) undertook an event study on the market reaction to a sample of eighty seasoned equity offers (SEOs) issued in the Brazilian stock market from 1992 to 2003. Initially, abnormal returns were calculated for each issue using the market model estimated by ordinary least squares (OLS). They found, however, the presence of ARCH (autoregressive conditional heteroscedasticity; see Engle 1982) processes in the regression residuals in 70 percent of the sample regressions. For these cases, they reestimated the regressions by ARCH or GARCH models, according to their best fit. Comparing the initial results to those taking into account the heteroscedastic (ARCH or GARCH) processes, they found that the original results indicated negative abnormal returns biased downwards with respect to the alternative estimation. However, reestimation, although leading to lower abnormal returns in absolute values, does not change the results in qualitative terms. Their results show that stockholders seem to be cautious about firms that issue stock to raise funds with the argument that they are investing in projects with positive net present value (NPV). When firms announce that they are raising funds through the issue of new stocks, it is inferred that these firms could be waiting for an opportunity window to issue. Thus, when the stock market reaches a given level in which the shares of these firms are overvalued, stockholders are satisfied to sell part of their investment at a profit. Actually, this shows that when companies announce the issue, there is a negative signaling to the market. As shown in De Medeiros and Matsumoto (2006), in the period from 1992 to 2003, Brazilian companies that raised funds through underwriting were unable to meet the benchmark, that is, the Brazilian stock market index. The results led them to infer that insiders sell their equity position about three weeks before the announcement, since they expect the announcement will convey negative information about the firm's true value. Actually, it is found that there are significant negative abnormal returns about three weeks before the announcement. Negative abnormal returns means falling prices, indicating the occurrence of inside information, which anticipates what should only occur on the announcement day. The results presented negative CARs (cumulative abnormal returns) of 4.6 percent thirty days around the announcement day, that is, fourteen days before and fifteen days after. On announcement day, a negative abnormal return of 2.4 percent was verified. The incidence of negative abnormal returns three weeks before the issue announcement may be interpreted as evidence of insider information, since it seems to be an anticipation of what should occur only on the announcement day. The negative abnormal returns related to the announcements are consistent with the asymmetric information hypothesis in which the management is better informed about a firm's value than outsider investors are.

Rochman and Eid (2006) performed a set of event studies on trades carried out with companies' stocks by insiders aiming at detecting abnormal returns, as a consequence of having access to inside information. Their sample is composed of trades performed by insiders of firms with stocks traded on Bovespa that are listed on superior levels of corporate governance (Levels 1 and 2, and the New Market). They found evidence of insider trading that resulted in statistically significant returns in excess of expected returns, such as in the purchase of ordinary stocks by controllers, families, and investment clubs or in sales of preferred stock by directors, board members, assessors, and board advisors. The authors show that insider traders, as defined by CVM Instruction No. 358 of 2002, are very active in the Brazilian stock market. The profile of trades involving stocks traded by insiders indicates that the directors sell more stocks than they buy, which is the opposite behavior with respect to controllers and board members, and that insiders of firms on Level 1, which require less governance requirements and restrictions, are more active in trading than their peers in firms on higher governance levels. They also show that purchases of ordinary stocks and sales of preferred stocks carried out by insiders show a significant cumulative average abnormal return, providing evidence of insider trading. This evidence is supported by the significant average cumulative abnormal returns obtained on the days following ordinary stock purchases by controllers, families, and investment clubs and by the sales of preferred stocks by directors, board members, assessors, and board advisors, where significant average cumulative abnormal returns were obtained ten days before the trades. As a consequence of these results, they reject the efficient market hypothesis in its strong form. The purchase of preferred stocks and the sale of common stocks did not present any significant average cumulative abnormal return. However, some of these operations presented significant average abnormal returns in some days near the event.

Beny (2006, 1) performs a study that "proposes three testable hypotheses regarding the relationship between insider trading laws and several measures of stock market performance. Using cross-country data, the paper finds that more stringent insider trading laws are generally associated with more dispersed equity ownership, greater stock price accuracy, and greater stock market liquidity." In her empirical study, Beny defines a series of variables associated with insider trading laws across countries. After analyzing these variables for a large sample of countries, she finds that Brazil presents the most restrictive insider trading prohibition, while, with respect to sanctions, the country shows the lowest expected sanctions. This result is very much aligned with the general idea that prevails in the country: Brazil has appropriate and sufficient laws, but the law enforcement fails, and criminals are not punished.

According to Bueno et al. (2000), one of the most visible economic phenomena in Brazil in the 1990s has been the mergers, acquisitions, separations, and restructuring of companies established in the country. As a direct consequence of globalization, economic openings, privatization, and monetary stabilization, the process seems to be far from ending its cycle. Because it is an effervescent market, the phenomenon, which is not entirely understood and assimilated, has been studied by several sciences, including economics, accounting, finance, marketing and strategy, and business policy. In the financial area, the effect of these operations on a capital market characterized by a high concentration of trades on a few securities, and low liquidity, has caused concerns. Now and then, the media report events involving the use of insider information, in stock, exchange rate, commodity, and derivative markets. Market participants argue that the origin of the phenomenon is a moral crisis in Brazilian society, combined with a sensation of impunity, supported by an inefficient legal system. Based on recent cases of mergers and acquisitions involving Brazilian publicly listed companies, Bueno, Braga, and Almeida (2000) selected a random sample containing sixteen cases involving fourteen stocks. The most liquid stocks from each firm were picked, making no distinction between stocks of acquiring/acquired or incorporating/incorporated or merged firms. Quotes were collected for one, five, ten, fifteen, and thirty trading days prior to the announcement of the facts relevant to legal dispositions or the media's disclosure of the event prior to the relevant announcement of the fact, even when, afterwards, they were unconfirmed or not concluded. The stocks included in the sample were issued by firms without regard to industry sector that were subject to merger or acquisition processes in the period from May 1995 to January 1998. The authors adopted Brown and Warner's (1985) comparison index (CI) method. Using this method, they calculated the ratio between the change of a stock's closing price and the Bovespa index (Ibovespa) change for the same period. The CI is given by $\log(CI) = \log[(P_t / P_{t-1}) / (I_t / I_{t-1})]$, where P_t is the stock price and I_t is the market index; log refers to the natural logarithm and t is the time subscript. The CI can be described as a test of abnormal returns as a function of private information unadjusted to risk. Based on the data collected, two types of analysis were performed: (a) Description of the frequency distribution of stock price changes with respect to the Ibovespa change. For such, the frequency distribution of the CI was used. (b) Testing the hypothesis that the stock price changes are different from the market index changes. For all cases the results reveal a mean higher than 1, with the mean return in the first trading day of 4.8 percent above the market. The variance tends to increase during longer periods. The same happens with the interval between minimum–maximum returns, and for the series of thirty days, with this measure being almost four times greater than the one-day series. With respect to the form and frequency of the series, with a warning about the small sample size, it can be noticed that, with the exception of the series with one and fifteen trading days, the others present negative skewness (inclined to the left) and also a negative kurtosis, meaning platikurtic distributions. This means that besides a greater dispersion, there is a higher probability for positive than negative abnormal returns (losses), given that there is a greater number of observations above the mean (negative skewness) than below it. The paper detects a market inefficiency in stock pricing performed with stock returns in one trading day before the announcement or disclosure. However, the authors point out that the method used has limitations for not taking into account the systematic risk of stocks and also because of the use of trading days instead of working days, which may imply that a different change of the market index might be caused by other factors, besides the use of inside information on mergers and acquisitions.

19.5 FINAL REMARKS

In this chapter, we attempted to present a picture of the Brazilian situation with respect to insider trading. We sought to discuss legal and economic aspects of insider trading in Brazil. We support the idea, raised by previous literature, that insider trading is a form of corruption. More specifically, a form of private or business corruption, although insider trading may also be practiced within the governmental arena.

We sustain the view that although an institutional framework exists in Brazil to fight insider trading practices in the stock market, the actual success and willingness of the authorities with respect to this form of corruption are still far from being efficient. One of the reasons for this is that Brazil is a country with a long tradition of corruption culture and corruption is still tolerated in the country.

Of course, we cannot deny the efforts applied by CVM, by Bovespa, and by the legislators toward increased transparency and ethical behavior, but the actual enforcement of the law against insider trading seems to require a much greater effort. More specifically, CVM lacks the proper material means and sufficient skilled personnel to monitor a large portion of stock market trades.

The struggle against insider trading in Brazil will probably be coincident with the struggle against corruption. Since corruption in this country seems to be largely a cultural phenomenon, it could take a very long time, possibly generations, until the country is able to see significant changes. As pointed out in many previous studies, corruption is inversely related to education. Therefore, the battle against corruption and insider trading will depend on whether the country is able to put forward sound and sustainable public educational policies in the future.

REFERENCES

Beny, L. N. 2006. Insider trading laws and stock markets around the world: An empirical contribution to the theoretical law and economics debate. Paper 04-004, John M. Olin Center for Law & Economics, University of Michigan. September.

Bezerra, M. O. 1995. *Corrupção: Um estudo sobre poder público e relações pessoais no Brasil*. Relume-Dumará, Rio de Janeiro.

Birdsall, N., and R. H. Sabot. 1996. *Opportunity foregone: Education in Brazil*. Baltimore, MD: Johns Hopkins University Press.

Black, F., M. C. Jensen, and M. Scholes. 1972. The capital asset pricing model: Some empirical tests. In *Studies in the Theory of Capital Markets*, ed. Michael C. Jensen. New York: Praeger.

Braguinsky, S. 1996. Corruption and Schumpeterian growth in different economic environments. *Contemporary Economic Policy* 14(3):14–25.

Brown, S. J., and J. B. Warner. 1985. Using daily stock returns: The case of event studies. *Journal of Financial Economics* 14(1):3–31.

Bueno, A. F., R. F. R. Braga, and R. J. Almeida. 2000. *Pesquisa sobre a eficiência informacional no mercado brasileiro nos casos de fusões e aquisições*. Florianópolis: Anais do Enanpad.

Da Costa, G. N. 2002. *A regulação das operações de compra e venda de valores mobiliários com base em informações privilegiadas: O caso Brasileiro 1989–1991*. Unpublished M.Sc. Dissertation in Economics, Fundacao Getulio Vargas, Rio de Janeiro.

De Araujo, G. V. (2005) Transparência e Educação no Combate à Corrupção no Brasil, I Concurso de Monografias e Redações, Controladoria Geral da União, Rio de Janeiro. Available at http://www.esaf.fazenda.gov.br/esafsite/premios/CGU/monografia/2_Gustavo_Viola_de_Araujo.pdf.

De Barros, R. P., and R. Mendonça. 1997. Investimentos em educação e desenvolvimento econômico. Discussion Paper 525, IPEA, Rio de Janeiro.

De Barros, R. P., R. Henriques, and R. Mendonça. 2002. Pelo fim das décadas perdidas: Educação e desenvolvimento sustentado no Brasil. Discussion Paper 857, IPEA, Rio de Janeiro.

De Medeiros, O. R., and A. S. Matsumoto. 2006. Market reaction to stock issues in Brazil: Insider trading, volatility effects and the new issues puzzle. *Investment Management and Financial Innovations* 3(1):142–50.

Du, J., and S. J. Wei. 2004. Does insider trading raise market volatility? *Economic Journal* 114:916–42.

Engle, R. F. 1982. Autoregressive conditional heteroskedasticity with estimates of the variance of the United Kingdom inflation. *Econometrica* 50(4):987–1007.

Filgueiras, F. B. 2006. A corrupção no Brasil e as instituições políticas. Working paper, Centro de Pequisas Estratégicas, Universidade Federal de Juiz de Fora.

Leal, R. P. C., and A. S. Amaral. 2000. Um momento para o insider trading: O período anterior ao anúncio de uma emissão pública de ações. In *Finanças Corporativas, Coleção COPPEAD de Administração,* ed. R. P. C. Leal, N. C. A. Costa Jr., and E. F. Lemgruber. São Paulo: Atlas.

MacKinlay, A. C. 1997. Event studies in economics and finance. *Journal of Economic Literature* 35(1):13–39.

Mauro, P. 1995. Corruption and growth. *Quarterly Journal of Economics* 110(3):681–712.

Pereira, J. M. 2004. Reforma do estado, transparência e democracia no Brasil. Revista Académica de Economia, n. 26. http://www.eumed.net/cursecon/ecolat/br/jmp-reforma.doc.

Power, T. J., and J. González. 2003. Cultura política, capital social e percepções sobre corrupção: Uma investigação quantitativa em nível mundial. *Revista de Sociologia e Política* 21:51–69.

Proença, J. M. M. 2005. *Insider trading—Regime jurídico do uso de informações privilegiadas no mercado de capitais*. São Paulo: Editora Quartier Latin.

Rochman, R., and W. Eid. 2006. *Insiders conseguem retornos anormais? Estudos de eventos sobre as operações de insiders das empresas de governança corporativa diferenciada da Bovespa*. Salvador: Anais do Enanpad.

Schwartzman, S. 2004. Educação: A nova geração de reformas. In *Reformas no Brasil: Balanço e Agenda,* organizers F. Giambiagi, J. G. Reis, and A. Urani. Rio de Janeiro: Editora Nova Fronteira.

Shleifer, A., and R. W. Vishny. 1993. Corruption. *Quarterly Journal of Economics* 108(3):599–617.

CHAPTER 20

Legal Insider Trading and Stock Market Reaction

Evidence from the Netherlands

Nihat Aktas, Eric de Bodt,
Ilham Riachi, and Jan de Smedt*

CONTENTS

* The authors are grateful to Katrijn Bergmans for editorial assistance. Jan de Smelt's contribution is made in his own name and the views expressed herein do not necessarily represent those of the CBFA.

20.1 INTRODUCTION

Insider trading regulation plays an important role in economies with developed stock markets. According to Battacharchya and Daouk (2002), 87 out of 103 countries with stock markets have insider trading laws, 38 of which have taken enforcement measures. One interesting aspect of these regulations is that they allow insiders to trade in their own companies' stocks, provided that certain conditions are fulfilled. Such transactions are referred to as *legal* insider trading. For example, under U.S. securities laws, legal insider trading occurs on a daily basis, as corporate insiders—officers, directors, or employees—buy or sell stock issued by their own companies. One constraint is that the insiders concerned have to report these trades to the Securities and Exchange Commission (SEC): once the trades are completed, filings have to be sent to the SEC which makes them public.

20.1.1 The Well-Known Debate

The question whether insider trading should be regulated has been deeply debated in the literature. This is certainly because corporate insiders are the persons most likely to possess privileged information regarding their company, and are therefore able to realize abnormal profits on the financial markets at the expense of outside investors. Critics of insider trading regulation mainly argue that restrictions are inefficient because insider trading allows new (private) information to be priced more quickly. As a result, stock prices reflect intrinsic firm value more accurately, promoting improved economic decision making and resource allocation (e.g., Manne 1966; Carlton and Fischel 1983; Leland 1992). On the other hand, those in favor of insider trading regulation essentially claim that prohibition promotes public confidence and participation in the stock market and allows outsiders to share in value-enhancing events on an equal footing (Ausubel 1990).

In recent decades, the academic literature, mainly within the U.S. context, has dealt extensively with the economic and financial analysis of legal insider trading. Without being exhaustive, topics such as the contribution of insider trades to market efficiency (e.g., Rozeff and Zaman 1988; Lakonishok and Lee 2001; Aktas, de Bodt, and Van Oppens 2007), the market-timing capacity of insiders, and their stock price predictive ability have attracted a great deal of attention (see Jenter 2005; or Piotroski and Roulstone 2005).

20.1.2 Short-Term Studies

A number of studies also appraise the impact of insider trading activities over a shorter period. Seyhun (1986), Lakonishok and Lee (2001), and more recently Aktas, de Bodt, and Van Oppens (2007) provide short-term event study results on U.S. legal insider trading. They observe statistically significant, but economically unimportant market movements around insider net purchase and net sale days. For example, Aktas, de Bodt, and Van Oppens (2007) report, using a sample of 59,244 aggregated daily insider trades, statistically significant five-day abnormal returns of 0.417 and 0.225 percent for net purchases and net sales, respectively. It is important to stress that these small returns could be considered economically significant, given that these trades contain transactions that are uninformative as well as others that do contain private information.

20.1.3 Long-Term Studies

There is overwhelming evidence in the literature that portfolios that are long on stocks purchased by insiders and short on stocks sold by insiders outperform the market over a time horizon ranging from one to several months (e.g., Jaffe 1974; Finnerty 1976; Seyhun 1986, 1998; Lin and Howe 1990; Jeng, Metrick, and Zeckhauser 2003). A notable exception is the study by Eckbo and Smith (1998), where the authors show that insiders on the Oslo Stock Exchange do not earn abnormal profits. However, it is important to note that the reported abnormal performance seems to be driven by latent risk factors such as size, earnings/price, or book-to-market (e.g., Rozeff and Zaman 1988; Lakonishok and Lee 2001).

As mentioned above, previous researches mainly focus on U.S. legal insider trading. They are based on the insider transactions notified to the

SEC. Insider trading in Europe has been the subject of only few comparable studies. One of the main reasons remains the lack of European data. Only until recently, the majority of European countries did not have a legal obligation of notification for insider transactions. As a consequence, these countries did not possess databases such as the register kept by the SEC in the United States. It was only after the promulgation of European Directive 2003/6 of February 28, 2003, that all European member states were conducted to adopt an obligation of notification in their national legal systems. However, some European countries such as the Netherlands and France already adopted a national obligation of notification before Directive 2003/6 came into existence. In the Netherlands, as we see later on, this obligation came into effect on January 1, 1999. Since then, insiders realizing transactions in their firm's own stock must notify these trades to the AFM, the Dutch financial markets authority.

The contribution of this chapter consists in an analysis of insider trading on the Euronext Amsterdam stock exchange by using data published in the register held by the AFM. More precisely, following the approach developed in Lakonishok and Lee (2001) and more recently in Aktas, de Bodt, and Van Oppens (2007), we provide market reactions on short event windows around insider trading days to test whether insider trades are information motivated. It is important to note that this test relies on the ability of financial markets to detect the presence of insiders in the market. We also present the abnormal returns over longer event windows to check whether insiders use long-term information and whether the notification of their transactions conveys valuable information to the other investors. Our sample encompasses 822 transactions executed by corporate insiders on the Euronext Amsterdam stock exchange between the beginning of January 1999 and the end of September 2005. Our analysis shows that the financial markets' response is not significant for purchases, and that the abnormal returns associated with the sales do not have the expected sign. However, over a longer time horizon, the average cumulated abnormal returns are positive for the stocks purchased, and negative for stocks sold by insiders. This result suggests either that insiders use long-term information for their trading activities or that they are able to time the market.

The chapter is organized as follows. We first describe the Dutch insider trading regulation, and compare it to the U.S. system. Then, to study the information content of insiders' trades (both insider net purchases and insider net sales are considered), we provide an analysis of market reactions around and following the transaction dates. The last section presents our conclusions.

20.2 ANALYSIS OF THE RELEVANT LEGISLATION

Before presenting the financial analysis of transactions notified by insiders to the Dutch AFM, we provide a brief overview of the Dutch law on insider trading. We mainly focus on the legal obligation for insiders to report their transactions in their own company's stock to the AFM, and we present some comparisons with the corresponding filing requirements under U.S. law.

The current Dutch law on insider trading is determined by the successive legal initiatives taken on the European level. The main piece of European legislation with respect to insider trading is Directive 2003/6/EC of January 28, 2003 on insider transactions and market abuse. As did its predecessor, Directive 89/592/EEC of November 13, 1989 coordinating regulations on insider trading, Directive 2003/6/EC prohibits persons who knowingly possess inside information from using that information by acquiring or disposing of financial securities to which that information relates. Further, the directive prohibits these persons from disclosing their inside information to any other person, as well as from recommending or inducing another person on the basis of that information to acquire or dispose of financial instruments to which that information relates.

For the purpose of these prohibitions, inside information must be interpreted as information of a precise nature which has not been made public, and which relates—directly or indirectly—to one or more issuers of financial instruments and which would have a significant effect on the prices of those financial instruments or on the price of related derivative financial instruments if it were made public.

To obtain transparency with regard to the transactions conducted by insiders, and to examine whether or not these transactions are conducted using inside information, Directive 2003/6/EC obliges certain insiders to notify the competent authorities of the transactions they conduct on their own account in the stock of the company to which they relate. The insiders subject to this obligation are the persons who discharge managerial responsibilities within an issuer of financial instruments as well as persons closely related to them.

It must be emphasized that this mechanism of notification in Directive 2003/6/EC does not imply an exception to the insider trading prohibition. Completing a transaction using inside information remains prohibited, and does not become authorized as a result of a notification of the transaction.

European Directives do not have direct effect in the legal order of the European member states. They establish the objectives that must be attained by the member states, leaving it to the states to decide on the means used to arrive at these objectives. Dutch national law on insider trading has been adapted several times in order to comply with the objectives prescribed in the successive European Directives (first Directive 89/592/EEC and then Directive 2003/6/EC). As a result, in imitation of the latter directive, current Dutch law on insider trading contains (1) a prohibition from using inside information by acquiring or disposing of financial instruments, (2) a prohibition from disclosing inside information to another person or from recommending or inducing another person to acquire or dispose of financial instruments to which that information relates, and (3) an obligation for certain insiders to notify the AFM, the Dutch authority surveying the financial markets, of their transactions. This last obligation is our main point of interest, and is the subject of the following paragraphs.

The Dutch legal obligation of notification in its current form—apart from some minor modifications inserted at the occasion of the introduction of the new Act on Financial Supervision (AFS) on September 26, 2006—came into effect on October 1, 2005, and translates the obligation of notification in Directive 2003/6/EC into Dutch national law. Prior to October 2005, Dutch law already contained an obligation of notification, as the Netherlands was one of the few countries that introduced such an obligation before it was generalized by Directive 2003/6/EC. This previous obligation, which came into effect in January 1999, was somewhat different from the current obligation. The transactions that serve as a basis for our analysis have been notified in the period running from January 1999 to September 2005, that is, under the previous legal regime. As a consequence, the following paragraphs discuss the obligation of notification under both the current and the previous state of Dutch law.

20.2.1 The Persons Subject to the Obligation of Notification

Current Dutch law imposes an obligation of notification upon the directors and commissioners of the issuing institution, as well as upon other persons who are not officially directors or commissioners, but who have similar authority and responsibilities. Furthermore, their relatives—

spouses and partners, children, and other relatives of the aforementioned insiders—are subject to the same obligation. Finally, corporate bodies, trusts, or personal companies that are controlled or managed by the aforementioned persons or that are set up for the benefit of these persons or the economic interests of which are equivalent to those of these persons are also subject to the obligation of notification.

Under the previous legal regime, the personal scope of the obligation of notification was somewhat different: apart from the directors and commissioners of the issuing institution and their relatives, the directors and commissioners of significant legal entities in which the issuing institution holds participation were also subject to the obligation. Further, persons holding more than 25 percent of the capital of the issuing institution—as well as their administrators and commissioners if this person was a company or a corporate body—were obliged to notify their transactions. Relatives of the aforementioned two categories were equally subject to the obligation, as were members of the works council of the issuer. Finally, the issuing institution itself was subject to the obligation of notification.

U.S. law also imposes an obligation of notification upon insiders. Without going into detail, this obligation principally regards officers and directors, as well as beneficial owners. The following, among others, must be considered "officers": the president, the principal financial and accounting officers, any vice president in charge of a principal business unit, division, or function, as well as any other officer or other person who performs policy-making functions. A "beneficial owner" is a person holding more than 10 percent of a class of registered equity securities, such as common stock or registered preferred stock (Hazen 2005).

20.2.2 The Transactions That Must Be Notified

Under current Dutch law, the aforementioned persons must notify transactions (1) in stocks that regard their own company and that are allowed to be traded on a regulated market, as well as (2) transactions in securities the value of which is determined by the value of the aforementioned stocks, that is, call and put options, warrants, and convertible debentures. Under the previous legal regime, the obligation of notification more generally envisaged transactions in securities that regarded the issuer.

U.S. law obliges officers, directors, and beneficial owners to file changes in their ownership of any class of any registered equity security of the issuer, as well as any purchase or sale of a security-based swap agreement involving such security.

20.2.3 The Delay of Notification

Under current law, the notification must be filed to the AFM *at the latest on the fifth working day after the transaction date.* A five-day period may be considered as rather long, certainly in comparison with the delay of two business days that is applicable in the United States since the Sarbanes–Oxley Act came into effect (Hazen 2005). In the period from January 1999 to September 2005, directors and commissioners of the issuing institution were obliged to notify the AFM *without delay*, whereas the other persons subject to the obligation had to notify their transactions *at the latest on the tenth day after the end of the calendar month during which the transaction was conducted or effected.*

20.2.4 Exceptions

Dutch law exempts from the obligation of notification transactions conducted or realized pursuant to a written mandate by a licensed portfolio manager, if that mandate stipulates that the principal shall not exert influence on transactions conducted or effected by the portfolio manager in his or her capacity of authorized representative. A similar exception existed under the previous legal regime.

The fact that a transaction does not come within the scope of application of the insider trading prohibition, that is, the prohibition from using inside information when conducting a transaction, does not imply that such a transaction is automatically exempted from notification to the AFM (see Grundmann-van de Krol 2004; Schutte 2006).

U.S. law provides for various exceptions to the reporting requirement. For instance, transactions effected in the framework of a distribution of securities where the insider acquires the securities for the purpose of distributing them are exempted from the reporting requirement. Further, there are exemptions from the reporting requirement for stock splits, stock dividends, and for rights issued pro rata (Hazen 2005).

20.2.5 Delay of Six Months under U.S. Law

The aforementioned U.S. reporting obligation for officers, directors, and beneficial owners is included in Section 16 of the 1934 Securities Exchange Act. Apart from the filing requirements, this section also imposes a short-swing prohibition: insiders must disgorge to the issuer any profit realized as a result of a purchase and sale of equity securities within a six-month period. In practice, this means that if an officer, director, or beneficial owner purchases relevant stock he or she must wait at least six months before reselling this stock in order not to incur liability (Hazen and Ratner 2006). No similar waiting period for insiders who have conducted a transaction in relevant financial instruments is provided for under European or Dutch law.

20.3 MARKET REACTIONS TO INSIDER TRADES ON EURONEXT AMSTERDAM

20.3.1 Data and Method

We use the database of the Dutch financial markets authority (AFM) to extract notified corporate insider purchases and sales. For each transaction, this database indicates the name of the insider, the transaction date, the type of the transaction (sale or purchase), the price at which the transaction was concluded, and the number of shares exchanged.

Our sample period ranges from January 1999 (when the obligation of notification came into effect in the Netherlands) to the end of September 2005. Because a new legal provision regarding the obligation of notification came into effect in October 2005, we limited the period under examination to the end of September 2005 for purposes of consistency. The number of transactions notified in the period running from January 1999 to September 2005 amounts to 11,970 transactions. To ensure the quality of the gathered data, we applied several filter rules to our initial sample. We kept only *stock* transactions realized by corporate insiders. We deleted transactions the prices of which were not reported in the AFM database or the prices of which were not in EUR. We eliminated any transaction in shares that are not listed on Euronext Amsterdam. The application of these filters reduced the sample size to 2,549 transactions. Moreover, we cross-checked the AFM price and volume information against that reported by the Datastream database. Doing so, we dropped from the sample records with a price outside the range of prices of that day, as well as records with a volume exceeding the number of shares exchanged on that day. As a last filter, we excluded transactions of less than 100 shares in order to focus only on the more meaningful events. Our final sample encompasses 822 transactions (Table 20.1).

TABLE 20.1. The sample: The filters used to obtain the final sample of 822 transactions

Initial sample	11,970
Deletion of	
Transactions in derivatives, or realized by the company, or with no price information, or in a currency different from EUR	9,158
Transactions in stock not listed on Euronext Amsterdam	263
Transactions with volume superior to total shares outstanding	227
Transactions with volume superior to the volume exchanged on the corresponding day	160
Transactions with price outside the range of prices of the corresponding day	1,320
Transactions with fewer than 100 shares	20
Final sample	822

Because several transactions for a given company in the sample were realized at the same date, we have computed the net transactions using the same method as in Fidrmuc, Goergen, and Renneboog (2006). For example, a purchase of 400 shares and a sale of 220 shares on a given day become a net purchase of 180 shares for that day, and a purchase of 210 shares and a sale of 400 shares become a net sale of 190 shares. Following this adjustment and the elimination of net transactions with a compensated volume or value of zero we reduced our sample from 822 to 600 transactions on the basis of the net volumes and to 602 on the basis of the net values. The remaining transactions include 163 net purchases (both in volume and in value), 439 net sales in value, and 437 net sales in volume.

20.3.2 Empirical Method

To measure the market reaction around insider transaction dates we perform a classic event study. We compute the daily abnormal returns (AR) as in Lakonishok and Lee (2001) and Aktas, de Bodt, and Van Oppens (2007) using the Beta-one model, which consists of subtracting the daily market portfolio return from the daily return for each company. We use the daily *All Shares* index of Euronext Amsterdam as a proxy for the market portfolio. The abnormal return for firm i is computed as follows:

$$\mathrm{AR}_{i,t} = R_{i,t} - R_{M,t},$$

where $R_{i,t}$ and $R_{M,t}$ are the observed return for stock i and for the market portfolio, respectively. The cumulated abnormal return (CAR) is simply the sum of the daily AR over the different considered event windows. The event windows are defined relative to the insider trading days (day 0).

20.3.3 Results

20.3.3.1 Summary Statistics

Table 20.2 shows some descriptive statistics for the insider transactions on the Amsterdam Euronext Stock Exchange between January 1999 and September 2005. The average number of insider net purchase days is 3.47 per firm. The corresponding average for the net sale is 8.09. On average, the companies that are subject to insider transactions in our sample have a market value of about EUR 5.3 billion.

The average number of stocks purchased is inferior to that of the sales. It is well known that corporate insiders are on average net sellers, probably for reasons of diversification (see Lakonishok and Lee 2001; Jenter 2005). On average, the number of stocks exchanged per purchase is 6,722. The median of the number of stocks purchased is 2,400. On the other hand, the average number of stocks sold per transaction is 14,975, the median being 4,063 stocks. Again, we observe that there is a large difference between the average value of the purchases and that of the sales. This value is EUR 116,665 for a purchase against EUR 216,010 for a sale. The medians are EUR 23,400 and EUR 45,662, respectively.

In Table 20.2 we also provide two ratios to analyze the relative size of the insider transactions. The first is the ratio of the insider net transaction to the volume of the corresponding day, and the second is the ratio of the net insider transaction to the market capitalization of the corresponding day. For the net purchases and net sales, the insider transactions amount on average to 12.09 and 13.38 percent of the daily volume exchanged, respectively. Relative to market capitalization of the firm, the average ratio is 0.04 percent for the purchases, and 0.08 percent for the sales.

20.3.3.2 Market Reactions

Table 20.3 displays market reactions to insider net purchases and sales around the transactions dates. For the entire sample, the two-day ($CAR_{-1,+1}$) and five-day ($CAR_{-2,+2}$) abnormal returns for the purchases are on average −1.20 and −0.92 percent, respectively. These CARs are not statistically significant. However, the CARs during the three days as of the transaction date

TABLE 20.2. Summary statistics: A number of summary statistics on legal insider transactions on Euronext Amsterdam in the period from January 1999 to the end of September 2005

	Mean	**Min**	**Q1**	**Median**	**Q3**	**Max**
Number of net purchases per firm	3.47	1	1	2	4.5	13
Number of net sales per firm	8.09	1	2	5	12	37
Market cap (in million €)	5,278	5	87	351	1,340	68,368
Number of stocks per net purchase	6,722	100	500	2,400	7,750	200,000
Number of stocks per net sale	14,975	100	1,000	4,063	13,036	315,513
Value of net purchases (in €)	116,666	45	4,263	23,400	93,236	5,000,000
Value of net sales (in €)	216,010	954	12,500	45,662	192,400	9,105,600
Net purchase over volume	12.09%	0.00%	0.16%	2.75%	15.63%	100.00%
Net sale over volume	13.38%	0.00%	0.54%	5.24%	16.57%	100.00%
Net purchase over market cap	0.04%	0.00%	0.00%	0.01%	0.03%	0.89%
Net sale over market cap	0.08%	0.00%	0.00%	0.01%	0.05%	5.16%

TABLE 20.3. Market reactions to insider trading activities: Average cumulative abnormal returns (CAR) around insider net purchases and insider net sales days

	Event windows			
	CAR-1,+1	CAR-2,+2	CAR0,+2	CAR+1,+2
	Panel A. Total sample			
Net purchases	-1.20%	-0.92%	0.39%	0.74%*
Net sales	0.47%	0.97%***	-0.19%	-0.31%
	Panel B. Split by trade size			
Net purchases				
Trade value ≤ Q1	-3.60%	-2.41%	-0.33%	2.29%**
Q1 < Trade value ≤ Q2	1.09%	1.12%	0.55%	0.43%
Q2 < Trade value ≤ Q3	-1.49%	-1.43%	0.75%	0.22%
Q3 < Trade value	-0.80%	-0.95%	0.61%	0.00%
Net sales				
Trade value ≤ Q1	-0.04%	0.40%	-0.57%	-0.70%
Q1 < Trade value ≤ Q2	0.17%	0.37%	0.00%	-0.46%
Q2 < Trade value ≤ Q3	0.80%	1.48%*	-0.19%	-0.24%
Q3 < Trade value	0.97%*	1.62%**	0.03%	0.10%

*The event windows are defined relative to the transaction day (day 0). Panel A deals with the total sample. Panel B provides a split of the sample by trade size in value. *, **, *** denote significance at the 10%, 5%, and 1% levels, respectively.*

($CAR_{0,+2}$) and during the two days after the transaction ($CAR_{+1,+2}$) are positive (0.39 and 0.74 percent, respectively) with the latter CAR being statistically significant at the 10 percent level. These results suggest that the market does not react on insider purchase days, but that the significant impact appears only during the subsequent two days. Since during the period under examination some categories of insiders (directors and commissioners of the issuing institution) were obliged to notify the AFM *without delay*, these ex post positive CARs could be explained by a buying pressure caused by other investors upon receipt of the information on the notification.

For the net insider sales, the average five-day CAR around the transaction date is 0.97 percent, and it is significant at the 1 percent level. Even if it is well known in literature that insider *sales* are more likely to be driven by other motives (such as diversification and liquidity reasons) than private information, this result is quite puzzling. One possible explanation is that insiders are more willing to sell stock when the market is dominated on the buy side, probably due to a positive (value-creating) public announcement. Consistent with this idea, Huddart, Ke, and Shi (2007) document in the U.S. context that insiders seem to sell after good news earnings announcements. On the other hand, the CARs observed within a period of three days as of the date of the sale (–0.19 percent) and during the two days following the transaction date (–0.31 percent) are negative, but without being statistically significant.

Overall, as do previous U.S. studies, our results show that the market reactions around the insider transaction date are too small to be economically significant.

In Panel B of Table 20.3 we have also explored the effect of transaction size by calculating the average CAR as a function of the trade size. Unlike the observations in some U.S. studies, the market impact does not seem to increase with trade size. For purchases, only the average CAR over the two days subsequent to the insider trading days is positive and statistically significant. The corresponding abnormal return is 2.29 percent. This result is consistent with the *stealth trading hypothesis* of Kraakman (1991) according to which insiders try not to alert the market (Friederich, Gregory, and Matatko 2002) by conducting, for example, several smaller transactions rather than one large transaction. The results regarding the net sales show positive and significant CARs during the five days around the dates of large value transactions—1.48 and 1.62 percent for trade size in value between the median (Q2 = EUR 46,662) and the third quartile (Q3 = EUR 192,400) and above the third quartile, respectively. The positive returns associated with large insider sales are again more likely to be explained by the announcement of positive corporate events (for example, positive earnings announcements).

We have also computed the average CAR over longer event windows to check whether insiders use long-term information. Figure 20.1 displays both for purchases (Panel A) and sales (Panel B) the average CAR from day +3 to day +200 relative to the insider trading day (which corresponds to day 0). For the purchase, we obtain an average CAR of 6.73 percent, while it is –14.91 percent for the sales. Both abnormal returns are statistically significant (Table 20.4). Moreover, the more informative insider trades seem to be, the smaller are the transactions (where the trade size in value is below the first quartile). The corresponding CARs are 7.90 and –20.20 percent, respectively, for the purchases and sales. These results suggest that on average insiders' transactions rely on long-term information and/or insiders have market timing skills. Moreover, since the maximum delay for the notification is 40 days after the

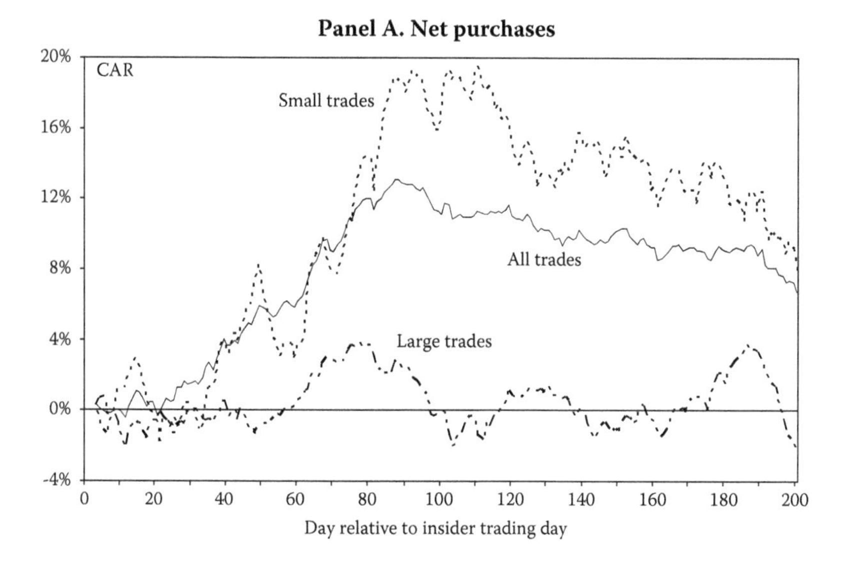

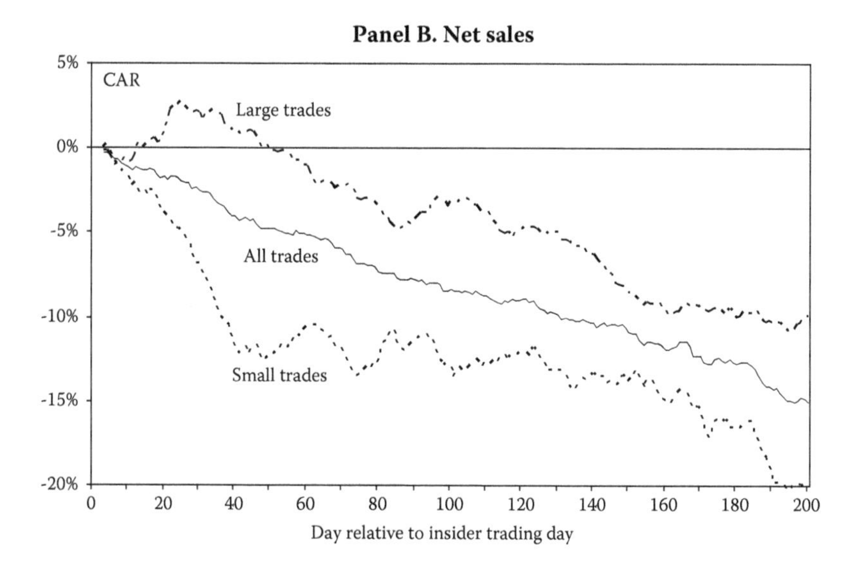

FIGURE 20.1. Cumulative abnormal returns (CAR) from day +3 until day +200 relative to the insider trading day, which is day 0. Small trades are trades whose size in value is below the first quartile. Large trades are the ones whose size in value is above the third quartile. Top, net purchases. Bottom, net sales.

TABLE 20.4. Market reactions over longer time horizons

	Event windows		
	CAR+3,200	**CAR+41,+100**	**CAR+41,+200**
Net purchases			
All trades	6.73%*	7.45%***	3.09%
Small trades	7.90%	12.80%**	4.59%
Large trades	-2.09%	-0.07%	-2.14%
Net sales			
All trades	-14.91%***	-4.40%***	-10.93%***
Small trades	-20.20%***	-1.18%	-8.51%**
Large trades	-9.93%**	-4.30%***	-11.08%***

*Average cumulative abnormal returns (CAR) for different event windows defined ex post to the insider trading day, which is day 0. Small trades are trades the size in value of which is below the first quartile. Large trades are the ones the size in value of which is above the third quartile. *, **, *** denote significance at the 10%, 5%, and 1% levels, respectively.*

transaction day and since an important proportion of the cumulative abnormal returns seems to be realized after day +40 (see Figure 20.1 and Table 20.4), outsiders mimicking insiders may also be able to realize an abnormal performance.

20.4 CONCLUSION

Public confidence in the integrity of the financial markets is crucial for the development of these markets. That is the reason developed countries constantly introduce and improve regulations that envisage the preservation of this public confidence and the protection of the financial markets against abuse and manipulation. The regulation of insider transactions forms part of these measures and envisages the prohibition of operations conducted by corporate insiders on the basis of private information.

In several countries, insider trading regulation implies an obligation to notify any transaction conducted by a corporate insider. This measure allows using stock price reactions an examination of the motives behind those transactions and provides the other investors in the market with a source of potential information. As a consequence, the information content of the transactions notified by insiders has been the subject of intense analysis in literature. Consistent to a large extent with previous literature, short-term abnormal returns associated with insider trades on Euronext Amsterdam are either nonsignificant or ambiguous. It is important to note that short-term abnormal returns are only a very noisy proxy for private information revelation in the context of insider trading. According to Aktas, de Bodt, and Van Oppens (2007) this is mainly due to two shortcomings. The first relates to the probable endogenous relation between abnormal returns and insider trading: insiders may decide to purchase on a specific day because they expect stock prices to increase on that day. The second shortcoming results from the fact that insiders can act strategically by timing the market, and can voluntarily choose a trading window in which they can hide their motivation for trading.

However, using longer event windows, we are able to show that the adjustment of the stock prices is notable and has the right direction. This suggests that insiders either have some market timing ability and/or use long-term information. The notification process seems to provide outsiders with an important source of information. However, to ensure that the excess returns are not simply a compensation for risk, more data are needed to perform a more sophisticated significance test while controlling for known priced factors in the market (e.g., such as beta, size, book-to-market, momentum). This is left for further research.

REFERENCES

Aktas, N., E. de Bodt, and H. Van Oppens. 2007. Evidence on the contribution of legal insider trading to market efficiency. Université catholique de Louvain, CORE discussion paper 2007/14.

Ausubel, L. M. 1990. Insider trading in a rational expectations economy. *American Economic Review* 80(5):1022–41.

Battacharya, U., and H. Daouk. 2002. The world price of insider trading. *Journal of Finance* 57(1):75–108.

Carlton, D. W., and D. R. Fischel. 1983. The regulation of insider trading. *Stanford Law Review* 35:857–99.

Eckbo, E. B., and D. C. Smith. 1998. The conditional performance of insider trades. *Journal of Finance* 53(2):467–98.

Fidrmuc, J., M. Goergen, and L. Renneboog. 2006. Insider trading, news releases and ownership concentration. *Journal of Finance* 61(6):2931–73.

Finnerty, J. E. 1976. Insiders and market efficiency. *Journal of Finance* 31(4):1141–48.

Friederich, S., A. Gregory, and J. Matatko. 2002. Short-term returns around the trades of corporate insiders on the London Stock Exchange. *European Financial Management* 8(1):7–30.

Grundmann-van de Krol, C. M. 2004. *Koersen door het effectenrecht. Beschouwingen omtrent Nederlands effectenrecht.* 5th ed. Den Haag: Boom Juridische Uitgevers.

Hazen, T. 2005. *The law of securities regulation.* 5th ed. St. Paul, MN: Thomson/West.

Hazen, T., and D. L. Ratner. 2006. *Securities regulation.* 9th ed. St. Paul, MN: Thomson/West.

Huddart, S., B. Ke, and C. Shi. 2007. Jeopardy, non-public information, and insider trading around SEC 10-K and 10-Q filings. *Journal of Accounting and Economics* 43(1):3–36.

Jaffe, J. F. 1974. Special information and insider trading. *Journal of Business* 47(3):410–28.

Jeng, L., A. Metrick, and R. Zeckhauser. 2003. The profits to insider trading: A performance-evaluation perspective. *Review of Economics and Statistics* 85(2):453–71.

Jenter, D. 2005. Market timing and managerial portfolio decisions. *Journal of Finance* 60(4):1903–49.

Kraakman, R. 1991. The legal theory of insider trading regulation of the United States. In *European insider dealing*, ed. K. E. Hopt and E. Wymeersch. London: Butterworths.

Lakonishok, J., and I. Lee. 2001. Are insider trades informative? *Review of Financial Studies* 14(1):79–111.

Leland, H. E. 1992. Insider trading: Should it be prohibited? *Journal of Political Economy* 100(4):859–87.

Lin, J., and J. Howe. 1990. Insider trading in the OTC market. *Journal of Finance* 45(4):1273–84.

Manne, H. G. 1966. *Insider trading and the stock market.* New York: Free Press.

Piotroski, J. D., and D. T. Roulstone. 2005. Do insider trades reflect both contrarian beliefs and superior knowledge about future cash flow realizations? *Journal of Accounting and Economics* 39(1):55–81.

Rozeff, M. S., and M. A. Zaman. 1988. Market efficiency and insider trading: New evidence. *Journal of Business* 61(1):25–44.

Schutte, J. 2006. Meldingsverplichtingen. In *Marktmisbruik, Series Onderneming en Recht,* D. R. Doorenbos, S. C. J. J. Kortmann, and M. P. Nieuwe Weme. Deventer: Kluwer.

Seyhun, N. H. 1986. Insiders' profits, costs of trading, and market efficiency. *Journal of Financial Economics* 16(2):189–212.

Seyhun, N. H. 1998. *Investment intelligence: From insider trading.* Cambridge, MA: MIT Press.

Index

B

C

D

E

J

K

L

M

N

O

P

T

U

V

W

X

Y

Z

For Product Safety Concerns and Information please contact our EU representative GPSR@taylorandfrancis.com
Taylor & Francis Verlag GmbH, Kaufingerstraße 24, 80331 München, Germany

www.ingramcontent.com/pod-product-compliance
Ingram Content Group UK Ltd.
Pitfield, Milton Keynes, MK11 3LW, UK
UKHW021624240425
457818UK00018B/722

* 9 7 8 0 3 6 7 3 8 6 9 9 3 *